Memoirs of the American Mathematical Society
Number 231

Steph
and G

Weil r

Ir ons
a

Published by
AMERICAN
Providence, Rhode Island, USA

May 1980 · Volume 25 · Number 231 (second of 3 numbers)

ABSTRACT

We set forth the foundations of the spectral decomposition of the Weil representation associated to a nondegenerate quadratic form Q over the field $\mathbb{R}$ of real numbers. The relevant intertwining distributions are constructed, and a complete analysis is made of the discrete spectrum.

AMS(MOS) SUBJECT CLASSIFICATION (1970):

Primary: 22E30, 22D10.

Secondary: 43A85.

Library of Congress Cataloging in Publication Data

Rallis, Stephen, 1942-
Weil representation I.

(Memoirs of the American Mathematical Society ; no.231
ISSN 0065-9266)
"Volume 25 ... (second of 3 numbers)"
Bibliography: p.
1. Lie groups. 2. Representations of groups.
3. Spectral theory (Mathematics) I. Schiffmann, Gerard, 1940- joint author. II. Title. III. Series: American Mathematical Society. Memoirs ; no. 231.
QA3.A57 no. 231 [QA387] 510s [512'.55] 80-12191
ISBN 0-8218-2231-4

TABLE OF CONTENTS

WEIL REPRESENTATION. I

INTERTWINING DISTRIBUTIONS AND DISCRETE SPECTRUM*

INTRODUCTION

Let F be a finite dimensional vector space over $\mathbb{R}$ and Q a non-degenerate quadratic form over F. One owes to A. Weil [26] the construction of a representation $\pi_{\mathfrak{m}}$ of the two fold covering $\widetilde{S\ell}_2(\mathbb{R})$ of $S\ell_2(\mathbb{R})$ on $L^2(F)$. The key property of this representation is the following one: given a lattice L in F such that $Q(L,L) \subset \mathbb{Z}$, there exists an "arithmetical" discrete subgroup Γ of $\widetilde{S\ell}_2(\mathbb{R})$ such that for any Schwartz function f on F,

$$G_f : g \mapsto \sum_{\xi \in L} \pi_{\mathfrak{m}}(g)^{-1}(f)(\xi)$$

is a C^∞ function on $\widetilde{S\ell}_2(\mathbb{R})$, left invariant under Γ. Now if Q is anisotropic, suitable choices of f give for G_f the classical θ series of Hecke. The goal of the authors is to study the series G_f for indefinite quadratic forms. More precisely one wants to choose for f a "K finite" function belonging to some irreducible component of the discrete spectrum of $\pi_{\mathfrak{m}}$. Unfortunately such functions are not of Schwartz type and rather serious technical problems arise. In any case to get significant results on G_f one needs a very detailed knowledge of $\pi_{\mathfrak{m}}$. The goal of this paper is to provide such a knowledge. The series G_f will be studied in another paper.

As it is clearly explained in [26], the whole problem can be put in a global adelic setup. However this would mean decomposing the Weil representation over an arbitrary local field. Some of the methods used in this paper remain valid in such a situation, and we hope to present in the near future some results for the general case.

Let us now describe in detail the contents of this paper.

*Received by the editors, March 1, 1976.

The general philosophy of horocycles in group representation theory can be put in the following general framework.

First let π be a unitary representation of a Lie group G on a Hilbert space H so that π has a cyclic vector in H. The spectral decomposition theory gives the following abstract equivalence. There is a locally compact space E and a Hilbert bundle E_π over E (see [25] for definition, etc.). Moreover on E there is a measure σ and a G intertwining unitary isomorphism ℓ_π between H and the space $L^2(E_\pi,\sigma)$ of square integrable sections of E_π relative to σ. Maurin has given in [18] a sharpening of this equivalence. We assume that there exists a G stable dense subspace $\tilde{H} \subseteq H$ provided with a linear convex Hausdorf topology τ so that $(\tilde{H},\tau)$ is a nuclear space and the embedding of $(\tilde{H},\tau)$ into H (with the usual Hilbert space topology) is continuous. We call $\tilde{\pi}$ the restriction of π to $\tilde{H}$. Then for almost all $\gamma \in E$ (relative to σ), there exists a linear G module injection $i_\gamma : E_\gamma \to \tilde{H}'$ (E_γ, the Hilbert space fiber over the point $\gamma \in E$ which carries an irreducible unitary representation π_γ of G, and $\tilde{H}'$, the continuous linear dual of H relative to the topology τ) so that relative to a basis $\{e_j(\gamma)\}_{j=1}^{\dim E_\gamma}$ of E_γ

$$(\ell_\pi(\varphi)(\gamma),\ e_j(\gamma))_\gamma = \langle \varphi,\ i_\gamma(e_j(\gamma))\rangle$$

for all $\varphi \in \tilde{H}$ ($(\ ,\ )_\gamma$, Hilbert inner product on E_γ and $\langle\ ,\ \rangle$, the pairing between $\tilde{H}$ and $\tilde{H}'$).

The importance of this sharpening is that we have a realization of E_γ in the dual $\tilde{H}'$ of $\tilde{H}$. It is thus important that one determines generally the space $\mathrm{Hom}_G(\pi_\gamma,\tilde{\pi}^c)$, the set of continuous G intertwining maps between π_γ and $\tilde{\pi}^c$ with $\tilde{\pi}^c$ the contragredient representation of $\tilde{\pi}$ on $\tilde{H}'$. Also by noting from [18] that the space $\tilde{H}$ with the topology τ is reflexive, we then have by duality that $\mathrm{Hom}_G(\pi_\gamma,\tilde{\pi}^c)$ is linearly isomorphic to $\mathrm{Hom}_G(\tilde{\pi},\tilde{\pi}_\gamma^c)$, where π_γ^c (the contragredient representation of π_γ) is again a unitary representation of G. The set $\mathrm{Hom}_G(\tilde{\pi},\pi_\gamma^c)$ can in general be determined more easily than $\mathrm{Hom}_G(\pi_{\gamma}, \tilde{\pi}^c)$.

At this point we are interested in reversing the spectral problem. Namely if we are given only the "cyclic" representation π above, we want to determine the space E , E_π , σ , etc. The starting point appears to be to find an "appropriate" nuclear G stable, dense subspace $\tilde{H}$ of H . Then given any unitary irreducible representation γ (any algebraically irreducible differentiable representation $\tilde{\gamma}$, respectively) of G on a Hilbert space H_γ (Frechet space $E_{\tilde{\gamma}}$ resp.), we are interested in determining the space $\mathrm{Hom}_G(\tilde{\pi},\gamma)$ ($\mathrm{Hom}_G(\tilde{\pi},\tilde{\gamma})$ resp.). The precise data required is first a completeness statement. Namely given a family $\{\ell_\gamma^\alpha\}_{\gamma\in W}$ where $\ell_\gamma^\alpha \in \mathrm{Hom}_G(\tilde{\pi},\gamma)$ and W any subset of $\hat{G}$ (the unitary dual space of G) , we want to know whether $\ell_\gamma^\alpha(e) = 0$ for all ℓ_γ^α implies $e \equiv 0$. Secondly fixing $\gamma \in \hat{G}$, we want to determine the relation between ℓ_γ^α and ℓ_γ^β when $\alpha \neq \beta$. For example a "multiplicity free" statement about a complete family $\{\ell_\gamma^\alpha\}_{\gamma\in W}$ is just that $\dim(\mathrm{Hom}_G(\tilde{\pi},\gamma)) \leq 1$ for all $\gamma \in W$. Indeed if $V_{\tilde{\pi}}$ is the algebra of all continuous operators on $\tilde{H}$ which commute with $\tilde{\pi}$ and $\{\ell_\gamma^\alpha\}$ has the "multiplicity free" property above, then $V_{\tilde{\pi}}$ is a commutative algebra.

With this problem at hand, we want to find methods to determine $\dim \mathrm{Hom}_G(\tilde{\pi},\gamma)$. The main tool for this is the theory of intertwining distributions developed by Bruhat in [3]. A special case is when we take a closed subgroup U of G and a continuous character χ of U . Then we form the induced representation σ_χ of G on the Frechet space (with Schwartz topology) $C^\infty(G/U,\chi) = \{f : G \to \mathbb{C} \mid f \in C^\infty(G)$ and $f(gu) = f(g)\chi(u)$ for $g \in G$, $u \in U\}$. Then it is easy to see that $\mathrm{Hom}_G(\tilde{\pi},\sigma_\chi)$ is linearly isomorphic to the space of vector distributions $\tilde{H}(\chi) = \{S \in \tilde{H}' \mid \tilde{\pi}^c(u)S = \chi(u)S$ for all $u \in U\}$. Thus if $\tilde{H}$ can be realized as a suitable space of functions on some manifold X which admits a U action, then to determine $\tilde{H}(\chi)$ we can bring into play the geometric properties of the U action on X with the support properties of elements of $\tilde{H}(\chi)$ (see section 3).

With this general setup, we consider the case of the Weil representation $\pi_{\mathfrak{M}}$ of the metaplectic group $\tilde{G}_2(Q) = \tilde{S\ell}_2(Q) \times O(Q)$ on the Hilbert space $L^2(\mathbb{R}^k)$ (where $O(Q)$ is the orthogonal group of a nondegenerate quadratic form Q on $\mathbb{R}^k$ and $\tilde{S\ell}_2(Q)$, the associated 2-fold covering of $\tilde{S\ell}_2(\mathbb{R})$). In this manuscript we lay the foundations of the spectral decomposition of $\pi_{\mathfrak{M}}$.

In section 1 we recall the properties of the metaplectic group $\tilde{G}_2(Q)$ and make precise the relation between Weil's construction of $\tilde{G}_2(Q)$ and the Kubota cocycle determining the covering group $\tilde{S\ell}_2(Q)$ (Theorem 1.1). Moreover we set forth functorial properties of the representation $\pi_{\mathfrak{M}}$ in relation to the addition of quadratic forms (Corollary to Theorem 1.1).

In section 2 we determine certain geometric and algebraic properties of the group $O(Q)$. Namely we set forth the symmetric space data on $O(Q)$ relative to a maximal compact subgroup K. Moreover we characterize the $O(Q)$ orbit structure in the Grassmann manifold of s-planes in $\mathbb{R}^k$ (Lemma 2.1). Then relative to the unipotent radical N_a of a maximal parabolic P_a in $O(Q)$, we define horocycles in $\mathbb{R}^k$ and show the relation to the geometric notion of horocycle of [10] (Lemmae 2.2 and 2.3). Namely there are 2 types of horocycles corresponding to the nondegeneracy of the Riemannian metric of $\mathbb{R}^k$ determined by Q restricted to the tangent spaces of the horocycles. In addition we determine the orbit structure of the group P_a operating on $\mathbb{R}^k$ (Proposition 2.4).

In section 3 we recall generalities of the theory of group representations. Specifically we set forth the theory of intertwining distributions of Bruhat. We apply this to construct the "conic" representations of $O(Q)$ (Propositions 3.4 and 3.5). Moreover we study the $O(Q)$ intertwining relations between such representations (Corollary to Proposition 3.5). Then again applying Bruhat's theory, we determine the space of $O(Q)$ intertwining maps between the standard representation π_t of $O(Q)$ on $C_c^\infty(\Gamma_t)$ (C^∞

functions of compact support on the hyperboloid $\Gamma_t = \{X \in \mathbb{R}^k | Q(X,X) = t\}$, $t \neq 0$) and the "conic" representations (Proposition 3.6). We conclude the section by showing the uniqueness of Whittaker models for the conic representations of the Lorentz group (Proposition 3.7).

In section 4 we study in detail the "conic" and Whittaker models of the representations of $\widetilde{s\ell}_2(Q)$. In particular taking the one parameter group $\underline{N}$ of $\widetilde{s\ell}_2(Q)$ isomorphic to $\left\{ \begin{bmatrix} 1 & x \\ 0 & 1 \end{bmatrix} \middle| x \in \mathbb{R} \right\}$ and any unitary character τ on $\underline{N}$, we study the unitarily induced representation π_τ of $\widetilde{s\ell}_2(Q)$ constructed from τ (τ trivial determining the "conic" representations and τ nontrivial determining the "Whittaker" models). We take the Casimir $\omega_{s\ell_2}$ of the enveloping algebra of $\widetilde{s\ell}_2(Q)$ and determine the generalized eigenspaces of $\omega_{s\ell_2}$ in $\mathscr{E}_\tau$ ($\mathscr{E}_\tau$ the dual space of $C_c^\infty(\tau)$, the C^∞ sections of compact support of the vector bundle determined by τ) (Propositions 4.6 and 4.9). Then the study of the possible intertwining maps between the various eigenspaces leads to (i) the classical picture of "$\widetilde{s\ell}_2(Q)$ intertwining" operators in the conic case and (ii) the unique embedding of the "conic" representations in the Whittaker model (Corollary 1 to Proposition 4.13 and Corollaries 1, 2, and 3 of Proposition 4.14). Finally we characterize completely the discrete spectrum of the representation π_τ (Proposition 4.15). We note at this point that the discussion in section 4 is self-contained, and none of the classical integral formulae of Whittaker functions, etc. are used. We need only the theory of asymptotic expansions of differential equations. (For another discussion of the representation theory of the universal cover of $s\ell_2(\mathbb{R})$, we refer the reader to [29] and [35]).

In section 5 we begin the study of intertwining distributions of the Weil representation $\pi_{\mathfrak{m}}$ of $\tilde{G}_2(Q)$. We take the parabolic subgroup $L = P \times P_a$ of $\tilde{G}_2(Q)$ (P isomorphic to the group $\left\{ \begin{bmatrix} r & s \\ 0 & r^{-1} \end{bmatrix} \middle| r \neq 0,\ s \in \mathbb{R} \right\}$ and recall that the continuous one dimensional representations of L are parametrized by a pair of quasicharacters (X_1, X_2) (X_1 (X_2 resp.) a character of P (P_a resp.)). Then as above we consider the family of induced representations $\sigma_{(\chi_1,\chi_2)}$ of $\tilde{G}_2(Q)$. The G stable nuclear space in this case is $S(\mathbb{R}^k)$, the space of Schwartz functions on $\mathbb{R}^k$ (e.g. $H = S(\mathbb{R}^k)$). Then Theorem 5.4 gives estimates on the numbers $a(\chi_1,\chi_2) = \dim \mathrm{Hom}_{\tilde{G}_2(Q)}(\tilde{\pi}_{\mathfrak{m}}, \sigma_{(\chi_1,\chi_2)})$. The important point here is that χ_1 and χ_2 must satisfy a linear relation in order that $a(\chi_1,\chi_2) \neq 0$. With these conditions we begin the study of the construction of the intertwining maps. Namely we first consider the case when $a(\chi_1,\chi_2) = 1$. Again Theorem 5.4 determines two families of such intertwining maps, each family being characterized by certain support properties; we start then with the distributions θ_f^χ and R_f^χ ($f \in S(\mathbb{R}^k)$, χ a quasicharacter) given by (5-30) and (5-31). However the first distribution θ_f^χ is defined in χ for a certain halfspace in $\mathbb{C}$. In Propositions 5.6 and 5.7 we exhibit certain integral and analytic expressions for θ_f^χ. This allows us by the famous trick of Tate's thesis to obtain Ξ_f^χ, the analytic continuation of θ_f^χ (Proposition 5.8). Then we obtain the "functional equation" of Θ_f^χ, a twisted version of Ξ_f^χ, in Theorem 5.9. This functional equation can be viewed as a generalization of the classical functional equation obtained by taking the Fourier transform of $e^{2\pi it\, Q(X,X)}$, $X \in \mathbb{R}^k$. Then we deduce by using the arguments of analytic continuation of section 4, the continuation of the second family of intertwining maps and the relation between the 2 families (Corollaries 3 and 4 to Theorem 5.9). As an interesting consequence of this work, we obtain the so-called functional equation of the "Intertwining Operator on $O(Q)$" (see (5-77)), generalizing the functional equation of the intertwining operator on $S\ell_2(\mathbb{R})$ given by Godement in [12] (see specifically formula (21) of [12], p. 217 and the relation of this functional equation to the study of Eisenstein series). Then in Theorem 5.10 we construct a complete family of $O(Q)$ intertwining maps $c_{\lambda,t}$ of $C_c^\infty(\Gamma_t)$ to the $O(Q)$ conic models constructed in section 3; we prove a multiplicity one statement about this family. In particular we deduce that the algebra of operators commuting with the standard representation of $O(Q)$ on $C_c^\infty(\Gamma_t)$ is commutative. Continuing with the general philosophy of horocycles outlined above, we consider the formal adjoint $T_{\lambda,t}^*$ (see (5-92)) of the map $c_{\lambda,t}$; whence $T_{\lambda,t}^*$ defines an $O(Q)$ intertwining map of the conic models to the distribution space $C_c^\infty(\Gamma_t)'$ (see Proposition 5.11). Finally in Proposition 5.12 we determine an analytic way to compute $T_{\lambda,t}^*$.

In section 6 we describe explicitly the discrete spectrum of the representation $\pi_{\mathfrak{M}}$ of $\tilde{G}_2(Q)$. First a trivial geometric characterization (Lemma 6.1) is given of the space $\mathbb{F}_Q$ of C^∞ vectors of $G_2(Q)$ in terms of the behavior of a function on the sets $\Omega_+ = \{X \in \mathbb{R}^k \mid Q(X,X) > 0\}$ and $\Omega_- = \{X \in \mathbb{R}^k \mid Q(X,X) < 0\}$. Then we let $\mathbb{F}_Q(\lambda) = \{\varphi \in \mathbb{F}_Q \mid \pi_{\mathfrak{M}}(\omega_{s\ell_2})\varphi = \lambda\varphi\}$, which is a $\tilde{G}_2(Q)$ submodule of $\mathbb{F}_Q$. In Lemma 6.2 we show that the spaces $\mathbb{F}_Q^+(\lambda) = \{\varphi \in \mathbb{F}_Q(\lambda) \mid \text{support } (\varphi) \subset \Omega_+\}$ and $\mathbb{F}_Q^-(\lambda) = \{\varphi \in \mathbb{F}_Q(\lambda) \mid \text{support } (\varphi) \subset \Omega_-\}$ determine $\tilde{G}_2(Q)$ stable submodules of $\mathbb{F}_Q$, and in fact, we have a direct sum decomposition $\mathbb{F}_Q(\lambda) = \mathbb{F}_Q^+(\lambda) \oplus F_Q^-(\lambda)$ as $\tilde{G}_2(Q)$ modules with $\mathbb{F}_Q^+(\lambda)$ and $\mathbb{F}_Q^-(\lambda)$ giving inequivalent $\tilde{G}_2(Q)$ representations. Then we determine explicitly the $s\tilde{\ell}_2$ action on $\mathbb{F}_Q^+(\lambda)$ and $\mathbb{F}_Q^-(\lambda)$ in Lemma 6.3. It is here that we note the importance of knowing the discrete spectrum of the "Whittaker" representation π_τ studied in section 4. Indeed we show in Theorem 6.4 that $\mathbb{F}_Q^+(\lambda)$ is $\tilde{G}_2(Q)$ equivalent to the tensor product representation $\left\{\begin{matrix}(D_{s-\frac{1}{2}}^A)_\infty \\ (H_s^A)_\infty\end{matrix}\right\} \hat{\otimes} \mathcal{A}_s^+$, where $s \in \frac{1}{2}\mathbb{Z}$, $D_{s-\frac{1}{2}}^A$ and H_s^A irreducible $s\tilde{\ell}_2$ modules occurring in L_τ^2 (see Proposition 4.15) and $\mathcal{A}_s^+$ is an $O(Q)$ invariant subspace (called "extreme vectors") of functions in $\mathbb{F}_Q$ determined by a family of differential equations (6-17).

The next problem is to characterize the space $\mathcal{A}_s^+$. For this purpose we use the Fock-Bargmann model of the metaplectic representation. The problem becomes purely an algebraic one of determining the infinitesimal representation of $\mathcal{U}(\mathfrak{S})$, the enveloping algebra of $O(Q)$, in the ring $\mathcal{F}_k$

of power series on $\mathbb{C}^k$. This is done in Lemma 6.6. Furthermore the algebraic method yields the algebraic irreducibility of the space of K finite vectors in $\mathcal{A}_s^+$. Hence we get the topological irreducibility of $\mathbb{F}_Q^+(\lambda)$ relative to the group $\tilde{G}_2(Q)$. Then the problem becomes one of showing for which $s \in \frac{1}{2}\mathbb{Z}$ that $\mathcal{A}_s^+ \neq 0$. However using Lemma 6.3 and the data of the K finite structure, we determine a nonvanishing criterion for $\mathcal{A}_s^+$ in Theorem 6.10. Thus we can completely determine $\mathrm{Spec}(\pi_{\mathfrak{m}}) = \{\lambda \mid \mathbb{F}_Q(\lambda) \neq 0\}$. As another consequence of Theorem 6.10 we deduce (Corollary 1 to Theorem 6.10) the rapid decrease and continuity properties of K finite functions in $\mathbb{F}_Q(\lambda)$. This will be important in later work for the construction of functions (which are not of Schwartz-Bruhat type) which satisfy the Poisson summation formula.

Moreover as an additional consequence of the K finite structure, we show the $O(Q)$ equivalence of $\mathcal{A}_s^+$ to a certain $O(Q)$ irreducible subspace in $L^2(\Gamma_{+1}, d\mu_{+1})$.

Then our work includes another method (for the case $b = 1$) to determine the space $\mathcal{A}_s^+$ (see Remark 6.5). Starting with the data in Lemma 6.6, we give an analytic method to pull back a function of the discrete spectrum in the Fock-Bargmann model to the Schrodinger model. In particular we show (Proposition 6.11) that starting from a function of the form $e^{-\mathfrak{h}(X)}$ $(X \in \mathbb{R}^k)$, where $\mathfrak{h}$ belongs to the space $\mathfrak{h}(Q)$ of Hermite majorants of Q (see [1]), we can obtain an "extreme vector" after (i) putting $\mathfrak{h}(Q)$ into polar coordinates, integrating against an appropriate spherical harmonic in the angular variables and then going to ∞ in the radial variable (see Proposition 6.12) and (ii) applying fractional differentiation in one of the coordinates (see (6-64) to (6-78)).

Finally our problem is to give a conic model for the representation of $O(Q)$ on $\mathcal{A}_s^+$ using Proposition 5.12. We thus determine explicitly in Proposition 6.15 the map $T^*_{\lambda,t}$ for a K finite isotypic component. From this we deduce the injectivity properties of $T^*_{\lambda,t}$ in <u>Theorem</u> 6.16. And in <u>Theorem</u> 6.17 we determine which "quotient" representations deduced from conic representations are $O(Q)$ equivalent to $\mathcal{A}^+_s$ via the map $T^*_{\lambda,t}$ or an appropriate residue map.

R. Strichhartz [36] has obtained in another framework some of the results on the discrete spectrum of the hyperboloid. His work contains an extensive reference on the harmonic analysis of $O(Q)$ on hyperboloids. We also refer the reader to [27] for a discussion of the $O(2,1)$ case in terms of the Weil representation.

SECOND INTRODUCTION

We have decided to update certain aspects of the problem in the light of recent applications of the Weil representation. The Weil representation of a dual reductive pair (à la Howe [37]) has proved invaluable for constructing automorphic forms (and representations) for certain classical reductive Lie groups. The philosophical point of view that a "pairing" occurs between representations of a dual reductive pair in a given Weil representation was more than apparent from the outset in the works of Siegel, Maass, and Weil. What is important in this context is to give an <u>explicit</u> description of such a pairing in relation to the spectral decomposition of the given Weil representation. This is the point of view adopted in this manuscript. We refer the reader to [30] for a summary of results which are an outgrowth of the ideas developed in this manuscript.

Using the results of §6, we have given in [31] new examples of automorphic forms on $\widetilde{S\ell}_2 \times O(Q)$ (which are cusp forms on $\widetilde{S\ell}_2$), generalizing earlier examples of Shimura, Shintani, Doi-Naganuma, and Zagier. One of the main consequences of §6 is an explicit study of the growth behavior of functions in the discrete spectrum. As a result we can apply the Poisson Summation Formula to a certain part of the discrete spectrum of $\pi_{\mathfrak{m}}$ (i.e. $\bigoplus_{s>k/2} \mathbb{F}_Q(s^2-2s)$); what is rather striking here is that precisely for the "cutoff eigenvalue" $s = k/2$, the $O(Q)$ irreducible component $\mathcal{A}^+_{k/2}$ in $\mathbb{F}_Q(\frac{k^2}{4} - k)$ has the property that it is cohomologically nontrivial, i.e. $H^b(\mathfrak{S},\mathfrak{K},\mathcal{A}^+_{k/2}) \neq 0$, where H^b is the b-th relative cohomology of $LA(O(Q)) = \mathfrak{S}$ with values in $\mathcal{A}^+_{k/2}$. This gives yet another candidate for a unitary irreducible representation in $\mathcal{A}^+_{k/2}$ (of $O(Q)$) which contributes to the b-th cohomology of a cocompact discrete subgroup Γ of $O(Q)$. Thus if we can find another method of summation to replace Poisson Summation (such as the Siegel-Maass idea of integration against a generalized θ- function), it should be possible to construct a nonzero $O(Q)$ intertwining operator from $\mathcal{A}^{+}_{k/2}$ to $L^2(O(Q)/\Gamma')$, where Γ' is some subgroup of finite index in Γ. This then would give another example of the nonvanishing of the b-th cohomology of Γ' at the critical point in the Borel-Wallach-Zuckermann theory. (We note here the work of Kazhdan [28], which gives the first example of the nonvanishing of the 1st cohomology of a cocompact discrete subgroup of SU(n,1).)

The ideas of §5 apply very easily to the general local field case. Indeed in [32], [33], and [34] we prove analogous statements to those given in §5. Moreover we interpret the "pairing" between representations occurring in the spectral decomposition of the dual reductive pair $(Sp_n, O(Q))$ in terms of "local" Langlands functoriality. On the other hand, we observe that such a pairing has a classical application in terms of the "Eichler Commutation Theorem", which arises in the theory of θ- functions.

§1. GENERALITIES OF THE WEIL REPRESENTATION

Let $\mathbb{R}^k$ be the vector space of k rows $\begin{pmatrix} x_1 \\ \vdots \\ x_k \end{pmatrix}$, $x_1 \in \mathbb{R}$.

Let τ: $\mathbb{R}^k \to S^1$ be the unitary character on $\mathbb{R}^k$ given by $\tau(x) = e^{2\pi ix}$.

Let $[\ ,\ ]$ be the bilinear form on $\mathbb{R}^k$ given by $[x,y] = \sum_{i=1}^{k} x_i y_i$.

Let $S(\mathbb{R}^k)$, $L^1(\mathbb{R}^k)$ be the space of Schwartz-Bruhat functions (L^1 functions) on $\mathbb{R}^k$. Then if $f \in S(\mathbb{R}^k)$ or $L^1(\mathbb{R}^k)$, we know that the Fourier transform of f exists and is given by the absolutely convergent integral.

$$(1 - 1)\quad \hat{f}(x) = \int_{\mathbb{R}^k} f(y)\tau(\ [y,x]\ )\ dy \quad .$$

Then if we define the inverse Fourier transform by

$$(1 - 2)\quad \check{f}(y) = \int_{\mathbb{R}^k} f(x)\tau(-[x,y]\ )\ dx \quad ,$$

we know that the Plancherel formula for Fourier transforms says that

$$(1 - 3)\quad \check{(\hat{f})}(x) = f(x) \qquad (f \in S(\mathbb{R}^k)) \quad .$$

We let $\langle\ ,\ \rangle$ be the bilinear pairing on $\mathbb{R}^k$ defined by $\langle z,w\rangle = \tau(\ [z,w]\ )$. This makes $\mathbb{R}^k$ self dual as a locally compact Abelian group. Then we consider the bicharacter on $\mathbb{R}^k \times \mathbb{R}^k$ defined by $B(z,w) = \langle z_1,w_2\rangle$ with $z = (z_1,z_2)$, $w = (w_1,w_2)$. From this we construct the Heisenberg group H_k as the topological product $\mathbb{R}^k \times \mathbb{R}^k \times S^1$ with the group law:

$$(1 - 4)\quad (z,t)\ (z',t') = (z + z',\ B(z,z')\ t\ t'),$$

with z, $z' \in \mathbb{R}^k \times \mathbb{R}^k$ and t, $t' \in S^1$.

We then consider the following unitary representation of H_k on the Hilbert space $L^2(\mathbb{R}^k)$. Namely define U_m: $H_k \to \text{Unit}\ (L^2(\mathbb{R}^k))$ by

$$(1 - 5)\quad U_m(z,t)\ f(x) = t^m\ f(x + z_1)\ \langle x,z_2\rangle\ , \quad z = (z_1,z_2).$$

Then the well known Stone-Von Neumann Theorem says that U_m is an irreducible unitary representation of H_k and that every unitary irreducible representation of H_k which is not finite dimensional is unitarily equivalent to U_m for some m.

If $B(H_k)$ is the Lie group of automorphisms which leave the center $\{ (0,t) \mid t \in S^1 \}$ pointwise fixed, then we know that $g \in B(H_k)$ has the form $g = (\sigma,f)$, where (σ,f) operates as $(z,t)^g = (\sigma(z),f(z)t)$ with σ a symplectic automorphism relative to the alternating bilinear form $A(z,w) = B(z,w)\,B(w,z)^{-1} = \tau\,(\,[z_1,w_2] - [w_1,z_2])$ and f a continuous quadratic character $\mathbb{R}^k \times \mathbb{R}^k \to S^1$ so that $f(z+w)\,f(z)^{-1}\,f(w)^{-1} = B(z,w)\,B(w,z)^{-1}$ (for fixed z a character in w, etc.). In the matrix form $\sigma = \left[\begin{array}{c|c} A & B \\ \hline C & D \end{array}\right]$ with $A^tD - C^tB = I$, the identity $k \times k$ matrix, and $A\cdot B^t$ and $C\cdot D^t$ symmetric. The group law on $B(H_k)$ is given by

$(\sigma_1,f_1)\,(\sigma_2,f_2) = (\sigma_1\sigma_2, f_1f_2^{\sigma_2})$ $(f_2^{\sigma_2}(w) = f(\sigma_2^{-1}(w))\,)$. The map $(\sigma,f)\underset{\psi}{\to}\sigma$ determines a Lie group homomorphism of $B(H_k)$ onto $Sp_k(\mathbb{R}^k)$, the symplectic group relative to A given above. The kernel of the map is the set of all (I,f) with f of the form $f(w) = \langle w_1,a_1\rangle\,\langle w_2,a_2\rangle$ for fixed $a_1,\,a_2 \in \mathbb{R}^k$ and all $w \in \mathbb{R}^k \times \mathbb{R}^k$. Hence Kernel (ψ) may be identified to the group of all inner automorphisms $\mathrm{Int}\,(H_k)$ $(\cong \mathbb{R}^k \times \mathbb{R}^k)$ of H_k. Moreover it is easy to determine that the map ψ splits so that in fact we have that $B(H_k)$ is the semidirect product $Sp_k(\mathbb{R}^k) \times \mathrm{Int}(H_k)$ (indeed the splitting section to ψ is given by $\sigma \overset{S_\psi}{\leadsto} (\sigma,f_\sigma)$, where $f_\sigma(w) = \langle w_1,\tfrac{1}{2}C^tAw_1\rangle\,\langle w_2,\tfrac{1}{2}B^tDw_2\rangle\,\langle Bw_2,Cw_1\rangle = \tau\,(\tfrac{1}{2}\,[w_1,C^tAw_1] + \tfrac{1}{2}\,[w_2,B^tDw_2] + [Bw_2,Cw_1]\,)$.

Then in the group of unitary operators $\mathrm{Unit}(L^2(\mathbb{R}^k))$, we consider the group $\beta(H_k) = \{g | g \text{ normalizes } U_m(H_k)\}$ endowed with the induced topology from the norm topology on $\mathrm{Unit}(L^2(\mathbb{R}^k))$. From [26] we recall that $g \in \beta(H_k)$ induces a topological group automorphism of H_k; thus we have a homomorphism $\rho:\beta(H_k)\to B(H_k)$. Conversely, starting with $x \in B(H_k)$ and defining the twisted representation $U_m{}^x$ as $U_m{}^x(z,t) = U_m\,[(z,t)^x]$ and by applying the Stone-Von Neumann Theorem above, we find a unitary operator $a(x) \in \beta(H_k)$ so that $a(x)U_m\, a(x)^{-1} = U_m{}^x$. Thus ρ is a surjective map. Morever by the irreducibility of U_m, we have that Kernel $(\rho) = \{\omega\cdot I\}$ with $\omega \in S^1$ and I the identity operator on $L^2(\mathbb{R}^k)$. However from [15] we recall that $\beta(H_k)$ is connected and locally compact and that the homomorphism ρ is continuous and open. Hence $\beta(H_k)$ is a central topological extension of $B(H_k)$; thus $\beta(H_k)$ is a Lie group. Moreover the unique simply connected Lie covering $\Omega(H_k)$ group of $\beta(H_k)$ is the semidirect product $\widetilde{Sp}_k(\mathbb{R}^k) \times \widetilde{H}_k$, where $\widetilde{Sp}_k$ and $\widetilde{H}_k$ are the unique simply connected covering groups of $Sp_k(\mathbb{R}^k)$ and H_k.

Following [26] we recall that every nonlinear quadratic character f on the space $\mathbb{R}^k$ is given by a unique symmetric matrix M_f on $\mathbb{R}^k$ so that $f(x) = \tau\,(\frac{1}{2}\,[x,M_f\, x])$ for all $x \in \mathbb{R}^k$. We then note that if determinant $(M_f) \neq 0$, then the map of multiplication $\varphi \to f\cdot\varphi$ for $\varphi \in S(\mathbb{R}^k)$ defines an automorphism of $S(\mathbb{R}^k)$, and we have the following formula:

$$(1-6)\quad (\varphi * f)\hat{}\,(z) = \int_{\mathbb{R}^k \times \mathbb{R}^k} \varphi(x)\, f(x-y)\tau([y,z])\, dx\, dy$$

$$= |\det(M_f)|^{-1/2}\, \theta\,(f)\, g_f\,(z)\, \hat{\varphi}\,(z)\,,$$

where $\theta(f)$ is a scalar factor of modulus one and $g_f(z) = [f(M_f{}^{-1}(z)\,)]^{-1}$ for all $z \in \mathbb{R}^k$. We know that $\theta(f)$ is an invariant of the signature of the quadratic form determined by M_f; explicitly if M_f is diagonalized to the form $D(\alpha_1,\dots,\alpha_k)$, then

$\theta(f) = \prod_{i=1}^{k} \gamma(\alpha_i)$, where $\gamma(\alpha) = \begin{cases} e^{\pi\sqrt{-1}/4} & \text{if } \alpha > 0. \\ e^{-\pi\sqrt{-1}/4} & \text{if } \alpha < 0. \end{cases}$ Then it is easy to see that $\theta(f)^2 = \left\{ \prod_{i=1}^{k} \left(\frac{\alpha_i}{-1}\right)_H \right\} \cdot \gamma(1)^{2k}$, where $\left(\frac{x}{y}\right)_H$ is the Hilbert symbol of the field $\mathbb{R}$.

Let $S(T) = \left[\begin{array}{c|c} 0 & -(T^t)^{-1} \\ \hline T & 0 \end{array}\right]$ for T any invertible symmetric matrix, and let $N(M) = \left[\begin{array}{c|c} I & M \\ \hline 0 & I \end{array}\right]$ with M any symmetric matrix. Then we recall that the group Sp_k is semisimple and that $P_k = \left\{ \left[\begin{array}{c|c} A & B \\ \hline 0 & (A^t)^{-1} \end{array}\right] \mid \det(A) \neq 0 \text{ and } A \cdot B^t \text{ symmetric} \right\}$ is a maximal parabolic subgroup of Sp_k. Then according to the Bruhat decomposition of Sp_k relative to P_k, we know that the unique "open cell" is the set $C_k = N(T)\left[\begin{array}{c|c} 0 & -I \\ \hline I & 0 \end{array}\right] P_k = \{N(T')S(A)N(T'')\}$ where T, T', and T'' range over all symmetric matrices, and A any invertible symmetric matrix. Indeed the set C_k can be characterized as the set of all $g = \left[\begin{array}{c|c} A & B \\ \hline C & D \end{array}\right]$ so that $\det(C) \neq 0$. In fact it is easy to see that C_k generates Sp_k with the condition $Sp_k = C_k^2 = \{g' \cdot g'' \mid g' \in C_k\}$ (see [15], p. 21). Moreover the Bruhat decomposition of $g = \left[\begin{array}{c|c} A & B \\ \hline C & D \end{array}\right] \in C_k$ is $N(AC^{-1})S(C)N(C^{-1}D)$.

We recall the construction of a local section for the map ρ above. We consider the subgroup $(P_k, 1)$ of $\beta(H_k)$ and the set $\tilde{C}_k = \{(g, f_g) \mid g \in C_k\}$; we use the identification of $\mathrm{Int}(H_k)$ to $\mathbb{R}^k \times \mathbb{R}^k$ above. Then we define the following unitary operators on $L^2(\mathbb{R}^k)$:

(i) $r((I,w))\varphi(x) = U_m(w,1)\varphi(x)$ for $w \in \mathrm{Int}(H_k)$,

(1 - 7)

(ii) $r\left(\left(\left[\begin{array}{c|c} A & 0 \\ \hline 0 & (A^t)^{-1} \end{array}\right], 1\right)\right)\varphi(x) = |\det(A)|^{1/2}\varphi(A(x)),$

(iii) $r((N(M),1))\varphi(x) = \tau(\frac{1}{2}[x,Mx])\varphi(x),$

(iv) $r((S(T),f_{S(T)}))\varphi(x) = |\det(T)|^{-1/2}\hat{\varphi}(-T^{-1}(x)).$

Then we can see that r can be extended to $(P_k,1)$ as an antigroup homomorphism of $(P_k,1)$ into $\beta(H_k)$. We then extend r to $\widetilde{C}_k$ in the following way. We take the Bruhat decomposition of the element g in C_k, transfer the above decomposition to $B(H_k)$ via the splitting section above, and then use the above formulae to define $r((g,f_g))$ as a composition of unitary operators. We thus have, with $g = \left[\begin{array}{c|c} A & B \\ \hline C & D \end{array}\right]$, $\det(C) \neq 0$, and $\varphi \in L^1(\mathbb{R}^k) \cap L^2(\mathbb{R}^k)$,

$$(1 - 8)\quad r((g,f_g))\varphi(w) = |\det(C)|^{-1/2} \int_{\mathbb{R}^k}\varphi(x)K(g,w,x)\,dx,$$

where the exponential kernel is

$$(1 - 9)\quad K(g,w,x) = \tau(\tfrac{1}{2}[C^{-1}Dx,x] + \tfrac{1}{2}[AC^{-1}w,w] - [C^{-1}x,w]).$$

We can then extend in the usual way $r((g,f_g))$ uniquely to a unitary operator on $L^2(\mathbb{R}^k)$.

Then we observe easily that

(1 - 10) $r((p,1))r((g,f_g)) = r((pg,f_{pg}))$ with $p \in P_k$ and $g \in C_k$ (observe that $pg \in C_k$),

(1 - 11) $r((g,f_g))r((p,1)) = r((gp,f_{gp}))$ with $p \in P_k$ and $g \in C_k$ (observe that $gp \in C_k$),

(1 - 12) $r((p,1))r((I,w)) = r((I,(p^t)^{-1}(w)))r((p,1))$ with $w \in \mathrm{Int}(H_k)$ and $p \in P_k$,

(1 - 13) $r((g,f_g))r((I,w)) = (f_{(g^t)^{-1}}(w))^m r((I,(g^t)^{-1}(w)))r((g,f_g))$ with $g \in C_k$ and $w \in \mathrm{Int}(H_k)$.

We then form the sets $X_k = \tilde{C}_k \cdot \mathrm{Int}(H_k)$ and $Y_k = (P_k,1)\cdot \mathrm{Int}(H_k)$ in $\beta(H_k)$. Noting that $\tilde{C}_k \cap \mathrm{Int}(H_k) = \{I\}$ and $(P_k,1) \cap \mathrm{Int}(H_k) = \{I\}$, we then can extend r to X_k and Y_k respectively by defining $r((g,f_g)(I,w)) = r((g,f_g))r((I,w))$ and $r((p,1)(I,w)) = r((p,1))r((I,w))$. Then we deduce from (1 - 12) and (1 - 13) that $r\circ\rho =$ the identity operator on X_k and Y_k.

At this point we recall from [15] that $r\colon X_k \to \beta(H_k)$ is a continuous map. We then let $g_1 = \left[\begin{array}{c|c} A_1 & B_1 \\ \hline C_1 & D_1 \end{array}\right]$, $g_2 = \left[\begin{array}{c|c} A_2 & B_2 \\ \hline C_2 & D_2 \end{array}\right] \in C_k$, so that

$g_1g_2 = \left[\begin{array}{c|c} A_3 & B_3 \\ \hline C_3 & D_3 \end{array}\right] \in C_k$. By computation, using the Bruhat decomposition of g_1, g_2 and $g_1\cdot g_2$, respectively, we derive the identity:

$$(1 - 14)\quad S(C_1)\cdot N(C_1^{-1}C_3C_2^{-1})\cdot S(C_2) = N(-A_1C_1^{-1})\{g_1g_2\}N(-C_2^{-1}D_2)$$
$$= N(A_3C_3^{-1} - A_1C_1^{-1})S(C_3)N(C_3^{-1}D_3 - C_2^{-1}D_2).$$

Then following [26] we embed the elements above in $B(H_k)$ via the splitting homomorphism and then apply r to both sides of the identity. We see that the unitary operator on the right hand side is equal to $\overline{\theta(h)}$ times the unitary operator on the left hand side, where h is the quadratic character $h(x) = \tau(\frac{1}{2}[x, C_1^{-1}C_3C_2^{-1}x])$. This gives

$$(1 - 15)\quad r((g_1g_2, f_{g_1g_2})) = \overline{\theta(h)}r((g_1,f_{g_1}))r((g_2,f_{g_2})).$$

Thus it follows that the map $R\colon \tilde{C}_k \times \tilde{C}_k \to S^1$ given by $R((g_1,f_{g_1}),(g_2,f_{g_2})) = \theta(h)$ satisfies the multiplier condition on $\tilde{C}_k$: if $g_i \in C_k$, $i = 1,2,3$, so that $g_ig_j \in C_k$ for $i \neq j$, then

$R((g_1,f_{g_1}),(g_2g_3,f_{g_2g_3}))R((g_2,f_{g_2}),(g_3,f_{g_3})) =$

$R((g_1,f_{g_1}),(g_2,f_{g_2}))R((g_1g_2,f_{g_1g_2}),(g_3,f_{g_3}))$.

Let S_k be the vector space of symmetric matrices in $\mathbb{R}^k$. It is easy to see that the map $T \to N(T)$ defines a vector space (and Lie group)

isomorphism of S_k to the unipotent group $\left\{\left[\begin{array}{c|c} I & S \\ \hline 0 & I \end{array}\right] \mid S \text{ symmetric}\right\}$.

Then for every $T \in S_k$ which is invertible, we associate to T two objects. First let $(Sp_k)_T = \{G \in Sp_k \mid GN(T)G^{-1} = N(T)\}$. Then $(Sp_k)_T$ is a closed Lie subgroup of Sp_k and by an easy matrix computation,

$(Sp_k)_T = \left\{\left[\begin{array}{c|c} A & B \\ \hline 0 & (A^t)^{-1} \end{array}\right] \mid A \in O(T),\ AB^t \text{ symmetric}\right\}$, with $O(T)$ the orthogonal group of T (i.e. all $A \in G\ell_k(\mathbb{R})$ so that $ATA^t = T$ and embedded in Sp_k as $\left[\begin{array}{c|c} A & 0 \\ \hline 0 & (A^t)^{-1} \end{array}\right]$). Moreover $(Sp_k)_T$ is the semidirect product $O(T) \times S_k$.

Secondly for $T \in S_k$ invertible, we consider the set $S\ell_2(T) =$ $\left\{\left[\begin{array}{c|c} \alpha\cdot I & \beta\cdot T \\ \hline \gamma\cdot T^{-1} & \delta\cdot I \end{array}\right] \mid \alpha,\beta,\gamma,\delta \in \mathbb{R} \text{ satisfying } \alpha\delta - \beta\gamma = 1\right\}$. Then we see that $S\ell_2(T)$ is a Lie subgroup of Sp_k and in fact is isomorphic to $S\ell_2(\mathbb{R})$ (thus the choice of notation here). The relation between $(Sp_k)_T$ and $S\ell_2(T)$ is given by

<u>Lemma 1.1</u> $O(T)' = \{X \in Sp_k \mid XZ = ZX, \forall Z \in O(T)\} = S\ell_2(T)$, <u>and</u> $S\ell_2(T)' = \{L \in Sp_k \mid LM = ML, \forall M \in S\ell_2(T)\} = O(T)$, <u>and finally</u> $S\ell_2(T) \cap O(T) = \{\pm I_{Sp_k}\}$.

Proof. By computation $\left[\begin{array}{c|c} A & B \\ \hline C & D \end{array}\right] \in O(T)'$ if and only if for all $X \in O(T)$, $AX = XA, D(X^t)^{-1} = (X^t)^{-1}, B(X^t)^{-1} = XB$, and $(X^t)^{-1}C = CX$. By the irreducibility of the standard representation of $O(T)$ in $\mathbb{C}^k$ (the complexification of $\mathbb{R}^k$) for all k, we have in the ring of complex matrices $M_k(\mathbb{C})$ that $A = \alpha\cdot I$, $D = \gamma\cdot I$, and $B = \beta\cdot T$, $C = \gamma\cdot T^{-1}$ with $\alpha,\beta,\gamma,\delta \in \mathbb{C}$. That $\left[\begin{array}{c|c} A & B \\ \hline C & D \end{array}\right] \in Sp_k$ determines both the relation $\alpha\delta - \beta\gamma = 1$ and that $\alpha,\beta,\gamma,\delta$ are real.

Conversely, if $\left[\begin{array}{c|c} A & B \\ \hline C & D \end{array}\right] \in S\ell_2(T)'$, then by computation, we have $A \cdot T = T \cdot D$ and $\gamma \cdot B \cdot T^{-1} = \beta \cdot T \cdot C$ for any pair of numbers γ, β. The second condition implies that $B = C = 0$, and hence $D = (A^t)^{-1}$; thus $A \in O(T)$.

Q.E.D.

Remark 1.1 It follows from Lemma 1.1 that the Lie subgroup $G_2(T)$ of Sp_k generated by $S\ell_2(T)$ and $O(T)$ is the product of $S\ell_2$ and $O(T)$, where the intersection $S\ell_2(T) \cap O(T) = \{\pm I\}$. We then write $G_2^{\natural}(T)$ as the direct product $S\ell_2(T) \times O(T)$ $((G,g), G \in S\ell_2(T), g \in O(T))$ and $G_2^{\natural}(T)$ covers $G_2(T)$ with Kernel equal to $\mathbb{Z}_2$.

We note that if T_1 and T_2 are invertible symmetric matrices, then $|\text{signature } (T_1)| = |\text{signature } (T_2)|$ if and only if $G_2(T_1)$ is Lie group isomorphic to $G_2(T_2)$.

We recall the theory of multipliers of locally compact groups (see [24] for relevant definitions and facts to be discussed below). Let S be any locally compact connected topological group and $\sigma\colon S \times S \to S^1$ a Borel multiplier on S. We form the group $S_\sigma = \{(x,\omega) \mid x \in S, \omega \in S^1\}$ where the group law is given by $(x,\omega)(x',\omega') = (xx', \sigma(x,x')\omega\omega')$. Then S_σ provided with the Weil topology becomes a locally compact connected topological group. We recall that the Weil topology is not the same as the product topology on $S \times S^1$; however if σ is a local C^∞ map at (e_s, e_s), the identity element of $S \times S$, then we know that there exists a neighborhood U_{e_s} of e_s in S so that the product topology on $U_{e_s} \times S^1$ coincides with the Weil topology on this set. Then noting that the map $(x,\omega) \to x$ of $S_\sigma \to S$ defines a central topological extension of S_σ by S^1, we have that if S is an analytic Lie group, then S_σ is an analytic Lie group (analytic structure compatible with the Weil group topology) and

$$\begin{array}{c} \omega \longrightarrow (I,\omega) \\ 1 \longrightarrow S^1 \longrightarrow S_\sigma \longrightarrow S \longrightarrow 1 \\ (x,\omega) \longrightarrow x \end{array} \tag{1 - 16}$$

is an exact sequence of analytic Lie groups and analytic group homomorphisms.

We then recall that if σ is locally C^∞ at (e_S, e_S), then the map $s \to (s,1)$ of $S \to S_\sigma$ is locally C^∞ at e_S of S. Under the condition that $\sigma(S \times S)$ generates a finite subgroup in S, we know that σ is locally C^∞ at (e_S, e_S) if and only if $\sigma = 1$ in a neighborhood of (e_S, e_S) of S. Thus under the latter condition there exists a "local homomorphism" of S to S_σ by $s \to (s,1)$ (see [4], p. 48). If S_σ' is the group generated by $(g,1)$ as $g \in S$, then S_σ' is a Lie subgroup of S_σ such that the identity component $(S_\sigma')_e$ of S_σ' is a finite to one topological covering of S and $[S_\sigma' : (S_\sigma')_e]$ is finite. We note in passing that set-theoretically $S_\sigma' = S \times T_\sigma$, with T_σ the group generated by $\sigma(S \times S)$, and that if T_σ is a group of prime order, then $S_\sigma' = (S_\sigma')_e$.

At this point we consider the group $S\ell_2(\mathbb{R}) = \left\{ \begin{bmatrix} a & b \\ c & d \end{bmatrix} \mid a,b,c,d \in \mathbb{R} \text{ and } ad - bc = 1 \right\}$. We recall the theory of n-fold coverings of $S\ell_2(\mathbb{R})$. Define

$$\theta_1(g) = \begin{cases} c(g) & \text{if } c(g) \neq 0 \\ d(g) & \text{if } c(g) = 0 \end{cases} \tag{1 - 17}$$

Then consider the map $A: S\ell_2(\mathbb{R}) \times S\ell_2(\mathbb{R}) \to \mathbb{Z}_2$, given by

$$A(g,h) = \left(\frac{\theta_1(g)}{\theta_1(h)} \right)_H \left(\frac{-\theta_1(g)^{-1}\theta_1(h)}{\theta_1(gh)} \right)_H \tag{1 - 18}$$

From [17] we know that A is a <u>nontrivial</u> Borel multiplier on $S\ell_2(\mathbb{R})$ and that $A = 1$ near the origin $(e_{S\ell_2}, e_{S\ell_2})$ in $S\ell_2(\mathbb{R}) \times S\ell_2(\mathbb{R})$.

We note that for any $g = \begin{bmatrix} a & b \\ c & d \end{bmatrix}$, $c \neq 0$ and $p = \begin{bmatrix} x & y \\ 0 & x^{-1} \end{bmatrix}$ with $x > 0$, that $A(g,p) = A(p,g) = 1$. Also if p_1 and $p_2 \in P_1$, then with $x(p_i) > 0$, $A(p_1 p_2) = A(p_2, p_1) = 1$.

Then we can apply the above considerations to $S = S\ell_2(T)$ and $\sigma = A$. We then consider the nontrivial two-fold covering $\widetilde{S\ell}_2(T)$ of $S\ell_2(T)$.

To save notation when the form T is fixed, we write the elements of $\widetilde{S\ell}_2(T)$ as $\left(\begin{pmatrix} a & b \\ c & d \end{pmatrix}, \epsilon\right)$ with $\epsilon = \pm 1$ with the group law defined as above.

Theorem 1.2 *For the elements* $\left(\begin{bmatrix} a & y \\ 0 & a^{-1} \end{bmatrix}, \epsilon\right)$ *and* $\left(\begin{bmatrix} a & b \\ c & d \end{bmatrix}, \epsilon\right)$ *with* $c \neq 0$, *we define for* $\varphi \in L^2(\mathbb{R}^k) \cap L^1(\mathbb{R}^k)$:

$$(1 - 19)\quad \pi_{\mathfrak{M}}\left(\left[\begin{matrix} a & y \\ 0 & a^{-1} \end{matrix}\right], \epsilon\right)\varphi(x) = (\mathrm{sgn}(\epsilon))^k f_k(a)|a|^{k/2} e^{\pi i y a T(x)} \varphi(a\cdot x)$$

$$(1 - 20)\quad \pi_{\mathfrak{M}}\left(\begin{pmatrix} a & b \\ c & d \end{pmatrix}, \epsilon\right)\varphi(x) = (\mathrm{sgn}(\epsilon))^k c_k(g)|\det(T)|^{-1/2}|c|^{-k/2}$$

$$\int_{\mathbb{R}^k} \varphi(w) e^{\frac{\pi i}{c}(aT^{-1}(x) + dT^{-1}(w) - 2[w, M_T^{-1}x])} dw$$

with

$$(1 - 21)\quad f_k(a) = \begin{cases} \left(\frac{a}{-1}\right)_H^{\delta(T)} & \text{if } k \text{ is even and } \delta(T) = \begin{cases} 0 & \text{if } \mathrm{sgn}(T) \equiv 0\ (4) \\ 1 & \text{if } \mathrm{sgn}(T) \equiv 2\ (4) \end{cases} \\ \begin{cases} 1 & \text{if } a > 0 \\ e^{\frac{\pi i}{2}\mathrm{sgn}(T)} & \text{if } a < 0 \end{cases} & \text{if } k \text{ is odd} \end{cases}$$

and $\mathrm{sgn}(T)$ = *the signature of* T. *Also*

$$(1-22)\quad c_k(g) = \begin{cases} e^{\frac{\pi i}{2}\operatorname{sgn}(T)}\,\theta(c\cdot T) & \text{if } k \text{ is even} \\ \left(\frac{c}{-1}\right)_H e^{-\frac{\pi i}{2}\operatorname{sgn}(T)}\,\theta(c\cdot T) & \text{if } k \text{ is odd}, \end{cases}$$

where $c = c(g)$. Then the map $\pi_{\mathfrak{M}}$ extends to a continuous unitary representation of the group $\widetilde{S\ell}_2(T)$ into Unit $(L^2(\mathbb{R}^k))$. Moreover

$\pi_{\mathfrak{M}}(\widetilde{S\ell}_2(T)) \subseteq \beta(H_k)$ and $\rho\circ\pi_{\mathfrak{M}}((g,\epsilon)) = S_\psi\left(\left[\begin{array}{c|c} \alpha\cdot I & \beta\cdot T \\ \hline \gamma T^{-1} & \delta I \end{array}\right]\right)$

with $g = \begin{pmatrix} \alpha & \beta \\ \gamma & \delta \end{pmatrix}$ any element of $S\ell_2(\mathbb{R})$ and S_ψ the splitting section to ψ given above.

Proof. We first observe that if $p = \begin{bmatrix} a & y \\ 0 & a^{-1} \end{bmatrix}$, then

$$(1-23)\quad \pi_{\mathfrak{M}}((p,\epsilon)) = (\operatorname{sgn}(\epsilon))^k f_k(a) r\left(S_\psi\left(\left[\begin{array}{c|c} a\cdot I & y\cdot T \\ \hline 0 & a^{-1}\cdot I \end{array}\right]\right)\right),$$

and if $g = \begin{bmatrix} a & b \\ c & d \end{bmatrix}$ with $c \neq 0$,

$$(1-24)\quad \pi_{\mathfrak{M}}((g,\epsilon)) = (\operatorname{sgn}(\epsilon))^k c_k(g) r\left(S_\psi\left(\left[\begin{array}{c|c} a\cdot I & b\cdot T \\ \hline c\cdot T^{-1} & d\cdot I \end{array}\right]\right)\right).$$

Thus $\pi_{\mathfrak{M}}((g,\epsilon))$ is unitary for any $g \in S\ell_2(\mathbb{R})$ and we have that

$\rho\circ\pi_{\mathfrak{M}}((g,\epsilon)) = r\left(S_\psi\left(\left[\begin{array}{c|c} a\cdot I & b\cdot T \\ \hline c\cdot T^{-1} & d\cdot I \end{array}\right]\right)\right)$ for arbitrary $g \in S\ell_2(\mathbb{R})$.

Thus it suffices to check that $\pi_{\mathfrak{M}}$ is a group homomorphism of $\widetilde{S\ell}_2(T)$ to $\beta(H_k)$. Since $S\ell_2(\mathbb{R}) = \left\{\begin{bmatrix} a & b \\ c & d \end{bmatrix} \mid c \neq 0\right\} \cup \left\{\begin{bmatrix} a & b \\ 0 & a^{-1} \end{bmatrix} \mid a \neq 0\right\}$

we must check all possible products of elements in both sets.

First assume that p_1, p_2, and $p_1 p_2 \in P_1$. Then noting that $r\colon (P_k, 1) \to \beta(H_k)$ is an antigroup homomorphism and using the relation between

r and $\pi_{\mathfrak{M}}$, we must have $f_k(a_1)f_k(a_2) = \{A(p_1,p_2)\}^k f_k(a_1 \cdot a_2)$ with $p_i = \begin{bmatrix} a_i & * \\ 0 & a_i^{-1} \end{bmatrix}$. By routine computation we have this relation.

Then let $p_1 \in P_1$ and $g_1 \in C_1$ so that p_1g_1 and $g_1p_1 \in C_1$. Using (1 - 11) above and the relation between r and $\pi_{\mathfrak{M}}$, we must have $f_k(a_1)c_k(g_1) = c_k(p_1g_1)\{A(p_1,g_1)\}^k$ and $c_k(g_1)f_k(a_1) = c_k(g_1p_1)\{A(g_1,p_1)\}^k$. This again follows by computation.

Then let g_1, g_2, and $g_1g_2 \in C_1$. From (1 - 15) we must have the relation $c_k(g_1)c_k(g_2) = \{A(g_1,g_2)\}^k c_k(g_1g_2)\theta(\frac{c(g_1g_2)}{c(g_1)c(g_2)}\, T)$, which again follows by checking from the definitions above.

Finally we let $g_1, g_2 \in C_1$ so that $g_1 \cdot g_2 \in P_1$. This implies that $c(g_1g_2) = 0$ and $d(g_1g_2) = -\frac{c(g_1)}{c(g_2)}$. Thus we must have $c_k(g_1)c_k(g_2) = \{A(g_1,g_2)\}^k f_k(-\frac{c(g_2)}{c(g_1)})$, which again follows by computation.

We recall that since C_1 is an open subset of $S\ell_2(\mathbb{R})$ and generates $S\ell_2(\mathbb{R})$, it follows that it suffices to prove the continuity of $\pi_{\mathfrak{M}}$ on the set (C_1,ϵ) with $\epsilon = \pm 1$. If $g \in C_1$ then the map $(g,\epsilon) \to r \circ \rho \circ \pi_{\mathfrak{M}}(g,\epsilon)$ is a continuous map of (C_1,ϵ) to $\beta(H_k)$ (since r is continuous). Then the map $(g,\epsilon) \to \{\mathrm{sgn}(\epsilon)\}^k c_k(g)$ with $g \in C_1$ is continuous. Indeed the set C_1 has two connected components in $S\ell_2(\mathbb{R})$ depending on the sign of $c(g)$, and the map C_1 to $S^1, g \to c_k(g)$ is <u>constant valued</u> on each component. Moreover noting that the map $(g,\omega) \to \omega^k$ (k integer) of $(S\ell_2)_A$ to S^1 is continuous in the Weil group topology on $(S\ell_2)_A$ and recalling that $\widetilde{S\ell}_2(T)(=\{(S\ell_2)_A\}')$ is a Lie subgroup of $(S\ell_2)_A$, we deduce that $(g,\epsilon) \to (\mathrm{sgn}(\epsilon))^k$ is continuous on the set (C_1,ϵ). Thus $\pi_{\mathfrak{M}}$ is a continuous map.

Q.E.D.

<u>Remark 1.2</u> If k is even then Kernel $(\pi_{\mathfrak{M}}(\widetilde{S\ell}_2(T))) = \{(I,\epsilon) \mid \epsilon = \pm 1\}$. Thus $\pi_{\mathfrak{M}}$ yields a continuous linear unitary representation of $S\ell_2(T)$ itself. However if k is odd then Kernel $(\pi_{\mathfrak{M}}) = \{(I,1)\}$ and hence $\pi_{\mathfrak{M}}$ does not in this case give a representation of $S\ell_2(T)$ but only of the covering $\widetilde{S\ell}_2(T)$.

Then we consider the continuous unitary representation of the Lie group $\widetilde{G}_2(T) = \widetilde{S\ell}_2(T) \times O(T)$ (a two-fold covering of $G_2^{\flat}(T)$) given by $\widetilde{\pi}_{\mathfrak{M}}((G,g)) = \pi_{\mathfrak{M}}(G)r((g,f_g)) = r((g,f_g))\pi_{\mathfrak{M}}(G)$ with $G \in \widetilde{S\ell}_2(T)$ and $g \in O(T)$.

We consider the symmetric matrix $T = \left[\begin{array}{c|c} T_1 & 0 \\ \hline 0 & T_2 \end{array}\right]$ where T_1 and T_2 are $r \times r$ and $(k-r) \times (k-r)$ symmetric matrices, respectively. T is invertible if and only if T_1 and T_2 are both invertible.

We let $F_1 = L^2(\mathbb{R}^r)$ and $F_2 = L^2(\mathbb{R}^{k-r})$. Then we can form the <u>Hilbert space tensor product</u> $F_1 \hat{\otimes} F_2$ (the completion of the algebraic tensor product $F_1 \otimes F_2$ with the scalar product $(f_1 \otimes f_2, g_1 \otimes g_2) = (f_1,g_1)_{F_1}(f_2,g_2)_{F_2}$). Then we can define the unitary representation of the direct product $\widetilde{S\ell}_2(T_1) \times \widetilde{S\ell}_2(T_2)$ on $F_1 \hat{\otimes} F_2$ by $\sigma(X,X') = \pi_{\mathfrak{M}}^r(X) \hat{\otimes} \pi_{\mathfrak{M}}^{k-r}(X')$ (defined on $F_1 \otimes F_2$ and then extended uniquely to the completion).

We let D_T be the diagonal subgroup of the direct product $\widetilde{S\ell}_2(T_1) \times \widetilde{S\ell}_2(T_2)$ given by $((x,\epsilon)_{T_1}, (x,\epsilon)_{T_2})$. Then D_T is a Lie group isomorphic to $S\ell_2(\mathbb{R})$. We then restrict the representation of σ to D_T.

<u>Corollary to Theorem 1.1</u> <u>The representation</u> $\sigma|_{D_T}$ <u>is unitarily equivalent to the representation</u> $\pi_{\mathfrak{M}}$ <u>of</u> $\widetilde{S\ell}_2(T)$ <u>in</u> $L^2(\mathbb{R}^k)$.

Proof. By elementary measure theory we know that $F_1 \hat{\otimes} F_2$ is unitarily equivalent to $L^2(\mathbb{R}^k)$ via the map $f \otimes g \to f \cdot g$ on the set of tensors. Thus it suffices to show "$\to$" intertwines $\sigma|_{D_T}$ and $\pi_{\mathfrak{M}}$. Then for $p =$

$\left(\begin{bmatrix} a & b \\ 0 & a^{-1} \end{bmatrix}, \epsilon \right)$ we have

$$(1-25) \quad \sigma(p)(f \otimes g) \to \left\{ \frac{f_r(a) f_{k-r}(a)}{f_k(a)} \right\} \pi_{\mathfrak{M}}(p)(f \cdot g) \quad ,$$

and if $g' = \left(\begin{bmatrix} a & b \\ c & d \end{bmatrix}, \epsilon \right)$ with $c \neq 0$ we have

$$(1-26) \quad \sigma(g')(f \otimes g) \to \left\{ \frac{c_r(g) c_{k-r}(g)}{c_k(g)} \right\} \pi_{\mathfrak{M}}(g')(f \cdot g) \quad .$$

By using the fact that $\theta(cT_1)\theta(cT_2) = \theta(cT)$ (θ a character on the Witt group of quadratic forms), we have that in both (1 - 25) and (1 - 26) the expressions $\{\ldots\ldots\} = 1$. Thus "$\to$" intertwines $\sigma|_{D_T}$ and $\pi_{\mathfrak{M}}$ of $\widetilde{S\ell}_2(T)$.

Q.E.D.

§2. GEOMETRIC AND ALGEBRAIC STRUCTURE OF ORTHOGONAL GROUP

Let Q be a nondegenerate quadratic form on $\mathbb{R}^k$. Then there exists an orthonormal basis of $\mathbb{R}^k$ relative to Q given by $\mathbb{R}^k \cong \langle e_1,\ldots,e_a\rangle \perp \langle e_{a+1},\ldots,e_k\rangle$ with $Q(e_i,e_j) = \delta_{ij}$ for $1 \leq i \leq a$, and $1 \leq j \leq a$, $Q(e_i,e_j) = -\delta_{ij}$ for $a+1 \leq i \leq k$ and $a+1 \leq j \leq k$; here $\perp$ means orthogonal relative to Q. Thus Q has signature (a,b) (with $a + b = k$, a = number of plus signs, b = number of negative signs). If we let $v_a = \frac{1}{\sqrt{2}}(e_1 - e_{a+1})$, $\tilde{v}_a = \frac{1}{\sqrt{2}}(e_1 + e_{a+1}),\ldots,$ $v_{k-1} = \frac{1}{\sqrt{2}}(e_b - e_k)$, $\tilde{v}_{k-1} = \frac{1}{\sqrt{2}}(e_b + e_k)$, then $\mathbb{R}^k \cong \langle v_a,\tilde{v}_a\rangle\perp\ldots\langle v_{k-1},\tilde{v}_{k-1}\rangle \perp \langle e_{b+1},\ldots,e_a\rangle$ where the $\langle v_i,\tilde{v}_i\rangle$, $a \leq i \leq k\text{-}1$ span a hyperbolic plane relative to Q, with $Q(v_i,v_i) = Q(\tilde{v}_i,\tilde{v}_i) = 0$ and $Q(v_i,\tilde{v}_i) = 1$.

Let $O(Q)$ be the orthogonal group of Q. Then by the Theorem of Witt the orbits of $O(Q)$ in $\mathbb{R}^k$ are exactly the sets $\Gamma_t = \{X \mid Q(X,X) = t\}$, $t \neq 0$ and $\Gamma_0 - \{0\} = \{X \neq 0 \mid Q(X,X) = 0\}$ and $\{0\}$. Then the isotropy group of a point $Z \in \Gamma_t$ $(t \neq 0)$ is $O(Q)_Z = \{g \in O(Q) \mid g(Z) = Z\}$, which is exactly isomorphic to the orthogonal group of the restriction of Q to the $k - 1$ dimensional subspace of $\mathbb{R}^k$ given by $(Z)^{\perp} = \{Y \mid Q(Y,Z) = 0\}$. Thus if $t > 0$ or $t < 0$ then $O(Q)_Z$ has signature $(a - 1,b)$ or $(a,b - 1)$. We say an element $X \in \mathbb{R}^k$ is elliptic, parabolic, or hyperbolic according as $Q(X,X) < 0$, $= 0$, or > 0.

We define $K = \{g \in O(Q) \mid g$ leaves stable the subspace $\langle e_1,\ldots,e_a\rangle\}$. Then K is a compact subgroup of $O(Q)$ isomorphic to the direct product $O(a) \times O(b)$ (where $O(s)$ is the orthogonal group of the standard positive definite quadratic form $\sum_{i=1}^{s} x_i^2$ on $\mathbb{R}^s$).

If $b > 0$ we recall that the group $O(Q)$ has four connected components, each component containing one component of the group K. Indeed the components of $O(Q)$ are constructed as follows. Let θ_1 and θ_2 be the involutions $\theta_1(e_1) = -e_1, \theta_1 = I$ on $(e_1)^\perp$ and $\theta_2(e_{a+1}) = -e_{a+1}$, $\theta_2 = I$ on $(e_{a+1})^\perp$. Thus let $S = \{I, \theta_1, \theta_2, \theta_1\theta_2\}$ be the Klein Viergruppe generated by θ_1 and θ_2. Then it is easy to see that the map $S \to O(Q)/O^+(Q)$ is an isomorphism where $O^+(Q)$ is the connected component of the identity in $O(Q)$.

Let $A_s(r)$, $r \in \mathbb{R}^*$ $(s = a, \ldots, k-1)$ be the one dimensional subgroup of $O(Q)$ given by

$$\begin{cases} v_s \to r v_s \\ \tilde{v}_s \to r^{-1} v_s \end{cases} \quad \text{and } I \text{ on } \langle v_s, \tilde{v}_s \rangle^\perp.$$

We let $A = \prod_{s=a}^{k-1} A_s(\mathbb{R}^*)$.

We let $N_s(\mathbb{R}^{k-2}) = N_s(s = a, \ldots, k-1)$ be the unipotent subgroup of $O(Q)$ given by

$$\begin{cases} v_s \to v_s \\ \tilde{v}_s \to \tilde{v}_s + Y - \frac{1}{2} Q(Y,Y) v_s \end{cases} \quad \text{and } Z \to Z - Q(Z,Y) v_s,$$

$Z \in \langle v_s, \tilde{v}_s \rangle^\perp$ for all $Y \in \langle v_s, \tilde{v}_s \rangle^\perp$. Then $N_s(\mathbb{R}^{k-2}) = N_s$ is a $k-2$ dimensional connected Abelian Lie group.

Then we observe that the isotropy group of the element $v_s \in \Gamma_o - \{0\}$ is given by the semidirect product $M_s \times N_s(\mathbb{R}^{k-2})$, where $M_s = O(Q)_{\langle v_s, \tilde{v}_s \rangle^\perp}$ is the orthogonal group of the form Q restricted to the $k-2$ dimensional subspace $\langle v_s, \tilde{v}_s \rangle^\perp$.

We recall that K is a maximal compact subgroup of $O(Q)$ and that all maximal compact subgroups of $O(Q)$ are conjugate by an inner automorphism of $O(Q)$. Also we recall that the homogeneous space $(O(Q)/K) = (O^+(Q)/K \cap O^+(Q))$ is connected and forms a Riemannian symmetric space with a realization to be given below.

The Grassmann manifold of all q dimensional subspaces of $\mathbb{R}^k$ is given by the set $G^q(\mathbb{R}^k) = \{(\ell_1 \wedge \ldots \wedge \ell_q) \in \mathbb{P}(\wedge^q(\mathbb{R}^k)) \mid (\ell_1, \ldots, \ell_q)$

is a linearly independent set of vectors in $\mathbb{R}^k\}$ where $\wedge^q(\mathbb{R}^k) =$ the q-th exterior product of $\mathbb{R}^k$ and $\mathbb{P}(\wedge^q(\mathbb{R}^k))$ is the real projective space attached to the real vector space $\wedge^q(\mathbb{R}^k)$. Then $\mathrm{Aut}(\mathbb{R}^k)$ acts transitively on $G^q(\mathbb{R}^k)$, and thus we put the differential and topological structure on $G^q(\mathbb{R}^k)$ viewed as a homogeneous space of the Lie group $\mathrm{Aut}(\mathbb{R}^k)$. Then $O(Q) \subseteq \mathrm{Aut}(\mathbb{R}^k)$ acts on $G^q(\mathbb{R}^k)$ and has a finite number of orbits. Indeed if Y is any q dimensional subspace of $\mathbb{R}^k$, then $Y \cong S_Y \oplus \mathrm{Rad}_Q(Y)$, where $\mathrm{Rad}_Q(Y) = \{Z \in Y \mid Q(Z,L) = 0$ for all $L \in Y\}$ and Q restricted to S_Y is nondegenerate. Thus again by Witt's Theorem, Z_1 and $Z_2 \in G^q(\mathbb{R}^k)$ are $O(Q)$ equivalent (conjugate) if and only if $\dim \mathrm{Rad}_Q(Z_1) = \dim \mathrm{Rad}_Q(Z_2)$ and $\mathrm{signature}_Q(S_{Z_1}) = \mathrm{signature}_Q(S_{Z_2})$.

Let $\Gamma_{(r,s,t)} = \{Z \in G^q(\mathbb{R}^k) \mid \dim \mathrm{Rad}_Q(Z) = t$ and $\mathrm{signature}_Q(S_Z) = (r,s)\}$. Then for $0 \le t \le b$, $r + s = q - t$, $r \le a$, and $s \le b$, the set of $\Gamma_{(r,s,t)}$ determine the set of distinct $O(Q)$ orbits in $G^q(\mathbb{R}^k)$. (Some $\Gamma_{(r,s,t)}$ may be empty.) Then relative to the basis $\langle e_1,\dots,e_k\rangle$ above, we see that any element Y of $\Gamma_{(r,s,t)}$ is $O(Q)$ conjugate to a subspace Y_o of the form:

(2 - 1)

(i) if $t = 0$ then $Y_o \cong \langle e_1,\dots,e_r\rangle \perp \langle e_{a+1},\dots,e_{a+s}\rangle$ if $r \ge 1$ and $s \ge 1$; $Y_o = \langle e_1,\dots,e_r\rangle$ if $r \ge 1$ and $s = 0$; $Y_o = \langle e_{a+1},\dots,e_{a+s}\rangle$ if $r = 0$ and $s \ge 1$.

(ii) if $t \ge 1$ then $Y_o = \langle v_a,\dots v_{a+t-1}\rangle \perp S_s \perp \tilde{S}_r$, where S_s is the subspace $\{0\}$ if $s = 0$ and $S_s \cong \langle e_{a+t+1},\dots,e_{a+t+s}\rangle$ if $s \ge 1$ and $\tilde{S}_r$ is the subspace $\{0\}$ if $r = 0$ and $\tilde{S}_r \cong \langle e_{t+1},\dots,e_{t+r}\rangle$ if $r \ge 1$.

We note that since $G^q(\mathbb{R}^k)$ is a homogeneous space of the group $\mathrm{Aut}(\mathbb{R}^k)$, that each $O(Q)$ orbit in $G^q(\mathbb{R}^k)$ is a submanifold of $G^q(\mathbb{R}^k)$.

Lemma 2.1 Relative to the differential structure given on $G^q(\mathbb{R}^k)$ *the sets* $\Gamma_{(r,s,0)}$ *as* $r + s = q$ *with* $r \leq a$ *and* $s \leq b$ *are open connected submanifolds of* $G^q(\mathbb{R}^k)$. *Moreover every open orbit of* $O(Q)$ *in* $G^q(\mathbb{R}^k)$ *is of the type* $\Gamma_{(r,s,0)}$.

Proof. Since each $\Gamma_{(r,s,t)}$ is a submanifold of $G^q(\mathbb{R}^k)$, we see that $\Gamma_{(r,s,t)}$ is open in $G^q(\mathbb{R}^k)$ if and only if $\dim \Gamma_{(r,s,t)} = \dim G^q(\mathbb{R}^k) = \frac{1}{2}k(k-1) - \frac{1}{2}\{q(q-1) + (k-q)(k-q-1)\}$. We then observe that if $Y = (\ell_1 \wedge \ldots \wedge \ell_q)$ is any q dimensional subspace where Q is nondegenerate, then $\mathbb{R}^k = Y \oplus Y^\perp$ and hence the isotropy group of Y is the direct product $O(Q)_Y \times O(Q)_{Y^\perp}$. Thus the orbit $O(Q)\cdot Y = \Gamma_{(r,s,0)}$ is open in $G^q(\mathbb{R}^k)$, where $(r,s) = \text{signature}_Q(Y)$. If $Y \in \Gamma_{(r,s,t)}$ with $t > 0$, then we let $S_Y = \langle x_1, \ldots, x_{r_1} \rangle$ and $\text{Rad}_Q(Y) = \langle y_1, \ldots, y_{r_2} \rangle$ be basis sets $(r + s = r_1,\ t = r_2)$. Then we can find sets $\langle \bar{y}_1, \ldots, \bar{y}_{r_2} \rangle$ and $\langle \bar{x}_1, \ldots, \bar{x}_{r_3} \rangle$ in $\mathbb{R}^k$ so that relative to Q, $\mathbb{R}^k = \langle x_1, \ldots, x_{r_1} \rangle \perp \langle y_1, \bar{y}_1 \rangle \perp \ldots \perp \langle y_{r_2}, \bar{y}_{r_2} \rangle \perp \langle \bar{x}_1, \ldots, \bar{x}_{r_3} \rangle$ with $\langle y_i, \bar{y}_i \rangle$ a hyperbolic plane relative to Q, and Q nondegenerate on $\langle x_1, \ldots, x_{r_1} \rangle$ and $\langle \bar{x}_1, \ldots, \bar{x}_{r_3} \rangle$. Then relative to this basis Q has the form $(2-2)$ and any element $g \in O(Q)_Y$ has the form $(2-3)$:

$$(2-2)\quad \begin{array}{c} r_1 \\ 2r_2 \\ r_3 \end{array} \left[\begin{array}{c|cc|c} M_1 & 0 & & 0 \\ \hline 0 & 0 & I & 0 \\ & I & 0 & \\ \hline 0 & 0 & & M_2 \end{array}\right] \qquad (2-3)\quad \begin{array}{cccc} r_1 & r_2 & r_2 & r_3 \end{array} \left[\begin{array}{c|c|c|c} \Omega_1 & 0 & \Omega_3 & 0 \\ \hline A_1 & A_2 & A_3 & A_4 \\ \hline & & (A_2^t)^{-1} & 0 \\ \cline{3-4} & 0 & T_3 & T_4 \end{array}\right]$$

with $\Omega_1 M_1 \Omega_1^t = M_1, T_4 M_2 T_4^t = M_2, A_2$ invertible, A_1 and A_4 any real matrices, and $\Omega_3 = -\Omega_1 M_1 A_1^t (A_2^t)^{-1}$, $T_3 = -T_4 M_2 A_4^t (A_2^t)^{-1}$ and A_3 satisfying $A_3 A_2^t + A_2 A_3^t = -(A_1 M_1 A_1^t + A_4 M_2 A_4^t)$. Hence we have $\dim (O(Q)\cdot Y) =$ $\frac{1}{2} r_2(r_2 - 1) + r_2(r_1 + r_3) + r_1 r_3 \leq r_2^2 + r_2(r_1 + r_3) + r_1 r_3 =$ $\dim G^q(\mathbb{R}^k)(q = r_1 + r_2)$. Then we have equality if and only if $r_2 = 0$.

By choosing $Y_o \in \Gamma_{(r,s,0)}$ as in (2 - 1), we see that θ_1 and θ_2 belong to the isotropy group $O(Q)_{Y_o}$, and hence the orbit $\Gamma_{(r,s,0)}$ is connected.

Q.E.D.

Remark 2.1 The symmetric space $O(Q)/K$ can be realized as either the open orbit $\Gamma_{(a,0,0)}$ in $G^a(\mathbb{R}^k)$ or the open orbit $\Gamma_{(0,b,0)}$ in $G^b(\mathbb{R}^k)$.

We record at this point the symmetric space data of the pair $O(Q)$, K. Relative to the basis $\langle v_a,\ldots,v_{k-1},\tilde{v}_a,\ldots,\tilde{v}_{k-1},e_{b+1},\ldots,e_a\rangle$ the group A, a maximal split torus, is given by (2 - 4):

$$(2-4)\qquad \begin{matrix} b \\ b \\ k-2b \end{matrix}\begin{bmatrix} D(r) & 0 & 0 \\ 0 & D(r)^{-1} & 0 \\ 0 & 0 & I \end{bmatrix}, \quad D(r) = \begin{bmatrix} r_1 & & 0 \\ & \ddots & \\ 0 & & r_b \end{bmatrix}.$$

Then $M = \mathrm{Ctz}_K(A) = \{\gamma \in K \mid \gamma a \gamma^{-1} = a \text{ for all } a \in A\}$ is the product

$$S \cdot \left\{ \begin{bmatrix} I & 0 & 0 \\ 0 & I & 0 \\ 0 & 0 & V \end{bmatrix} \;\middle|\; V \in O(k-2b) \right\}.$$ The Iwasawa U^+ unipotent group is given by the set of all

$$\begin{matrix} b \\ b \\ k-2b \end{matrix}\begin{bmatrix} M_1 & A & B \\ 0 & (M_1^t)^{-1} & 0 \\ 0 & C & I \end{bmatrix},$$ with M_1 upper triangular

with 1 on the diagonal, B an arbitrary $b \times (k-2b)$ real matrix, $C = -B^t \cdot (M_1^t)^{-1}$, and A satisfying $AM_1^t + M_1A^t = -BB^t$. Then we recall the Iwasawa decomposition $O(Q) = K \cdot A \cdot U^+$ with $K \cap AU^+ = \{I\}$ and $M \cdot AU^+ = \text{Normalizer}_{O(Q)}(U^+)$. In fact the map $K \times A \times U^+ \to O(Q)$ given by $(k,a,n) \to k \cdot a \cdot n$ is a diffeomorphism of differentiable manifolds.

From above we note that the group $M_s \cdot A_s$ normalizes N_s, and hence we can form the Lie group $P_s = M_sA_sN_s$.

Let $\mathfrak{G}$, $\mathfrak{K}$, $\mathfrak{A}$, $\mathfrak{U}^+$, $\mathfrak{M}$, $\mathfrak{A}_a$, $\mathfrak{N}_a$, $\mathfrak{M}_a$, and $\mathfrak{P}_a$ be the Lie algebras of $O(Q)$, K, A, U^+, M, A_a, N_a, M_a, and P_a respectively. Let Δ be the set of restricted roots of $\mathfrak{A}$ on $\mathfrak{G}$. By easy and straight-forward computation, we see that the set Δ_+ of positive roots has three different types:

I. $\lambda_i - \lambda_j$, $1 \leq i < j \leq b$, where the corresponding root subgroup

$$U^+_{\lambda_i - \lambda_j} \text{ is } \begin{bmatrix} I+se_{ij} & 0 & \\ & I-se_{ji} & \\ 0 & & I \end{bmatrix} \begin{matrix} \updownarrow b \\ \updownarrow b \\ \updownarrow k-2b \end{matrix}, \quad s \in \mathbb{R}.$$

II. λ_i, $1 \leq i \leq b$, where the corresponding root subgroup $U^+_{\lambda_i}$

$$\text{is} \quad \begin{matrix} b \updownarrow \\ b \updownarrow \\ k-2b \updownarrow \end{matrix} \begin{bmatrix} I & A_i & B_i \\ 0 & I & 0 \\ 0 & C_i & I \end{bmatrix} \quad \text{where } B_i = \begin{bmatrix} & 0 & \\ * & \cdots\cdots & * \\ & 0 & \end{bmatrix}, \text{ the}$$

i-th row being a $(k - 2b)$ dimensional real vector space and

$$A_i = \frac{1}{2} \begin{bmatrix} d_1^i & & 0 \\ & \ddots & \\ 0 & & d_b^i \end{bmatrix} \quad \text{with } d_j^i = \delta_{ij} \text{ trace } (B_iB_i^t) \text{ and}$$

$C_i = -B_i^t$.

III. $\lambda_i + \lambda_j$, $1 \leq i < j \leq b$, where the corresponding root subgroup

$$U^+_{\lambda_i + \lambda_j} \quad \text{is} \quad \begin{matrix} b \updownarrow \\ b \updownarrow \\ k\text{-}2b \updownarrow \end{matrix} \begin{bmatrix} I & s(e_{ij}-e_{ji}) & 0 \\ 0 & I & 0 \\ 0 & 0 & I \end{bmatrix}, \quad s \in \mathbb{R}.$$

Then we recall that $\mathfrak{G} = \mathfrak{K} \oplus \mathfrak{A} \oplus \mathfrak{U}^+$; $\mathfrak{U}^+ = \sum_{1\leq i<j\leq b} \mathfrak{U}^+_{\lambda_i-\lambda_j} \oplus$ $\sum_{1\leq i<j\leq b} \mathfrak{U}^+_{\lambda_i+\lambda_j} \oplus \sum_{i=1}^{b} \mathfrak{U}^+_{\lambda_i}$. Then we know that $\mathfrak{P}_a$ is a parabolic algebra of $\mathfrak{G}$ where $\mathfrak{P}_a = \widetilde{\mathfrak{M}}_a \oplus U(\mathfrak{P}_a)$, with $U(\mathfrak{P}_a) = \mathfrak{N}_a$, the unipotent radical of $\mathfrak{P}_a$ given by $\sum_{1<j\leq b} \mathfrak{U}^+_{\lambda_1-\lambda_j} \oplus \sum_{1<j\leq b} \mathfrak{U}^+_{\lambda_1+\lambda_j} \oplus \mathfrak{U}^+_{\lambda_1}$ and $\widetilde{\mathfrak{M}}_a$ a reductive subalgebra of $\mathfrak{P}_a$ so that $\widetilde{\mathfrak{M}}_a = \mathfrak{M}_a \oplus \mathfrak{A}_a$.

The simple roots of Δ_+ are $\{\alpha_1,\ldots,\alpha_b\}$ with $\alpha_i = \lambda_i - \lambda_{i+1}$, $i\leq b-1$ (when $b > 1$) and $\alpha_b = \begin{cases} \lambda_b & \text{if } k \neq 2b \\ \lambda_{b-1} + \lambda_b & \text{if } k = 2b \end{cases}$. Let T be any subset of $\{\alpha_1,\ldots,\alpha_b\}$ and Σ_T be the $\mathbb{Z}$ linear span of the elements of T. Let $T = \emptyset$ if $b = 1$ or if $k = 4$ with $b = 2$; otherwise let $T = \{\alpha_2,\ldots,\alpha_b\}$. Then we see that $\widetilde{\mathfrak{M}}_a = \mathfrak{A} \oplus \mathfrak{M} \oplus \sum_{\beta\in\Sigma_T \cap \Delta} \mathfrak{U}_\beta$ ($\mathfrak{U}_\beta$ the root space corresponding to β). We use here the standard construction of a parabolic subalgebra of $\mathfrak{G}$ given in [26].

Let $c(\ell,m) \in O(Q)$ be the transformation of $\mathbb{R}^k$ which interchanges changes $v_\ell \longleftrightarrow v_m$ and $\tilde{v}_\ell \longleftrightarrow \tilde{v}_m$ ($\ell \neq m$ and $a \leq \ell \leq k-1$ and $a \leq m \leq k-1$) and fixes everything else. Let S_b be the subgroup $O(Q)$ generated by the family of $c(\ell,m)$ above. We see easily that S_b is isomorphic to the symmetric group on b letters $\{a,\ldots,k-1\}$ (hence the choice of notation).

Let $d(j)(a \leq j \leq k-1)$ be the transformation on $\mathbb{R}^k$ which interchanges $v_j \longleftrightarrow \tilde{v}_j$ and fixes everything else. Then we let $(\mathbb{Z}_2)^b$ be the group generated by the family of $d(j)$ (direct product b times of the group $\mathbb{Z}_2$ given by $\{d(j),I\}$).

It is easy to see that S_b normalizes $(\mathbb{Z}_2)^b$ and that $S_b \cap (\mathbb{Z}_2)^b = \{I\}$. Thus we can form the semidirect product $S_b \times (\mathbb{Z}_2)^b = W$ (the group generated by $c(\ell,m)$ and $d(j)$).

First we note that $W \subseteq \text{Normalizer}_K(A)$ and by the theory of symmetric spaces in [13], the group W under the adjoint representation $Ad_{O(Q)}(W)$ restricted to $\mathfrak{A}$ determines completely the "Weyl group" of $\mathfrak{A}$. Indeed $Ad(d(\ell))|_{\mathfrak{A}}$ is the reflection τ_{λ_ℓ} determined by the root λ_ℓ, and $Ad(c(j,j+1))|_{\mathfrak{A}}$ is the reflection $\tau_{\lambda_j-\lambda_{j+1}}$ determined by the root $\lambda_j - \lambda_{j+1}$. Then noting the relation $\tau_{\lambda_i}\tau_{\lambda_i-\lambda_j}\tau_{\lambda_i} = \tau_{\lambda_i+\lambda_j}$ and the fact that the reflections determined by the simple positive roots generate the Weyl group, it follows that $Ad_{O(Q)}(W)$ restricted to $\mathfrak{A}$ is exactly the Weyl group of $\mathfrak{A}$.

Let W_j be the subgroup of W given by $S_b^j \times (\mathbb{Z}_2)^{b-1}$, where S_b^j is generated by $c(\ell,m)$ with $\ell \neq j$ and $m \neq j$, <u>and</u> $(\mathbb{Z}_2)^{b-1}$ is generated by $d(s)$ where $s \neq j$. Then it is elementary to see that the <u>double coset</u> decomposition of W relative to W_j is given as $\{W_j\cdot W_j, W_j d(j) W_j, W_j c(s,j) W_j\}$ with $s \neq j$. Observe that $[W_j\backslash W/W_j] = 2$ if $b = 1$ and $= 3$ if $b > 1$.

W then consider $N_j \subseteq P_j$ and for $x \in O(Q)$ all possible conjugates $x\cdot N_j\cdot x^{-1}$. Then we consider the family $\mathcal{S}$ of <u>all</u> orbits $x\cdot N_j\cdot x^{-1}(\xi)$ as $\xi \in \mathbb{R}^k$ and $x \in O(Q)$. Call the set $x\cdot N_j\cdot x^{-1}(\xi)$ a <u>horocycle</u>. It is easy to see that the family $\mathcal{S}$ is independent of the choice of N_j (any N_s will work for $s = a,\ldots,k-1$). Then $Y \in \mathcal{S}$ has the form $x\cdot N_j\cdot(\xi')$ with $\xi' \in \mathbb{R}^k(\xi' = x^{-1}(\xi))$. The group $O(Q)$ acts on the set of all conjugates $x\cdot N_j\cdot x^{-1}$ by: $x\cdot N_j\cdot x^{-1} \underset{g}{\rightarrow} (gx)N_j(gx)^{-1}$; hence the group $O(Q)$ has an action on the family $\mathcal{S}$.

From above since the group N_a is unipotent, the orbit $N_a(\xi)$ for any $\xi \in \mathbb{R}^k$ is a closed subset of $\mathbb{R}^k$ and is given explicitly by $N_a(Y)(rv_a + s\tilde{v}_a + Z) = (r - \frac{1}{2}sQ(Y,Y) - Q(Z,Y))v_a + s\tilde{v}_a + (Z + sY)$ with $Z \in \langle v_a, \tilde{v}_a\rangle^{\perp}$. If $Q(\xi, v_a) \neq 0$, we observe by simple computation that for any $X = uv_a + s\tilde{v}_a + \tilde{Z}$ with the property that $Q(X,X) = Q(\xi,\xi)$, there exists a unique $Y \in \langle v_a, \tilde{v}_a\rangle^{\perp}$ so that $\tilde{Z} = Z + sY$ and $u = r - \frac{1}{2}sQ(Y,Y) - Q(Z,Y)$. Thus if $Q(\xi, x^{-1}v_a) \neq 0$, we have $x \cdot N_a(Y)(\xi) = \{Z \in \Gamma_t - \{0\} \mid Q(Z, x^{-1}v_a) = Q(\xi, v_a)\}$ with $Q(\xi,\xi) = t$. On the other hand $N_a(Y)(\xi)$ is a <u>line</u> through the point Z in the direction of v_a if and only if $Q(\xi, v_a) = 0$ $(s = 0)$. Thus $x \cdot N_a(Y)(\xi) \in \mathcal{S}$ is a line through the point $x \cdot Z$ in the direction of $x^{-1}v_a$ if and only if $Q(\xi, x^{-1}v_a) = 0$.

We say a horocycle $Y \in \mathcal{S}$ is <u>singular</u> if Y is a line in $\mathbb{R}^k$. Otherwise we say $Y \in \mathcal{S}$ is <u>nonsingular</u>. Thus we divide the set $\mathcal{S} = \mathcal{S}_s \cup \mathcal{S}_n$ where $\mathcal{S}_s$ is the set of singular horocycles and $\mathcal{S}_n$ is the set of nonsingular horocycles; the union is disjoint with the sets $\mathcal{S}_s$ and $\mathcal{S}_n$ both $O(Q)$ stable.

We note at this point that it is easy to see that the group P_a can be characterised as the set $\{g \in O(Q) \mid g$ leaves stable the line $(v_a)\}$ (i.e. $\Gamma_{(0,0,1)} \cong O(Q)/P_a$).

<u>Lemma 2.2</u> <u>There is a bijective correspondence between the set</u> $\Gamma_o - \{0\}$ <u>and the set</u> $\mathcal{S}_n \cap (\Gamma_t - \{0\})$ <u>of all nonsingular horocycles on</u> $\Gamma_t - \{0\}$ <u>given by</u> $\lambda \in \Gamma_o - \{0\}$ <u>determining the nonsingular horocycle</u> $B_{\lambda,t} = \{Z \in \Gamma_t - \{0\} \mid Q(Z,\lambda) = 1\}$.

<u>Proof</u>. From above we consider the nonsingular horocycle $x \cdot N_a(Y)(\xi)$ and let $\lambda = x^{-1}(v_a)/Q(\xi, v_a)$. Thus the map $\lambda \to B_\lambda$ is surjective. However since this map commutes with the $O(Q)$ actions on $\Gamma_o - \{0\}$ and $\mathcal{S}_n \cap (\Gamma_t - \{0\})$,

it suffices to determine the isotropy group of $B_{\tilde{v}_a,t} = N_a(Y)(\tilde{v}_a + \frac{t}{2}v_a) = \frac{1}{2}(t - Q(Y,Y))v_a + \tilde{v}_a + Y$. Then if $x \in O(Q)$ has the property $xB_{\tilde{v}_a,t} = B_{\tilde{v}_a,t}$, then $Q(x \cdot N_a(Y)(\tilde{v}_a + \frac{t}{2}v_a), v_a) = 1$ for all Y. Then taking $s \cdot Y$ and differentiating with respect to s we get $-sQ(Y,Y)Q(v_a, x^{-1}v_a) + Q(Y, x^{-1}v_a) = 0$ for all s and Y. Thus $x \cdot v_a \in (v_a)$ and hence $x \in P_a$. But clearly $B_{\tilde{v}_a,t} \cap B_{s\tilde{v}_a,t} = \emptyset$ for any $s \neq 1$. Thus $x \in M_a \cdot N_a$ and hence the map above is injective.

Q.E.D.

<u>Remark 2.2</u> We note that Lemma 2.2 is true for every orbit $\Gamma_t - \{0\}$ including $t = 0$.

If L is any s dimensional subspace of $\mathbb{R}^k$, we denote the L hyperplane passing through the point $Z \in \mathbb{R}^k$ as $Z + (L)$. We let $\langle \ell_1, \ldots, \ell_s \rangle$ be a basis of (L). Then for $Z + (L) \in \Gamma_t$ with $Q(Z,Z) = t$, it follows that $Q(Z,Z) = Q(Z + \sum_{i=1}^{s} t_i\ell_i, Z + \sum_{i=1}^{s} t_i\ell_i) = Q(Z,Z) + \sum_{i=1}^{s} 2t_iQ(Z,\ell_i) +$

$\sum_{i<j} 2t_it_jQ(\ell_i,\ell_j) + \sum_{i=1}^{s} t_i^2Q(\ell_i,\ell_i)$. Since the t_i are arbitrary, we have that $Q(\ell_i,\ell_j) = 0$ for all i and j <u>and</u> $Q(Z,u) = 0$ for all $u \in L$.

<u>Lemma 2.3</u> <u>The set $\Gamma_t - \{0\}$ contains a line (a hyperplane of dimension 1) for all t except when $t \leq 0$ with $b = 1$. The set of all lines on $\Gamma_t - \{0\}$ is exactly the same as the set of all singular horocycles $\mathcal{S}_s \cap (\Gamma_t - \{0\})$ on $\Gamma_t - \{0\}$. The set $\mathcal{S}_s \cap (\Gamma_t - \{0\})$ forms a two to one surjective covering of $\Gamma_{(1,0,1)}(\Gamma_{(0,1,1)})$ for $t > 0$ $(t < 0)$ via the map ψ which sends the line $x + (y)(Q(x,x) = t, Q(y,y) = 0)$ to the two plane $(x \wedge y)$. Thus the group $O(Q)$ operates transitively on $\mathcal{S}_s \cap (\Gamma_t - \{0\})$ for all $t \neq 0$, and in fact $\mathcal{S}_s \cap (\Gamma_t - \{0\})$ is isomorphic</u>

to the homogeneous space $O(Q)/M_a^X A_a N_a$ *with* $M_a^X = \{m \in M_a \mid m(x) = X\}$ *with* $X \in \langle v_a, \tilde{v}_a\rangle^\perp$ *so that* $Q(X,X) = t$.

Proof. Let $\eta \in \Gamma_o - \{0\}$. Then there exists $\zeta \in \Gamma_o - \{0\}$ so that $\langle \zeta, \eta\rangle$ spans a hyperbolic plane. Then $\Gamma_t \cap \langle \zeta, \eta\rangle^\perp \neq \emptyset$ for all t with $b \geq 2$ and for $t > 0$ with $b = 1$. Hence if $t \leq 0$ with $b = 1$, then $Q(\xi, X) = 0$ with $\xi \in \Gamma_o - \{0\}$ and $X \in \mathbb{R}^k$ implies that $\xi_k x_k = \sum_{i=1}^{k-1} \xi_i x_i$.

By Cauchy inequality $\xi_k^2 x_k^2 \leq (\sum_{i=1}^{k-1} \xi_i^2)(\sum_{i=1}^{k-1} x_i^2)$. However $\xi \in \Gamma_o - \{0\}$ means that $\xi_k^2 = \sum_{i=1}^{k-1} \xi_i^2$ and hence $x_k^2 \leq \sum_{i=1}^{k-1} x_i^2$ or $Q(X,X) \geq 0$. Thus either X is hyperbolic or $X = s\xi$. Thus the first statement of the Lemma follows.

It is easy to see that the map ψ defined in the Lemma is well-defined and surjective. Let $a + (b)$ and $x + (y)$ be lines in $\Gamma_t - \{0\}$ satisfying $Q(a,a) = Q(x,x) = t$ and $Q(b,b) = Q(y,y) = 0$. If $\psi(a + (b)) = \psi(x + (y))$, then we may express $a = \alpha x + \beta y$ and $b = \gamma x + \delta y$ $(\alpha, \beta, \gamma, \delta \in \mathbb{R})$. Then using the properties of a and b above, we deduce that $\gamma = 0$ and $\alpha^2 = 1$. Thus $\psi^{-1}(x \wedge y) = \{x + (y), -x + (y)\}$. Since $O(Q)$ operates transitively on $\Gamma_{(1,0,1)}$ $(\Gamma_{(0,1,1)})$, we deduce that $O(Q)$ operates transitively on $\mathcal{S}_s \cap (\Gamma_t - \{0\})$ for $t \neq 0$.

We must determine the isotropy group of the line $X + (v_a)$ with $X \in \langle v_a, \tilde{v}_a\rangle^\perp \cap \Gamma_t$. Then if $g \in O(Q)$ stabilizes the line $X + (v_a)$, we deduce that $Q(gX, z) + tQ(gv_a, z) = 0$ for all $z \in \langle X, v_a\rangle^\perp$ and all t. Thus gX and $gv_a \in \langle X, v_a\rangle^{\perp\perp} = \langle X, v_a\rangle$. Then using the same argument as in the paragraph above, we deduce that $g \in P_a$. We then note that $A_a N_a$ stabilizes the line $X + (v_a)$ and hence we deduce our result.

Q.E.D.

We are next interested in the action of the parabolic P_a on the orbit Γ_t.

Proposition 2.4

(1) *Let* $t \neq 0$. *If* $b > 1$ *then there is the disjoint decomposition*

$$(2-5) \quad \Gamma_t = P_a(\tilde{v}_a + \tfrac{1}{2}tv_a) \cup P_a c(a,a+1)(\tilde{v}_a + \tfrac{1}{2}tv_a),$$

where $P_a(\tilde{v}_a + \frac{1}{2}tv_a) = A_a(r)N_a(Y)(\tilde{v}_a + \frac{1}{2}tv_a)$ *and* $P_a \cdot c(a,a+1)(\tilde{v}_a + \frac{1}{2}tv_a) = N_a(Y)M_a c(a,a+1)(\tilde{v}_a + \frac{1}{2}tv_a)$.

If $b = 1$ *and* $t < 0$ *then*

$$(2-6) \quad \Gamma_t = P_a(\tilde{v}_a + \tfrac{1}{2}tv_a)$$

where $P_a(\tilde{v}_a + \frac{1}{2}tv_a) = A_a(r)N_a(Y)(\tilde{v}_a + \frac{1}{2}tv_a)$.

If $b = 1$ *and* $t > 0$ *then*

$$(2-7) \quad \Gamma_t = P_a(\tilde{v}_a + \tfrac{1}{2}tv_a) \cup P_a \cdot \gamma_t \cdot (\tilde{v}_a + \tfrac{1}{2}tv_a)$$

where $\gamma_t \in O(Q)$ *has the property* $\gamma_t(\tilde{v}_a + \frac{1}{2}tv_a) = t^{1/2} \cdot e_2$ *and* $P_a(\tilde{v}_a + \frac{1}{2}tv_a) = A_a(r)N_a(Y)(v_a + \frac{1}{2}tv_a)$ *and* $P_a\gamma_t(\tilde{v}_a + \frac{1}{2}tv_a) = N_a(Y)M_a\gamma_t(\tilde{v}_a + \frac{1}{2}tv_a)$.

(2) *Let* $t = 0$. *If* $b > 1$ *there is the disjoint decomposition*

$$(2-8) \quad \Gamma_o - \{0\} = P_a d(a)v_a \cup P_a c(a,a+1)v_a \cup P_a \cdot v_a$$

where $P_a d(a)v_a = A_a(r)N_a(Y)d(a)v_a$, $P_a c(a,a+1)v_a = N_a(Y)M_a c(a,a+1)v_a$, *and* $P_a v_a = A_a(r)v_a$.

If $b = 1$ *then*

$$(2-9) \quad \Gamma_o - \{0\} = P_a d(a)v_a \cup P_a v_a$$

where $P_a d(a)v_a = A_a(r)N_a(Y)d(a)v_a$ *and* $P_a v_a = A_a(r)v_a$.

Proof. Consider the point $\tilde{v}_a + \frac{1}{2}tv_a$. Then from above

$A_a(r)N_a(Y)(\tilde{v}_a + \frac{1}{2}tv_a) = (\frac{1}{2}t - \frac{1}{2}Q(Y,Y))rv_a + r^{-1}\tilde{v}_a + Y = T_a^t$.

Then it follows from Lemma 2.2 that $T_a^t \cap (\Gamma_t - \{0\}) = \{X \in \Gamma_t - \{0\} |$ $Q(X,v_a) \neq 0\}$; then $X \in {}^cT_a^t = \Gamma_t - T_a^t$ has the property that $X =$ $sv_a + \xi$, $\xi \in \langle v_a, \tilde{v}_a\rangle^{\perp}$ with $Q(X,X) = Q(\xi,\xi) = t$. Then let $b > 1$ and $t \neq 0$. There exists $c(a,a+1)$ so that $c(a,a+1)(\tilde{v}_a + \frac{1}{2}tv_a) = \tilde{v}_{a+1} + \frac{1}{2}tv_{a+1}$. Since the group M_a operates transitively on the set $\Gamma_t \cap \langle v_a, \tilde{v}_a\rangle^{\perp}$, we have $M_a(\tilde{v}_{a+1} + \frac{1}{2}tv_{a+1}) = \xi$ for $Q(\xi,\xi) = t$. Then $N_a(Y)\xi =$ $\xi + sv_a$, $s \in \mathbb{R}$, and hence $N_a(Y)M_ac(a,a+1)(\tilde{v}_a + \frac{1}{2}tv_a) = {}^cT_a^t$. If $b = 1$ and $t < 0$, then $T_a^t = \emptyset$. If $b = 1$ and $t > 0$, then by the same argument $sv_a + \xi = {}^cT_a^t$ clearly has the form $N_a(Y)M_a\gamma_t(\tilde{v}_a + \frac{1}{2}tv_a)$.

Let $t = 0$. If $b > 1$ we use the same arguments and obtain that ${}^cT_a^0 = \{sv_a + \xi | Q(\xi,\xi) = 0\} \cup \{sv_a | s \neq 0\}$ with $sv_a + \xi = N_a(Y)M_ac(a,a+1)v_a$ and $(s\cdot v_a) = A_a(r)v_a$. If $b = 1$ then $\xi = 0$, and we just have ${}^cT_a^0 = \{sv_a | s \neq 0\}$.

Q.E.D.

Corollary 1 to Proposition 2.4. ("The Bruhat decomposition") The double coset decomposition of $O(Q)$ relative to P_j is $P_jd(j)P_j \cup P_jc(j,j+1)P_j \cup P_j$ if $b > 1$ and $P_jd(j)P_j \cup P_j$ if $b = 1$. Thus there is a bijective correspondence between $W_j\backslash W/W_j$ and $P_j\backslash O(Q)/P_j$.

Remark 2.3 In the right coset decomposition of $O(Q)/P_a$, let $\tilde{P}_a$ be the isotropy group of the coset $c(a,a+1)P_a$ when $b > 1$. Then $\tilde{P}_a$

$= P_a \cap c(a,a+1)P_a c(a,a+1) = P_a \cap P_{a+1}$. Then by an easy argument $\widetilde{P}_a = (M_a \cap P_{a+1})\cdot A_a \cdot \{N_a(Z) \mid Z \in \langle v_a, \widetilde{v}_a, v_{a+1}\rangle^{\perp}\}$. And since $\widetilde{P}_a$ is a closed subgroup of $O(Q)$, it follows that the Lie algebra of $\widetilde{P}_a$ is

$$\mathfrak{P}_a \cap \mathrm{Ad}\, c(a,a+1)\mathfrak{P}_a = \mathfrak{M} \oplus \mathfrak{A} \oplus \sum_{\substack{\beta>0 \\ \beta\neq\alpha_1}} \mathfrak{U}_\beta \oplus \sum_{2<i<j\leq b} \mathfrak{U}_{-(\lambda_i-\lambda_j)} \oplus$$

$$\sum_{2<i<j\leq b} \mathfrak{U}_{-(\lambda_i+\lambda_j)} \oplus \sum_{2<i\leq b} \mathfrak{U}_{-\lambda_i}.$$

§3. REPRESENTATION THEORY OF ORTHOGONAL GROUP

We recall here general facts of representation theory and the theory of induced representations.

If E and F are locally convex complete Hausdorf topological vector spaces, we form the <u>projective</u> tensor product $E \hat{\otimes} F$ which is again a locally convex complete Hausdorf topological vector space (see [25] for relevant definitions).

Let S be an arbitrary real C^∞ Lie group with a fixed left Haar measure $d_S(x)$. Let δ_S be the <u>modular</u> function on S relative to d_S. Then we know that $\delta_S(x) = |\det Ad_S(x^{-1})|$, where $x \in S$ and Ad_S is the adjoint representation of S on its Lie algebra $\mathcal{S}$.

Let E be a locally convex complete Hausdorf topological vector space and $C^\infty(S,E)$ ($C_c^\infty(S,E)$), the space of f: $S \to E$ which are C^∞ (C^∞ of compact support). We recall that $f \in C^\infty(S,E)$ <u>if and only if</u> f is weakly C^∞, i.e. $x \to \langle T, f(x)\rangle \in C^\infty(S)$ for all $T \in E'$, the continuous linear dual space of E. For any $X \in \mathcal{S}$ we consider the left invariant differential operator X^* on $C^\infty(S,E)$ given by

$$(3-1) \quad X^*f(P) = \frac{d}{dt} f(\exp(-tX)\cdot P)\Big|_{t=0}, \; P \in S.$$

In this way for any $D \in \mathcal{U}(\mathcal{S})$, the universal enveloping algebra of $\mathcal{S}$, we can define D^*f. Then we topologize $C^\infty(S,E)$ as follows. Let $\{|\ |_\alpha\}$ be any family of continuous seminorms on the space E, and ω any compact subset of S. Then define

$$(3-2) \quad |f|^\alpha_{D,\omega} = \sup_\omega |D^*f|_\alpha .$$

This set of seminorms puts a locally convex complete Hausdorf topological vector space structure on $C^\infty(S,E)$ (the <u>Schwartz</u> topology). We then note that $C^\infty(S,E)$ is a Frechet space if E is a Frechet space.

Let $C_T^\infty(S,E) = \{f \in C^\infty(S,E) \mid \text{support } (f) \subseteq T\}$, where T is a compact subset of S. Consider the topology on $C_T^\infty(S,E)$ as that induced by the Schwartz topology of $C^\infty(S,E)$. Then $C_T^\infty(S,E)$ is again a locally convex complete Hausdorf topological vector space, and we topologize $C_c^\infty(S,E)$ as the inductive limit of $C_T^\infty(S,E)$ as T ranges over all compact subsets of S. In this way $C_c^\infty(S,E)$ is also a locally convex complete Hausdorf topological vector space.

We note at this point the following identifications: $C^\infty(S,E) = C^\infty(S) \hat{\otimes} E$ and $C_T^\infty(S,E) = C_T^\infty(S) \hat{\otimes} E$ where T is compact.

Let π be a continuous representation of S on E (with structure as above). We recall that a vector $a \in E$ is <u>differentiable</u> for π if the function $a^\natural : g \to \pi(g)a \in C^\infty(S,E)$. Then if E_∞ is the linear space of all differentiable vectors for π we know from [25] that E_∞ is stable by $\pi(g)$ as $g \in S$ (call the restriction of π to E_∞ by π_∞) and that E_∞ is dense in E. The map $a \to a^\natural$ of E_∞ to $C^\infty(S,E)$ is injective and identifies E_∞ to a closed subspace of $C^\infty(S,E)$. We equip E_∞ with the induced topology from the Schwartz topology on $C^\infty(S,E)$ so that E_∞ becomes a locally convex complete Hausdorf topological vector space. Then we observe that π_∞ gives a continuous representation of S on E_∞. Moreover we say that a representation is differentiable if $a \to a^\natural$ is a topological isomorphism into (i.e. $E = E_\infty$ with the topologies coinciding). Also if E is a Frechet space then E_∞ is a Frechet space.

If $Z \in \mathscr{S}$ then we know that for all $a \in E_\infty$ that

$$(3-3) \quad \pi(Z)a = \lim_{t \to 0} \frac{1}{t}(\pi(\exp(tZ)) - I)(a)$$

exists, and hence we obtain a linear representation of $\mathscr{S}$ on the space E_∞. Then the representation extends to one of $\mathcal{U}(\mathscr{S})$ on E_∞.

We recall that the involution $X \to -X$ on $\mathcal{S}$ extends uniquely to an anti-algebra automorphism $*$ on $\mathcal{U}(\mathcal{S})$ ($(\ell_1 \ell_2)* = \ell_2^* \ell_1^*$ for all $\ell_i \in \mathcal{U}(\mathcal{S})$). Moreover we say an element $\ell \in \mathcal{U}(\mathcal{S})$ is <u>symmetric</u> if $\ell^* = \ell$. An element $\ell \in \mathcal{U}(\mathcal{S})$ is <u>central</u> if $\beta \cdot \ell = \ell \cdot \beta$ for all $\beta \in \mathcal{U}(\mathcal{S})$. Then the set $\mathfrak{z}_{\mathcal{S}}$ of central elements of $\mathcal{U}(\mathcal{S})$ forms a commutative subalgebra of $\mathcal{U}(\mathcal{S})$, and the set Q of central symmetric elements of $\mathcal{U}(\mathcal{S})$ has the following remarkable property [11]. Let π be a unitary representation of S on a Hilbert space H and V, a dense subspace of H contained in H_∞ and stable by $\pi(x)$ for all $x \in S$. Let $\ell \in Q$ and $\pi_\infty(\ell)|_V$ be the restriction of $\pi_\infty(\ell)$ to V. Then $\pi_\infty(\ell)|_V$ has <u>one and only one self adjoint extension to</u> H.

We recall certain functorial properties of differentiable representations. If π_1 and π_2 are differentiable representations on spaces E_1 and E_2, then $\pi_1 \hat{\otimes} \pi_2$ (given on $E_1 \otimes E_2$ by $\pi_1 \otimes \pi_2(g)(v \otimes w) = \pi_1(g)(v) \otimes \pi_2(g)(w)$) defines a differentiable representation on $E_1 \hat{\otimes} E_2$.

If E' is the continuous dual space of E, the contragredient representation π^c on E' is given by $\langle \pi^c(g)T, f \rangle = \langle T, \pi(g)^{-1} f \rangle$ for all $f \in E$ and $T \in E'$. However in general π^c is not continuous in the various topologies on E'. But if we consider the space $E^\vee = \{v \in E' \mid$ the map $g \to \pi^c(g)v$ for all g is continuous relative to the <u>strong topology</u> on $E'\}$, then $E^\vee$ is stable under $\pi^c(S)$ and $E^\vee$ is weakly dense in E'. And if E is <u>complete</u> relative to the <u>strong</u> topology, then $E^\vee = E'$; hence if π is differentiable on E, and E' satisfies the latter condition, then π^c is a differentiable representation of S on E' (relative to the strong dual topology on E').

We note at this point that if E is a Banach space and V any dense subspace of E contained in E_∞ and stable by $\pi(x)$ as $x \in S$, then V is dense in E_∞ (in the E_∞ topology) (see [20], p. 94).

We recall the notions of intertwining operator and intertwining form. If π_1 and π_2 are continuous representations on spaces E and F which are locally convex complete Hausdorf topological vector spaces, we say a continuous linear operator $U: E \to F$ is an intertwining operator if $U \circ \pi_1(g) = \pi_2(g) \circ U$ for all $g \in S$. Let $\mathrm{Hom}_S(\pi_1, \pi_2)$ be the vector space of all continuous intertwining operators.

Moreover if E and F are also barreled spaces, then we recall that a separately continuous bilinear form B on $E \times F$ is an intertwining form of π_1 to π_2 if for every $g \in S$, $B(\pi_1(g)e, \pi_2(g)f) = B(e,f)$ for all $e,f \in E,F$. Let $I(\pi_1, \pi_2)$ be the vector space of intertwining forms.

The relation between intertwining operator and form is given as follows. We have a linear bijection between the sets $I(\pi_1, \pi_2)$ and $\mathrm{Hom}_S(\pi_1, \pi_2^c)$, the correspondence being $B \to$ the linear operator $\varphi_B: E \to F^{\vee}$ (relative to induced strong topology) given by $\langle b, \varphi_B(a) \rangle = B(a,b)$ for $a \in E$ and $b \in F$.

Then with the hypotheses on E and F as above (barreled, etc. ..), we know that $\dim I(\pi_1, \pi_2) \leq \dim I((\pi_1)_\infty, (\pi_2)_\infty)$ and $\dim \mathrm{Hom}_S(\pi_1, \pi_2) \leq \dim \mathrm{Hom}_S((\pi_1)_\infty, (\pi_2)_\infty)$. Moreover if π is a unitary representation of S on a Hilbert space E, then $\dim I(\pi, \pi^c) = 1$ if and only if π is a unitary irreducible representation (i.e. the only $\pi(S)$ closed invariant subspaces of E are E and $\{0\}$).

Let K be any compact Lie group and π any continuous representation of K on a locally convex complete Hausdorf topological vector space. Let $\hat{K}$ = equivalence classes of unitary irreducible representations of K.

If $\lambda \in \hat{K}$ let χ_λ be the character of λ and $d(\lambda)$ the degree of λ. Then for any $w \in E$ we can define

$$(3-4) \qquad P(\lambda)(w) = d(\lambda) \int_K \overline{\chi_\lambda(k)} \pi(k)(w) dk \ .$$

We know that $P(\lambda)$ is a projection operator on E $(P(\lambda)^2 = P(\lambda))$ and that Image $P(\lambda)(E) = E(\lambda)$ is a closed subspace of E. Moreover $E(\lambda)$ is stable by $\pi(K)$ and the space $E(\lambda)$ is isotypic (relative to K) of type λ. If $\lambda_1 \neq \lambda_2$ then $P(\lambda_1)P(\lambda_2) = P(\lambda_2)P(\lambda_1) = 0$ $(E(\lambda_1) \cap E(\lambda_2) = 0)$. And if E_K is the algebraic direct sum of $E(\lambda)$, then E_K is dense in E and $E_K = \{v \in E \mid \text{the span of } \pi(k)v \text{ (as } k \in K) \text{ is finite dimensional}\}$. And if $w \in E_\infty$ (differentiable vectors relative to π), we know that the series $\sum_{\lambda \in \hat{K}} P(\lambda)(w)$ converges absolutely to w (in the topology of E_∞ defined above) and that $E_K \cap E_\infty =$ (the algebraic direct sum $E(\lambda) \cap E_\infty$) is dense in E.

We let U be a closed subgroup of S. Then the homogeneous space S/U exists as a C^∞ manifold and admits a quasi invariant measure. Indeed if ρ is any rho function on S (any positive Borel function f on S bounded above and below on compact subsets of S and satisfying $f(x \cdot \xi) = \frac{\delta_U(\xi)}{\delta_S(\xi)} f(x)$ for all $\xi \in U$), then there exists a quasi invariant measure $d\dot{\mu}_\rho$ on S/U so that

$$(3-5) \qquad \int_S f(g)\rho(g)d_S(g) = \oint_{S/U} \left\{ \int_U f(g\xi)d_U(\xi) \right\} d\dot{\mu}_\rho(\dot{g})$$

for all $f \in C_c(S)$.

The first notion of induced representation is with the assumption that σ is a differentiable representation of U on a Frechet space E. Then we define

$$(3-6) \qquad \begin{aligned} C^\infty_\sigma(S,E) = \{ f \in C^\infty(S,E) \mid\ & f \text{ has compact support mod } U \text{ and} \\ & f(g\xi) = \left[\frac{\delta_U(\xi)}{\delta_S(\xi)}\right]^{1/2} \sigma(\xi)^{-1} f(g) \text{ for all } g \in S \text{ and } \xi \in U\}. \end{aligned}$$

Then for any $z \in S$ we define $\pi_\sigma(z)f(y) = f(z^{-1}y)$ for $f \in C^\infty_\sigma(S,E)$ and hence we have a linear representation π_σ of S on $C^\infty_\sigma(S,E)$. For a compact subset ω of S we consider $C_\sigma^\infty(S,E,\omega) = \{f \in C_\sigma^\infty(S,E) \mid \text{support } (f) \subseteq \omega \cdot U\}$ and put the induced topology of $C^\infty(S,E)$ on $C_\sigma^\infty(S,E,\omega)$, which makes $C_\sigma^\infty(S,E,\omega)$ into a Frechet space. Then equipping $C_\sigma^\infty(S,E)$ with the inductive limit topology of the $C_\sigma^\infty(S,E,\omega)$ as ω ranges over compact subsets of S, we know that π_σ is a differentiable representation of S on $C_\sigma^\infty(S,E)$.

Remark 3.1 By the structure of the topology on $C_\sigma^\infty(S,E)$ (an LF space), we see that $C_\sigma^\infty(S,E)'$ is complete in the strong dual topology and hence the contragredient representation π_σ^c of S on $C_\sigma^\infty(S,E)'$ is differentiable.

Remark 3.2 If the homogeneous space S/U is compact, then $C_\sigma^\infty(S,E)$ is a Frechet space. Indeed $C_\sigma^\infty(S,E)$ is a closed subspace of $C^\infty(S,E)$ and the inductive limit topology on $C^\infty(S,E)$ is the same as that induced on $C_\sigma^\infty(S,E)$ by $C^\infty(S,E)$. The topology on $C_\sigma^\infty(S,E)$ can be described by the family of seminorms: $|f|_{D,\omega} = \sup_\omega |\pi_\sigma(D)*f|$ as ω ranges over compact subsets of S and $D \in \mathcal{U}(\mathscr{S})$.

At this point we recall that the linear map

$$(3-7) \qquad f \xrightarrow[S_\sigma]{} f^\sigma(x) = \int_U \left[\frac{\delta_U(\xi)}{\delta_S(\xi)}\right]^{1/2} \sigma(\xi) f(x\xi) d_U(\xi)$$

from $C_c^\infty(S,E)$ to $C_\sigma^\infty(S,E)$ is a continuous surjective open linear mapping; hence if $L_\sigma = \{f \in C^\infty(S,E) \mid f^\sigma \equiv 0\}$, then $C^\infty(S,E)/L_\sigma$ is linearly and topologically isomorphic to $C_\sigma^\infty(S,E)$. Moreover the map S_σ intertwines the usual left action of S on $C_c^\infty(S,E)$ with π_σ given above.

The second notion of induced representation is with the assumption that σ is a unitary representation of U on a Hilbert space E with norm $(\ |\ )_E$. We start with a rho function ρ on S so that $\rho(e_S) = 1$; let $d\mu_\rho$ be the quasi invariant measure on S/U relative to ρ. Then

we define

$$E_\sigma^m = \{f\colon\ S\to E \mid \text{(i) for every } a\in E \text{ the function } a\to(f(x)\mid a)_E \text{ is a Borel measurable function on } S;\ \text{(ii) for all } \xi\in U,\ x\in S,$$

$$(3-8)\qquad f(x\xi) = \left[\frac{\delta_U(\xi)}{\delta_S(\xi)}\right]^{1/2} \sigma(\xi)^{-1}f(x);\ \text{and}$$

$$\text{(iii)} \oint \rho(x)^{-1}\|f(x)\|_E^2\, d\dot{\mu}_\rho(\dot{x}) < \infty\}.$$

Then we consider on E_σ^m the Hermitian form $(f\mid g)_\rho = \oint \rho(x)^{-1}(f(x),g(x))_E\, d\dot{\mu}_\rho(\dot{x})$ for f and $g \in E_\sigma$. We let E_σ be the space of functions in E_σ^m identified modulo Borel null sets of S. Then we know that $(\ \mid\ )_\rho$ induces a Hilbert space structure on E_σ. Moreover if $y \in S$ and $f \in E_\sigma$, then we define $\tilde{\pi}_\sigma(y)f(x) = f(y^{-1}x)$; then $\tilde{\pi}_\sigma$ defines a unitary representation of S on E_σ. We note that the representation is independent of the choice of ρ in the sense that another choice of ρ gives a unitarily equivalent representation.

The relation between the two notions of induced representation is given in the following way. Let σ be a unitary representation of U on a Hilbert space E; from above we have a differentiable representation σ_∞ of U on the Frechet space E_∞. Then we observe that $C_\sigma^\infty(S,E_\infty) \subseteq E_\sigma$ and the injection of $C_\sigma^\infty(S,E_\infty)$ to E_σ is continuous. Moreover $C_\sigma^\infty(S,E_\infty)$ is dense in E_σ.

<u>Remark 3.3</u> We note that $(E_\sigma)_\infty \supseteq C_\sigma^\infty(S,E_\infty)$. Moreover we know from [25] that $(E_\sigma)_\infty$ can be characterized as all $f \in C^\infty(S,E)$ such that $\tilde{\pi}_\sigma(D)f \in E_\sigma$ for <u>all</u> $D \in \mathcal{U}(\mathcal{S})$. Thus if S/U is compact, we have that $C_\sigma^\infty(S,E_\infty) = (E_\sigma)_\infty$ (indeed every function $f \in C^\infty(S,E) \cap E_\sigma$ takes values in E_∞; for every $x \in S$ the map $\xi\to\sigma(\xi)f(x) = \left[\frac{\delta_U(\xi)}{\delta_S(\xi)}\right]^{1/2} f(x\xi^{-1})$ is clearly an element of $C^\infty(U,E)$).

Remark 3.4 If λ is any finite dimensional continuous representation of U on E, we consider the total topological space $S \times_\lambda E$ consisting of all equivalence classes of elements $(s,v) \in S \times E$ under the relation $(s\xi,v) \sim (s,\lambda^{-1}(\xi)v)$ for all $\xi \in U$. The projection map $[s,v] \to s\cdot U$ coset in S/U defines a S homogeneous vector bundle $S \times_\lambda E$ over S/U (with fiber $\{[s,v] \mid v \in E\} \cong E$ over the point sU). Thus taking

$$\lambda = \left[\frac{\delta_S}{\delta_U}\right]^{1/2} \sigma,$$

where σ is a differentiable finite dimensional representation of U on E, then $C^\infty_\sigma(S,E)$ is identified to the set of all C^∞ sections of compact support of the vector bundle $S \times_\lambda E$ over S/U.

We then recall the following well known fact. If L is a closed subgroup of S so that $S = L\cdot U$ and $L \cap U$ is compact, then with given left Haar measures d_L and d_U on L and U, it is possible to normalize d_S so that

$$(3-9) \qquad \int_S f(x)d_S(x) = \int_{L\times U} f(\ell\cdot\xi)\frac{\delta_S(\xi)}{\delta_U(\xi)} d_L(\ell)d_U(\xi)$$

for all $f \in C_c(S)$.

We let S act differentiably on any C^∞ manifold X. We recall that for any orbit $\mathfrak{O} = S\cdot p$ equipped with the induced topology from X, that the canonical map $S/S_p \to \mathfrak{O}$ is a homeomorphism of S/S_p onto $\mathfrak{O}$(S_p the isotropy group of the point $p \in X$). On the other hand putting the differential structure of S/S_p on the set $\mathfrak{O}$ makes $\mathfrak{O}$ into a submanifold of X. And hence if $\mathfrak{O}$ is closed in X, then the latter differential structure on $\mathfrak{O}$ coincides with the induced differential structure on $\mathfrak{O}$ from X.

Let π be any differentiable representation of S on a Frechet space E. Put the usual Schwartz topology on the space $C^\infty_c(X,E)$. Then consider the two representations of S on $C_c^\infty(X,E)$ given by: (i) I, the "coinduced representation", $I(g)\varphi(m) = \varphi(g^{-1}\cdot m)$ and (ii) $\underset{\sim}{\pi}(g)\varphi(m) = \pi(g)(\varphi(m))$ for all $g \in S$ and $m \in X$.

We define the following spaces:

$$(3-10) \quad T_\pi(X) = \{T \in C_c^\infty(X,E)' \mid I^c(g)T = \underset{\sim}{\pi}^c(g^{-1})T \text{ for all } g \in S\}$$

$$(3-11) \quad T_{\pi,\mathcal{O}}(X) = \{T \in T_\pi(X) \mid \text{support } (T) \subseteq \mathcal{O}\}$$

where $\mathcal{O}$ is an orbit in X.

The problem of Bruhat in [3] is to find the dimension of these spaces.

We know that since S_p leaves the point p fixed, we have a linear representation ν_p: $S_p \to \mathrm{Aut}\, T_p(X)$, where $T_p(X)$ is the tangent space of X at the point p and ν_p is the differential of the map $\tau_p(g)$: $x \to g\,x$ $(x \in X)$ on the tangent space $T_p(X)$. Moreover since $\mathcal{O}$ is a submanifold of X, we can consider $T_p(\mathcal{O})$ (the tangent space of $\mathcal{O}$ at p) as a subspace of $T_p(X)$. Then the restriction of the map ν_p to $T_p(\mathcal{O})$ is the same as the differential of the map $\tau_p(g)$ on the tangent space $T_p(\mathcal{O})$.

We complexify $T_p^{\mathbb{C}}(X) = \mathbb{C} \otimes_{\mathbb{R}} T_p(X)$, $T_p^{\mathbb{C}}(\mathcal{O}) = \mathbb{C} \otimes_{\mathbb{R}} T_p(\mathcal{O})$, and let R_p be the quotient space $T_p^{\mathbb{C}}(X)/T_p^{\mathbb{C}}(\mathcal{O})$. Let $\tilde{\nu}_p$ be the linear representation of S_p constructed on R_p in the obvious way from ν_p. If $S(R_p) = \bigoplus_{\ell \geq 0} S^\ell(R_p)$ is the symmetric algebra of R_p (S^ℓ the homogeneous component of degree ℓ), we consider the corresponding representation $\tilde{\nu}_p^\ell$ of S_p on $S^\ell(R_p)$ constructed from $\tilde{\nu}_p$.

Then from [25], p. 398 we know that if $\mathcal{O}$ is a closed orbit in X, then

$$(3-12) \qquad \dim T_{\pi,\mathcal{O}}(X) \leq \sum_{\ell \geq 0} \dim \mathrm{Hom}_{S_p}\left(\pi|_{S_p}, \left(\frac{\delta_S}{\delta_{S_p}}\right)(\tilde{\nu}_p^\ell)\right).$$

We note that $\dim \mathrm{Hom}_{S_p}\left(\pi|_{S_p}, \left(\frac{\delta_S}{\delta_{S_p}}\right)(\tilde{\nu}_p^\ell)\right)$ is independent of the choice of the point p in the orbit $\mathcal{O}$.

Then to determine $\dim T_\pi(X)$ in general, we assume that X has (at most) countably many orbits. Then from [25] it is possible to define a decreasing sequence X_α (as α runs through all ordinals) of closed S stable subsets constructed by: (i) $X_0 = X$; (ii) for each ordinal α let $X_{\alpha+1}$ be the complement of the union of the family of all S open orbits contained in X_α; and (iii) if β is a limit ordinal, then $X_\beta = \bigcap_{\alpha<\beta} X_\alpha$. Thus $X_\alpha = \emptyset$ for some countable ordinal α. Moreover for any orbit $\mathcal{O}$ there is a unique ordinal α so that $\mathcal{O} \subseteq X_\alpha$ and $\mathcal{O}$ is open in X_α. Then define $\Omega_{\mathcal{O}} = \mathcal{O} \cup (X - X_\alpha)$. We have that $\Omega_{\mathcal{O}}$ is an open S stable subset of X and $\mathcal{O}$ is a closed subset of $\Omega_{\mathcal{O}}$.

Then from [25], p. 401 we have if X has (at most) countably many S orbits that

$$(3-13) \quad \dim T_\pi(X) \leq \sum_{\mathcal{O}} \dim T_{\pi,\mathcal{O}}(\Omega_{\mathcal{O}})$$

as $\mathcal{O}$ runs over the set of (at most) countably many orbits of S in X.

Let U_1 and U_2 be two closed subgroups of S and σ_1, σ_2 differentiable representations of U_i on Frechet spaces E_i, $i = 1, 2$. The problem in Bruhat's thesis is to determine $I(\pi_{\sigma_1}, \pi_{\sigma_2})$, the space of intertwining forms of $C^\infty_{\sigma_1}(S,E_1)$ and $C^\infty_{\sigma_2}(S,E_2)$.

Remark 3.5 The importance of solving this problem comes from the following observation. Namely if $U_1 = U_2$ and $\sigma_1 = (\beta)_\infty$ and $\sigma_2 = (\beta^c)_\infty$, where β is a unitary representation of U_1 on a Hilbert space E, then $\dim I((\pi_{\sigma_1})_\infty, (\pi_{\sigma_2})_\infty) = 1$ implies that $\tilde{\pi}_\beta$ (the unitary representation of S on E_β constructed above) is irreducible (recall that $C^\infty_\beta(S,E_\infty)$ is dense in E_β).

We consider the representation of the product group $U_2 \times U_1$ on $E_1 \hat{\otimes} E_2$ given by $U(\xi_2,\xi_1) = [\delta_S(\xi_1\xi_2^{-1})\delta_{U_1}(\xi_1)\delta_{U_2}(\xi_2)]^{1/2}\sigma_1(\xi_1) \hat{\otimes} \sigma_2(\xi_2)$.

Then it is easy to see that U is a differentiable representation. We then consider the differentiable action of $U_2 \times U_1$ on S given by $y \xrightarrow[(\xi_2,\xi_1)]{} \xi_2 y \xi_1^{-1}$. The set of orbits is exactly the set of U_2, U_1 double cosets in S. Then the isotropy group $(U_2 \times U_1)_y$ of $y \in S$ is the group $\{(g_2,g_1) \in U_2 \times U_1 \mid y^{-1}g_2 y = g_1\}$.

We first recall from [25] that there is a linear bijective correspondence between the space $I(\pi_{\sigma_1},\pi_{\sigma_2})$ and the space $I'(\pi_{\sigma_1},\pi_{\sigma_2})$ of all separately continuous bilinear forms B' on $C_c^\infty(S,E_1) \times C_c^\infty(S,E_2)$ having the properties

(a) $B'(f,g) = 0$ if and only if $f \in L_{\sigma_1}$ or $g \in L_{\sigma_2}$;

$$(3-14)\quad \text{(b)}\ B'(f(*\xi_1),g(*\xi_2)) = [\delta_S(\xi_1\xi_2)\delta_{U_1}(\xi_1)\delta_{U_2}(\xi_2)]^{1/2} B'(\sigma_1(\xi_1^{-1})f,\sigma_2(\xi_2^{-1})g)$$

for all ξ_1, $\xi_2 \in U_1$, U_2 and all f, $g \in C_c^\infty(S,E_1), C_c^\infty(S,E_2)$. The correspondence is given by $B \to B'$ where $B'(f,g) = B(f^{\sigma_1},g^{\sigma_2})$.

Then we know from [25] that $I'(\pi_{\sigma_1},\pi_{\sigma_2})$ is linearly isomorphic to the space of distributions $T_U(S) = \{T \in C_c^\infty(S,E_1 \hat{\otimes} E_2)' \mid I^c(\xi_2,\xi_1)T = U^c(\xi_2,\xi_1)T$ for all $\xi_2,\xi_1 \in U_2,U_1\}$ (here using the notation of (3 - 10) and (3 - 11)). The correspondence $B' \to T_{B'}$ is

$$(3-15)\quad B'(f,g) = \int_S \langle T_{B'}, f(y*) \otimes g(y)\rangle d_S(y)$$

for $f \in C_c^\infty(S,E_1)$, $g \in C_c^\infty(S,E_2)$.

We then consider a simple case and minor adaption of the above theory.

<u>Lemma 3.1</u> <u>Let</u> σ_1 <u>be any differentiable representation of</u> U_1 <u>on a Frechet space</u> E_1 <u>and</u> σ_2 <u>any continuous character on the group</u> U_2. <u>There exists a bijective linear correspondence between</u> $I'(\pi_{\sigma_1},\pi_{\sigma_2})$ <u>and the space</u>

$$F^{\sigma_1}_{\sigma_2} = \{T \in C^{\infty}_{\sigma_1}(S,E_1)' \mid \pi^{c}_{\sigma_1}(\xi_2)T = \left[\frac{\delta_{U_2}(\xi)}{\delta_S(\xi)}\right]^{1/2} \sigma_2(\xi_2)T \quad \text{for all} \quad \xi_2 \in U_2\}$$

with the equivalence given by $B' \to T'$

$$(3 - 16) \quad B'(f,g) = \int_S \langle T', f^{\sigma_1}(y*)g(y)\rangle d_S(y).$$

Proof. In the notation above $E_2 = \mathbb{C}$ and hence $E_1 \hat{\otimes}_{\mathbb{C}} E_2 = E_1$. Thus $C^{\infty}_c(S,E_1)$ is identified to $C^{\infty}_c(S,E_1 \hat{\otimes} E_2)$. Since the map $f \xrightarrow[S_{\sigma_1}]{} f^{\sigma_1}$ of $C^{\infty}_c(S,E_1)$ to $C^{\infty}_{\sigma_1}(S,E_1)$ is surjective and open, then the transpose $(S_{\sigma_1})^t\colon C^{\infty}_{\sigma_1}(S,E_1)' \to C^{\infty}_c(S,E_1)'$ is a bijection of $C^{\infty}_{\sigma_1}(S,E_1)'$ to the space of all $T \in C^{\infty}_c(S,E_1)'$, so that $T(L_{\sigma_1}) \equiv 0$. Then it is formal to check that $T \in F^{\sigma_1}_{\sigma_2}$ if and only if $(S_{\sigma_1})^t(T) \in T_U(S)$ defined above.

Q.E.D.

Remark 3.6 From Lemma 3.1 it follows that $I(\pi_{\sigma_1},\pi_{\sigma_2})$ is linearly isomorphic to $F^{\sigma_1}_{\sigma_2}$ under the hypotheses of the Lemma.

Let $D_{\sigma_2}(S) = \{f\colon S \to \mathbb{C} \mid f \in C^{\infty}(S)$ and $f(x\xi_2) = \left[\frac{\delta_{U_2}(\xi_2)}{\delta_S(\xi_2)}\right]^{1/2}$ $\sigma_2(\xi_2)f(x)$ for all $x \in S$ and $\xi_2 \in U_2\}$. Then since σ_2 is continuous (and hence differentiable), $D_{\sigma_2}(S)$ is a closed subspace of $C^{\infty}(S)$ and thus $D_{\sigma_2}(S)$ (with the induced Schwartz topology) is a locally convex complete Hausdorff topological vector space, and the representation $\underset{\sim}{\pi}_{\sigma_2}\colon f \to \underset{\sim}{\pi}_{\sigma_2}(z)f(y) = f(z^{-1}y)$ of S on $D_{\sigma_2}(S)$ is clearly differentiable.

Remark 3.7 If S/U_2 is compact, then $D_{\sigma_2}(S)$ with the induced Schwartz topology coincides with $C^{\infty}_{\sigma_2^{-1}}(S)$ constructed above.

<u>Proposition 3.2</u> *If* σ_1 *and* σ_2 *satisfy the hypotheses of Lemma* 3.1, *then the space* $F_{\sigma_2}^{\sigma_1}$ *is linearly isomorphic to the set* $\mathrm{Hom}_S(\pi_{\sigma_1}, \underset{\sim}{\pi}_{\sigma_2})$ *of all continuous* S *intertwining maps from* $C_{\sigma_1}^{\infty}(S,E_1)$ *to* $D_{\sigma_2}(S)$. *The equivalence is given by the map*

$$(3-17) \quad T \underset{\Delta}{\leadsto} (f \to \phi_f^T(x) = \langle \pi_{\sigma_1}^c(x)T, f\rangle).$$

<u>Proof</u>. It is clear by definition that $\phi_f^T \in D_{\sigma_2}(S)$ (T is continuous). Then noting that $D*\phi_f^T(x) = \langle T, \underset{\sim}{\pi}_{\sigma_2}(x)\underset{\sim}{\pi}_{\sigma_2}(D)f\rangle$, it is easy to see that the map Δ is continuous. It then suffices to show that Δ is surjective. Indeed if $L \in \mathrm{Hom}_S(\pi_{\sigma_1}, \underset{\sim}{\pi}_{\sigma_2})$ then consider the distribution $L_1 \in F_{\sigma_2}^{\sigma_1}$ so that $\langle L_1, f\rangle = L(f)(e_S)$. Then $\Delta(L_1) = L$.

Q.E.D.

<u>Remark 3.8</u> Noting that $\lambda = \left[\dfrac{\delta_{U_2}}{\delta_S}\right]^{1/2} \sigma_2$ is a differentiable character on U_2, and that the representation $\pi_{\sigma_1}^c$ of S on $C_{\sigma_1}^{\infty}(S,E_1)'$ is differentiable, we have that $F_{\sigma_2}^{\sigma_1} \subseteq \tilde{F}_{\sigma_2}^{\sigma_1} = \{T \in C_{\sigma_1}^{\infty}(S,E_1)' \mid \pi_{\sigma_1}^c(X)T = \lambda_*(X)T$ for all $X \in \mathfrak{U}_2\}$ with $\mathfrak{U}_2$ the Lie algebra of U_2 and λ_* the infinitesimal character of $\mathfrak{U}_2$ by taking the differential of λ. By an easy argument if U_2 is <u>connected</u> then $F_{\sigma_2}^{\sigma_1} = \tilde{F}_{\sigma_2}^{\sigma_1}$.

<u>Corollary 1 to Proposition 3.2</u> *Let* $U_1 = U_2$ *with* U_1 *connected and* σ_1 *a unitary character of* U_1 *so that* $\sigma_2 = \sigma_1^{-1} = \sigma_1^c$. *Then the unitary representation* $\tilde{\pi}_{\sigma_1}$ *of* S *on* E_{σ_1} *is irreducible if* $\dim(\tilde{F}_{\sigma_1^{-1}}^{\sigma_1} = \{T \in C_{\sigma_1}^{\infty}(S)' \mid \pi_{\sigma_1}^c(X)T = \lambda_{\sigma_2}(X)T$ for all $X \in \mathfrak{U}_1\})$ *is equal to* 1 (*where* λ_{σ_2} *is the differential of the character* $\left[\dfrac{\delta_{U_1}}{\delta_S}\right]^{1/2} \sigma_1^{-1}$ *on* U_1).

We return to the determination of the space $T_U(S)$ relative to the methods used above. For the double coset $Q_y = U_2\, y\, U_1$ we consider the corresponding space $T_{U,Q_y}(S)$. If L_g $(g \in S)$ is left translation on S, we consider the induced representation L_g^* on the space $C_c^\infty(S, E_1 \hat{\otimes} E_2)$. An easy argument shows that $(L^*_{y^{-1}})^c(T_{U,Q_y}(S)) = \{T \in C_c^\infty(S, E_1 \hat{\otimes} E_2) \mid I^c(\xi_2, \xi_1)T = V^c(\xi_2, \xi_1)T$ for all $\xi_2 \in y^{-1}U_2 y$ and $\xi_1 \in U_1$ and support $(T) \subseteq y^{-1}Q_y\}$, where V is the differentiable representation of $(y^{-1}U_2 y) \times U_1$ on $E_1 \hat{\otimes} E_2$ given by $[\delta_S(\xi_1\xi_2^{-1})\delta_{U_1}(\xi_1)\delta_{U_2}(y\xi_2 y^{-1})]^{1/2}\sigma_1(\xi_1) \hat{\otimes} \sigma_2(y\xi_2 y^{-1})$ for $\xi_2 \in y^{-1}U_2 y$ and $\xi_1 \in U_1$. Thus we apply the methods above relative to the group $(y^{-1}U_2 y) \times U_1$ (the action of $(y^{-1}U_2 y) \times U_1$ on S given as $u \mapsto rus^{-1}$ with $r \in y^{-1}U_2 y$ and $s \in U_1$) and the representation V. We note that the isotropy group of $y^{-1}Q_y$ (= the orbit of $(y^{-1}U_2 y) \times U_1$ of the identity element) is $\tilde{U}_y = \{(\zeta, \zeta) \in (y^{-1}U_2 y) \times U_1 \mid \zeta \in (y^{-1}U_2 y) \cap U_1\}$; we observe that $\tilde{U}_y$ is Lie group isomorphic to $U_y = (y^{-1}U_2 y) \cap U_1$.

We observe that the tangent space of S at e_S is $\mathscr{S}$ and that the tangent space of $y^{-1}Q_y$ at e_S is $\mathfrak{U}_1 + \mathrm{Ad}(y^{-1})\mathfrak{U}_2 = \mathfrak{U}_y$, where $\mathfrak{U}_i$ is the Lie algebra of U_i, $i = 1, 2$. And since the U_i are closed, it follows that the Lie algebra of U_y is $[\mathrm{Ad}(y^{-1})(\mathfrak{U}_2)] \cap \mathfrak{U}_1$. If $\mathscr{S}_{\mathbb{C}}$ and $(\mathfrak{U}_i)_{\mathbb{C}}$ are the complexifications of $\mathscr{S}$ and $\mathfrak{U}_i$, $i = 1, 2$, respectively, then $(\mathfrak{U}_1)_{\mathbb{C}} + \mathrm{Ad}(y^{-1})(\mathfrak{U}_2)_{\mathbb{C}} = ({}_y\mathscr{S})_{\mathbb{C}}$ is stable under $\mathrm{Ad}_S(U_y)$, and as a result $t_y = \mathscr{S}_{\mathbb{C}}/({}_y\mathscr{S})_{\mathbb{C}}$ is stable under $\mathrm{Ad}_S(U_y)$. Let $S^\ell(t_y)$ be the ℓ-th component of the symmetric algebra on t_y and Λ_ℓ the corresponding canonical representation of $\mathrm{Ad}_S(U_y)$ on $S^\ell(t_y)$. Then taking the appropriate modular functions and using the identification between $\tilde{U}_y$ and U_y, we see that the "representation $\left(\dfrac{\delta_S}{\delta_{S_p}}\right) \tilde{\nu}_p^{\,\ell}$" (see notation in

(3 - 12)) of U_y is equivalent to the representation $A_\ell^*(\xi) = [\delta_{U_1}(\xi)\delta_{U_2}(y\xi y^{-1})]\delta_{U_y}(\xi^{-1})\Lambda_\ell(\xi)$ for $\xi \in U_y$.

On the other hand the representation V restricted to $\tilde{U}_y$ is equivalent to the representation of U_y on $E_1 \hat{\otimes} E_2$ given by $\psi_y^*(\xi) = [\delta_{U_1}(\xi)\delta_{U_2}(y\xi y^{-1})]^{1/2}\sigma_1(\xi) \hat{\otimes} \sigma_2(y\xi y^{-1})$. Let $I(\sigma_1,\sigma_2,y,\ell)$ be the space of all continuous U_y intertwining maps between ψ_y^* and A_ℓ^*. We note that if $\tilde{A}_\ell(\xi) = [\delta_{U_1}(\xi)\delta_{U_2}(y\xi y^{-1})]^{-1/2}A_\ell^*(\xi)$ and $\psi_y(\xi) = [\delta_{U_1}(\xi)\delta_{U_2}(y\xi y^{-1})]^{-1/2}\psi_y^*(\xi)$, then it is easy to see that $\dim I(\sigma_1,\sigma_2,y,\ell) = \dim \mathrm{Hom}_{U_y}(\psi_y^*,A_\ell^*) = \dim \mathrm{Hom}_{U_y}(\psi_y,\tilde{A}_\ell)$. Then with the assumption that S has (at most) countably many U_1, U_2 double cosets and by use of the criteria above

$$(3 - 18) \quad \dim I(\pi_{\sigma_1},\pi_{\sigma_2}) \leq \sum \dim I(\sigma_1,\sigma_2,y,\ell),$$

where the sum ranges over all $\ell \geq 0$ and all distinct representatives of the U_1, U_2 double cosets in S.

Remark 3.9 The results of Bruhat's thesis are in fact more specific in this case. That is, if there are countably many U_1, U_2 double cosets in S and if Σ_1 is a set of representatives of those cosets $\{U_2 \, z \, U_1\}$ which are closed in S, then $\dim I(\pi_{\sigma_1},\pi_{\sigma_2}) \geq \sum_{z \in \Sigma_1} \dim I(\sigma_1,\sigma_2,z,0)$.

Another point in Bruhat's thesis is the treatment of the case when $U_1 = U_2$. In such case if $\dim I(\sigma_1,\sigma_2,y,m) = 0$ for all m and all y so that $y \notin U_1$, then $\dim I(\pi_{\sigma_1},\pi_{\sigma_2}) = \dim I(\sigma_1,\sigma_2)$ (see [25], p. 422).

We now give several applications of the above theory.

We assume here for the rest of this section that $k \geq 3$.

Let $S = O(Q)$. Then $\delta_{O(Q)} = 1$ since $O(Q)$ is unimodular. Let $U_1 = U_2 = P_a$. Then by an easy argument $\delta_{P_a}(m) = \delta_{P_a}(n) = 1$ for

$m \in M_a$ and $n \in N_a(Y)$ and $\delta_{P_a}(A_a(r)) = |r|^{2-k}$ for $r \in \mathbb{R}^*$.

We recall the Bruhat decomposition of $O(Q)$ relative to P_a (see Corollary 1 to Proposition 2.4).

We let $y = d(a)$. Then $U_{d(a)} = P_a \cap d(a)P_a d(a) = M_a \cdot A_a$, which is unimodular. Then $\mathfrak{S}_{d(a)} = \mathfrak{P}_a + Ad(d(a))\mathfrak{P}_a = \mathfrak{S}$, and hence $t_{d(a)} = \{0\}$. Thus $S^{\ell}(t_{d(a)}) = \mathbb{C}$ if $\ell = 0$ and $= 0$ if $\ell > 0$. As a result $\tilde{A}_{\ell}(mA_a(r)) = [\delta_{P_a}(mA_a(r))\delta_{P_a}(md(a)A_a(r)d(a))]^{1/2}\wedge_{\ell}(mA_a(r)) = I$ if $\ell = 0$ and $= 0$ if $\ell > 0$.

We then let $y = c(a, a+1)$ (with $b > 1$). From Remark 2.3 we have that $U_{c(a,a+1)} = P_a \cap c(a, a+1)P_a c(a, a+1) = \tilde{P}_a$. Then $A_a(r_1)A_{a+1}(r_2) \subseteq \tilde{P}_a$ and using the determination of the Lie algebra of $\tilde{P}_a$ in Remark 2.3, we see that $\delta_{\tilde{P}_a}(A_a(r_1)\cdot A_{a+1}(r_2)) = |r_1|^{3-k}|r_2|^{3-k}$. Moreover $\mathfrak{S}_{c(a,a+1)} = \mathfrak{P}_a + Ad(c(a, a+1))(\mathfrak{P}_a) = \mathfrak{P}_a + \sum_{2\leq j\leq b} \mathfrak{U}_{\lambda_j-\lambda_1} + \sum_{2<j\leq b} \mathfrak{U}_{-(\lambda_j+\lambda_1)} + \mathfrak{U}_{-\lambda_1}$. Thus

$t_{c(a,a+1)} \cong \mathfrak{U}_{-(\lambda_1+\lambda_2)}$ and $\wedge_{\ell}(A_a(r_1)A_{a+1}(r_2)) = (r_1r_2)^{-\ell}$ for $\ell \geq 0$. Hence $\tilde{A}_{\ell}(A_a(r_1)A_{a+1}(r_2)) = [\delta_{P_a}(A_a(r_1)A_{a+1}(r_2))\delta_{P_a}(A_{a+1}(r_1)A_a(r_2))]^{1/2}$

$\delta_{\tilde{P}_a}(A_a(r_1^{-1})A_{a+1}(r_2^{-1}))(r_1r_2)^{-\ell} = |r_1r_2|^{\frac{k-4}{2}}(r_1r_2)^{-\ell}$.

Let $y = e_{O(Q)}$. Then $U_e = P_a$ and $t_e = \mathfrak{S}_{\mathbb{C}}/(\mathfrak{P}_a)_{\mathbb{C}} \cong \sum_{\beta\in\Delta_+-(\Delta_+\cap\Sigma_T)} \mathfrak{U}_{-\beta}$ (Σ_T defined in section 2). Then $\wedge_{\ell}(A_a(r)) = \bigoplus_1^{v_\ell} r^{-\ell}$ where v_ℓ is $\dim S^{\ell}(t_e)$. The representation $Ad(M_a)$ on t_e is obtained by taking the complexification of the standard orthogonal representation of $M_a(\cong O(a-1, b-1))$ on $\mathbb{R}^{k-2}$. The decomposition of M_a on $S^{\ell}(t_e)$ is well known. Indeed we have for every integer $s \geq 0$ a distinguished subspace $(s) \cong H^s(\mathbb{R}^{k-2})$ of $S^s(\mathbb{R}^{k-2})$ (called the harmonic polynomials) which is irreducible under M_a (if $k = 3$, then $H^0 = \mathbb{C}$, $H^1 = \mathbb{C}\cdot t$ and $H^s = 0$ for $s \geq 2$). Thus the representation of M_a in $S^\ell(\mathbb{R}^{k-2})$ is equivalent to (i) the direct sum $(\ell) + (\ell-2) + \dots + (\ell-2[\frac{\ell}{2}])$ if $k > 3$ (where $[\ ]$ = "greatest integer function") or to (ii) (u_ℓ) where $\ell \equiv u_\ell \bmod 2$ with $u = 0, 1$, if $k = 3$. We note that (ℓ_1) and (ℓ_2) are inequivalent M_a representations if $\ell_1 \neq \ell_2$ when $k > 3$, and (ℓ) is a self dual representation of M_a for all ℓ. Then noting by a simple computation that $[\mathfrak{N}_a, \mathfrak{S}] \subseteq \mathfrak{P}_a$ (see section 2, $\mathfrak{S}$ = Lie algebra of $O(Q)$), we have that N_a acts as the identity representation on $S^1(t_e)$ (and hence acts as the identity representation on all $S^\ell(t_e)$). Thus we have that $\widetilde{\Lambda}_\ell(mA_a(r)N_a(Y)) = [\delta_{P_a}(A_a(r))^2]^{1/2}\delta_{P_a}(A_a(r^{-1}))\Lambda_\ell(mA_a(r)N_a(Y)) = r^{-\ell}\{(\ell) + (\ell-2) + \dots + (\ell - 2[\frac{\ell}{2}])\}(m)$ for $k > 3$ <u>and</u> $= r^{-\ell}(u_\ell)(m)$ for $k = 3$.

We let $\lambda\colon \mathbb{R}^* \to \mathbb{C}$ be a quasicharacter on R^*. Then λ is given by the pair (λ_0, ϵ), $\lambda_0 \in \mathbb{C}$, $\epsilon = \begin{cases} 1 & \text{if } \tilde{\epsilon} = 0 \\ -1 & \text{if } \tilde{\epsilon} = 1 \end{cases}$ with $\lambda(x) = |x|^{\lambda_0}(\mathrm{sgn}(x))^{\tilde{\epsilon}}$. Moreover we denote by $\underset{\sim}{\ell}$ (ℓ integer) the quasicharacter $x \to x^\ell$; hence $\underset{\sim}{\ell} = (\ell, \epsilon(\ell))$ where $\epsilon(\ell) = \begin{cases} 1 & \text{if } \ell \text{ even} \\ -1 & \text{if } \ell \text{ odd} \end{cases}$. By convention we extend λ to a function on $\mathbb{R}$ by setting at $r = 0$, $\lambda(r) = 0$. Then we observe that for every fixed $r \in \mathbb{R}$, the map $\lambda_0 \to \lambda(r)$ is an analytic function of $\lambda_0 \in \mathbb{C}$.

We then consider a quasicharacter λ on $A_a(r)$: $A_a(r) \to \lambda(r)$. Let β be any differentiable representation of M_a on a Frechet space F_β. Then we consider the differentiable representation $\beta \otimes \lambda$: $M_aA_a(r)N_a(Y) \to \beta(m)\lambda(r)$ of P_a on the space $F_\beta \otimes_{\mathbb{C}} \mathbb{C} \cong F_\beta$. Then for β_1, β_2 two such representations of M_a and λ_1, λ_2 two quasicharacters, we determine an estimate for $I(\pi_{\beta_1 \otimes \lambda_1}, \pi_{\beta_2 \otimes \lambda_2})$.

First let $y = d(a)$. Then $\psi_{d(a)}(mA_a(r)) = \lambda_1(r)\lambda_2(r)\beta_1(m) \hat{\otimes} \beta_2(m)$, and hence we have that $\dim I(\beta_1 \otimes \lambda_1, \beta_2 \otimes \lambda_2, d(a), s) = 0$ if either

$s > 0$ or $\lambda_1 \neq \lambda_2$. On the other hand if $s = 0$ and $\lambda_1 = \lambda_2$, then $\dim I(\beta_1 \otimes \lambda, \beta_2 \otimes \lambda, d(a), 0) \leq \dim \mathrm{Hom}_{M_a}(\beta_1 \hat{\otimes} \beta_2, I)$, where I is the one dimensional identity representation of M_a.

If $y = c(a, a+1)(b > 1)$, then $\psi_{c(a,a+1)}(A_a(r_1)A_{a+1}(r_2)) = \lambda_1(r_1)\lambda_2(r_2)\beta_1(A_{a+1}(r_2)) \hat{\otimes} \beta_2(A_{a+1}(r_1))$. We define $L_{(\beta_1,\beta_2)}$ as the representation of the group $\mathbb{R}^* \times \mathbb{R}^*$ on $F_{\beta_1} \hat{\otimes} F_{\beta_2}$ given by $(r_1,r_2) \to \beta_1(A_{a+1}(r_2)) \hat{\otimes} \beta_2(A_{a+1}(r_1))$. Thus $\dim I(\beta_1 \otimes \lambda_1, \beta_2 \otimes \lambda_2, c(a, a+1), \ell) \leq \dim \mathrm{Hom}_{\mathbb{R}^* \times \mathbb{R}^*}(L_{(\beta_1,\beta_2)}, (\lambda_1^{-1}\gamma_\ell, \lambda_2^{-1}\gamma_\ell))$, where $(\lambda_1^{-1}\gamma_\ell, \lambda_2^{-1}\gamma_\ell)$ is the quasicharacter on $\mathbb{R}^* \times \mathbb{R}^*$ given by $(r_1,r_2) \to \lambda_1^{-1}(r_1)|r_1|^{k/2} r_1^{-(\ell+2)} \lambda_2^{-1}(r_2)|r_2|^{k/2} r_2^{-(\ell+2)}$.

Finally let $y = e_{0(Q)}$. Then $\psi_e(A_a(r)) = \lambda_1(r)\lambda_2(r) I_{\beta_1 \otimes \beta_2}$ with $I_{\beta_1 \otimes \beta_2}$ the identity operator on $F_{\beta_1} \hat{\otimes} F_{\beta_2}$. Thus we have that if $\lambda_1\lambda_2 \neq -\underset{\sim}{\ell}$ (with ℓ a nonnegative integer), then $\dim I(\beta_1 \otimes \lambda_1, \beta_2 \otimes \lambda_2, e, \ell) = 0$. However if $\lambda_1\lambda_2 = -\underset{\sim}{\ell}$ and $k > 3$, then $\dim I(\beta_1 \otimes \lambda_1, \beta_2 \otimes \lambda_2, e, \ell) \leq \sum_{j=0}^{2\left[\frac{\ell}{2}\right]} \dim \mathrm{Hom}_{M_a}(\beta_1 \hat{\otimes} \beta_2, (\ell - 2j))$; if $k = 3$, then $\dim I(\beta_1 \otimes \lambda_1, \beta_2 \otimes \lambda_2, e, \ell) \leq \dim \mathrm{Hom}_{M_a}(\beta_1 \hat{\otimes} \beta_2, (u_\ell))$.

<u>Remark 3.10</u> If S is an arbitrary Lie group, σ a differentiable representation of S on a Frechet space E, and γ a finite dimensional, continuous (and hence differentiable) representation of S, we say S has the <u>null (σ, γ) property</u> if $\dim \mathrm{Hom}_S(\sigma, \gamma) = 0$. We note that if γ is a continuous character of S, then S has the <u>null</u> (σ, γ) <u>property</u> if and only if $\dim \{f \in E^\vee \mid \sigma^c(s)(f) = \gamma^{-1}(s)\cdot f \text{ for all } s \in S\} = 0$. And we also observe that S has the <u>null</u> $(\beta_1 \hat{\otimes} \beta_2, I)$ <u>property</u> (where β_i is a differentiable representation of S on a Frechet space F_i, $i = 1, 2$) if and

only if $\dim I(\beta_1,\beta_2) = 0$.

Proposition 3.3 *Consider the following conditions:*

(i) *either* $\lambda_1 \neq \lambda_2$ *or* M_a *has the null* $(\beta_1 \hat{\otimes} \beta_2, I)$ *property;*

(ii) *if* $b > 1$ *then* $\mathbb{R}^* \times \mathbb{R}^*$ *has the null* $(L_{(\beta_1,\beta_2)}, (\lambda_1^{-1}\gamma_\ell, \lambda_2^{-1}\gamma_\ell))$ *property for all integers* $\ell \geq 0$;

(iii) *for every integer* $\ell \geq 0$ *either* $\lambda_1\lambda_2 \neq -\ell$ *or* M_a *has the null* $(\beta_1 \hat{\otimes} \beta_2, [s])$ *property for* $s = \ell, \ell - 2, \ldots, \ell - 2\left[\frac{\ell}{2}\right]$ *if* $k > 3$, and for $s = (u_\ell)$ if $k = 3$.

If conditions (i) *and* (ii) *hold then* $\dim I(\pi_{\beta_1} \otimes \lambda_1, \pi_{\beta_2} \otimes \lambda_2) = \dim I(\beta_1 \otimes \lambda_1, \beta_2 \otimes \lambda_2)$. *If conditions* (ii) *and* (iii) *hold, then* $\dim I(\pi_{\beta_1} \otimes \lambda_1, \pi_{\beta_2} \otimes \lambda_2) \leq \dim I(\beta_1,\beta_2)$.

Proof. The proof follows by what has been done above and Remark 3.9.

Q.E.D.

We consider a special case of Proposition 3.3. Namely we let $\beta_1 = \beta_2 = I$, the identity representation of M_a.

Corollary 1 to Proposition 3.3 *In general if* $\lambda_2 \neq \lambda_1$ *and* $\lambda_2 \neq \lambda_1^{-1}$, *then* $\dim \mathrm{Hom}_{O(Q)}(\pi_{I \otimes \lambda_1}, \pi_{I \otimes \lambda_2}) = 0$. *Moreover if* $\lambda \neq (0, 1)$ *and* $\lambda \neq (0, -1)$, *then* $\dim \mathrm{Hom}_{O(Q)}(\pi_{I \otimes \lambda}, \pi_{I \otimes \lambda}) = 1$. *Thus if* $\lambda = (it, \epsilon)$ *with* $t \neq 0$, *then the unitary induced representation* $\tilde{\pi}_{I \otimes \lambda}$ *of* $O(Q)$ *on the Hilbert space* $E_{I \otimes \lambda}$ *is irreducible.*

Corollary 2 to Proposition 3.3

(i) *If* $b = 1$ *and* $\lambda \notin \{(j,\epsilon) \mid j \leq 0$ *and* j *integer*$\}$ *then* $\dim \mathrm{Hom}_{O(Q)}(\pi_{I \otimes \lambda}, \pi_{I \otimes \lambda^{-1}}) \leq 1$. *However if* $\lambda = (j,\epsilon)$ *with* j

integer <u>and</u> $j \leq 0$ <u>then</u> $\dim \mathrm{Hom}_{O(Q)}(\pi_{I \otimes \lambda}, \pi_{I \otimes \lambda^{-1}}) \leq 2$.

(ii) <u>If</u> $b > 1$ <u>and</u> $\lambda \notin \{(j,\epsilon) \mid j$ integer <u>and</u> $j \leq 0\} \cup \{(\frac{1}{2}k - s, \epsilon(s)) \mid s \geq 2\}$, <u>then</u> $\dim \mathrm{Hom}_{O(Q)}(\pi_{I \otimes \lambda}, \pi_{I \otimes \lambda^{-1}}) \leq 1$.

<u>If</u> $b > 1$ <u>and</u> k <u>is odd and</u> $\lambda \notin \{(j,\epsilon) \mid j$ integer and $j \leq 0\} \cup \{(\frac{1}{2}k - s, \epsilon(s)) \mid s \geq 2\}$, <u>then</u> $\dim \mathrm{Hom}_{O(Q)}(\pi_{I \otimes \lambda}, \pi_{I \otimes \lambda^{-1}}) \leq 2$.

<u>If</u> $b > 1$ <u>and</u> k <u>is even and either</u> $\lambda = (j, \epsilon')$ (j integer and $j \leq 0$), <u>so that</u> $\epsilon' \neq \epsilon(\frac{1}{2}k - j)$, <u>or</u> $\lambda = (\frac{1}{2}k - s, \epsilon(s))$ <u>so that</u> $s \geq 2$ <u>and</u> $\frac{1}{2}k - s > 0$, <u>we have</u> $\dim \mathrm{Hom}_{O(Q)}(\pi_{I \otimes \lambda}, \pi_{I \otimes \lambda^{-1}}) \leq 2$.

<u>If</u> $b > 1$ <u>and</u> k <u>is even and</u> $\lambda = (j, \epsilon')$ (j integer and $j \leq 0$) <u>so that</u> $\epsilon' = \epsilon(\frac{1}{2}k - j)$, <u>then</u> $\dim \mathrm{Hom}_{O(Q)}(\pi_{I \otimes \lambda}, \pi_{I \otimes \lambda^{-1}}) \leq 3$.

Q.E.D.

Recalling that $O(Q)/P_a$ is compact, we know that $C^\infty_{I \otimes \lambda}(O(Q)) = C^\infty_\lambda(O(Q)) = \{f\colon O(Q) \to \mathbb{C} \mid f \in C^\infty(O(Q))$ and $f(gmA_a(r)N_a(Y)) = f(g)|r|^{\frac{2-k}{2}} \lambda^{-1}(r)$ for all $g \in O(Q)$, $m \in M_a$, $r \in \mathbb{R}^*$ and $Y \in \mathbb{R}^{k-2}\}$. Noting that the parabolic set $\Gamma_o - \{0\} \cong O(Q)/M_a \cdot N_a$, we consider the linear space $S_\lambda(\Gamma_o) = \{\tilde{f}\colon \Gamma_o - \{0\} \to \mathbb{C} \mid \tilde{f} \in C^\infty(\Gamma_o - \{0\})$ and $\tilde{f}(t\xi) = |t|^{\frac{2-k}{2}} \lambda^{-1}(t)\tilde{f}(\xi)$ for all $t \in \mathbb{R}^*$ and $\xi \in \Gamma_o - \{0\}\}$. The map $f \to \tilde{f}(C^\infty_\lambda(O(Q)) \to S_\lambda(\Gamma_o))$ given by $f \to \tilde{f}(\xi) = f(g \cdot v_a)$ with $\xi = gv_a$ defines a linear isomorphism between $C^\infty_\lambda(O(Q))$ and $S_\lambda(\Gamma_o)$. We then transport the topology of $C^\infty_\lambda(O(Q))$ to $S_\lambda(\Gamma_o)$ via this identification.

We note that every element $X \in \Gamma_o - \{0\}$ has the unique form $X = t(\omega_1 + \omega_2)$, $\omega_1 \in S^{a-1}$, $\omega_2 \in S^{b-1}$, with $t > 0$; here $S^{a-1} = \{Z \in \mathbb{R}e_1 \oplus \dots \oplus \mathbb{R}e_a \mid Q(X,X) = 1\}$ ($\cong$ the unit sphere in $\mathbb{R}^a$) while

$S^{b-1} = \{Z \in \mathbb{R}e_{a+1} \oplus \dots \oplus \mathbb{R}e_k \mid Q(X,X) = -1\}$ ($\cong$ the unit sphere in $\mathbb{R}^b$).

Then $\tilde{f} \in S_\lambda(\Gamma_o)$ can be written as $\tilde{f}(t(\omega_1 + \omega_2)) = |t|^{\frac{2-k}{2} - \lambda_o} \varphi_{\tilde{f}}(\omega_1 + \omega_2)$, where $\varphi_{\tilde{f}} \in C^\infty(S^{a-1} \times S^{b-1})$ and $\varphi_{\tilde{f}}$ is even (odd) as $\epsilon = 1$ ($= -1$). For $g \in O(Q)$ then $g^{-1}\xi = r(g^{-1}\cdot\xi)(\omega_1(g^{-1}\cdot\xi) + \omega_2(g^{-1}\cdot\xi))$ relative to the decomposition above. For any $s \in \mathbb{C}$ consider the function $\phi_s(g) = [r(g\cdot v_a)]^s$ for $g \in O(Q)$. Then $\phi_s(gmA_a(r)N_a(Y)) = |r|^s\phi_s(g)$ and $\phi_s(g) > 0$ for $s \in \mathbb{R}$. Moreover $\phi_s \in C^\infty(O(Q))$ and for $s \in \mathbb{R}$, ϕ_s is bounded above and below on any compact set ω in $O(Q)$. Indeed the projection ω' of ω in $\Gamma_o - \{0\}$ is compact and hence $\omega' \subseteq \{S_1(S^{a-1} \times S^{b-1}) \mid S_1$ a compact subset of $\mathbb{R}^*\}$. If $s = 2 - k$ then ϕ_{2-k} is a <u>rho</u> function for $O(Q)/P_a$ with $\phi_{2-k}(e_{O(Q)}) = 1$. Let $d\mu_k$ be the $O(Q)$ quasi-invariant measure on $O(Q)/P_a$ chosen relative to ϕ_{2-k} as above.

The representation σ_λ of $O(Q)$ transported to $S_\lambda(\Gamma_o)$ is given by

$$(3 - 19) \quad \sigma_\lambda(g)(\tilde{f})(g'\cdot v_a) = \phi_{\frac{2-k}{2} - \lambda_o}(g^{-1}g')\varphi_{\tilde{f}}(\omega_1(g^{-1}g'\cdot v_a) + \omega_2(g^{-1}g'\cdot v_a))$$

for $g, g' \in O(Q)$.

We recall from [2] that the measure $d\mu_k$ defines on the space $C^\infty_{(\frac{k-2}{2},1)}(O(Q))$ a <u>positive continuous linear form</u>

$$(3 - 20) \quad \nu(\varphi) = \oint_{O(Q)/P_a} \varphi(x)d\mu_k(\dot{x}), \quad (\varphi \in C^\infty_{(\frac{k-2}{2},1)}(O(Q))),$$

which is invariant by left translation by $O(Q)$. Then we note that for $\varphi_1 \in C^\infty_\lambda(O(Q))$ and $\varphi_2 \in C^\infty_\beta(O(Q))$, so that $\lambda_o = -\beta_o$ and $\epsilon(\lambda) = \epsilon(\beta)$, we have $\varphi_1\cdot\varphi_2 \in C^\infty_{(\frac{k-2}{2},1)}(O(Q))$. Thus

$$(3 - 21) \quad (\varphi_1 \mid \varphi_2) = \nu(\varphi_1\cdot\varphi_2)$$

defines a continuous bilinear form on $C^\infty_\lambda(O(Q))$ and $C^\infty_\beta(O(Q))$ (so that $\lambda_o = -\beta_o$ and $\varepsilon(\lambda) = \varepsilon(\beta)$). By the invariance of ν, we see that $(\sigma_\lambda(g)\varphi_1 | \sigma_\beta(g)\varphi_2) = (\varphi_1 | \varphi_2)$ for all $g \in O(Q)$.

Moreover recalling that $O(Q) = K \cdot P_a$ and then using (3 - 9), we deduce that relative to some choice of Haar measure dk on K

$$(3 - 22) \quad (\varphi_1 | \varphi_2) = c \int_K \varphi_1(k)\varphi_2(k)dk, \ (c \neq 0).$$

For any s dimensional Euclidean space $\mathbb{R}^s$, we normalize the Haar measure $d\sigma_s$ on the compact orthogonal group $O(s)$ so that $\int_{O(s)} d\sigma_s = 1$; this then implies that for $f \in L^1(O(s), d\sigma_s)$ we have

$$(3 - 23) \quad \int_{O(s)} f(z)d\sigma_s(z) = \int_{S^{s-1}} \{\int_{O(s-1)} f(zy)d\sigma_{s-1}(y)\}d\omega_s,$$

where $d\omega_s$ is the unique $O(s)$ invariant measure on S^{s-1}, the s-1 sphere in $\mathbb{R}^s$, so that for the usual Euclidean measure dX on $\mathbb{R}^s$, we have $dX = \text{volume}(S^{s-1})|r|^{s-1}dr d\omega_s$ (here we have the convention that $O(o) = \{1\}$ and $O(1) \cong \mathbb{Z}_2$, the 2-element group).

Proposition 3.4 The bilinear form $(\ |\ )$ defined in (3 - 21) is, up to scalars, the unique $O(Q)$ intertwining form between $S_\lambda(\Gamma_o)$ and $S_\beta(\Gamma_o)$ with $\lambda_o + \beta_o = 0$ and $\varepsilon(\lambda) = \varepsilon(\beta)$. Moreover for $\tilde{f}_1 \in S_\lambda(\Gamma_o)$, $\tilde{f}_2 \in S_\beta(\Gamma_o)$ with the above relation between λ and β, we have

$$(3 - 24) \quad (f_1 | f_2) = c_1 \int_{S^{a-1} \times S^{b-1}} \varphi_{\tilde{f}_1}(\omega_1 + \omega_2)\varphi_{\tilde{f}_2}(\omega_1 + \omega_2)d\omega_a(\omega_1)d\omega_b(\omega_2)$$

(c_1, a nonzero constant). And if for $\tilde{f} \in S_\lambda(\Gamma_o)$ and for any λ, we define the "norm" of f as

$$(3 - 25) \quad \|\tilde{f}\|^2 = \int_{S^{a-1} \times S^{b-1}} |\varphi_{\tilde{f}}(\omega_1 + \omega_2)|^2 d\omega_a(\omega_1)d\omega_b(\omega_2),$$

then σ_λ defines a continuous representation of $O(Q)$ on $S_\lambda(\Gamma_o)$, equipped with the topology given by $\|\ \|$.

Proof. The proof follows by using Corollary 1 to Proposition 3.3 and the discussion preceding Proposition 3.4. The continuity of σ_λ in the $\| \ \|$ topology is a standard argument similar to that of [25], p. 445.

Q.E.D.

Remark 3.11 For any λ we define $B_\lambda(\Gamma_o) = \{f\colon \Gamma_o - \{0\} \to \mathbb{C} \mid f$ is Borel measurable, $f(t\xi) = |t|^{\frac{2-k}{2}} \lambda^{-1}(t) f(\xi)$, $t \in \mathbb{R}^*$, $\xi \in \Gamma_o - \{0\}$, and $\|f\| < \infty\}$. As above we let $B_\lambda(\Gamma_o)$ be the function $f \in B_\lambda(\Gamma_o)$ identified modulo Borel null sets of $\Gamma_o - \{0\}$. Then we note that the map $\tilde{f} \mapsto \varphi_{\tilde{f}}$ ($\varphi_{\tilde{f}}$ = the restriction of $\tilde{f}$ to $S^{a-1} \times S^{b-1}$) of $(B_\lambda(\Gamma_o), \| \ \|)$ to $L^2(S^{a-1} \times S^{b-1}, \begin{cases} \text{even if } \varepsilon(\lambda) = 1 \\ \text{odd if } \varepsilon(\lambda) = -1 \end{cases}, d\omega_a \times d\omega_b)$ is a bijective linear isometry; thus $B_\lambda(\Gamma_o)$ is a Hilbert space relative to $\| \ \|$. Moreover by the same reasoning, $S_\lambda(\Gamma_o)$ is dense in $B_\lambda(\Gamma_o)$. Then by Proposition 3.4 σ_λ extends uniquely to a continuous representation $\tilde{\sigma}_\lambda$ of $O(Q)$ on $B_\lambda(\Gamma_o)$. And if $\lambda = (it, \varepsilon)$, then the representation $\tilde{\sigma}_\lambda$ is unitary and is unitarily equivalent to the representation $\tilde{\pi}_{I\otimes\lambda}$ of $O(Q)$ on E_λ constructed above.

Proposition 3.5 *For all $\lambda \in \mathbb{C}$, $S_\lambda(\Gamma_o)$ is exactly the space $(B_\lambda(\Gamma_o))_\infty$ of C^∞ vectors of the representation $\tilde{\sigma}_\lambda$ of $O(Q)$ on $B_\lambda(\Gamma_o)$.*

Proof. We follow the same argument as in the proof of Lemma 5.2 in [20]. Using the identification $C^\infty_\lambda(O(Q))$ to $S_\lambda(\Gamma_o)$ (via the map $\sim$ above), we see that, by simple application of Sobolev's Lemma, there exists for any compact set T of $O(Q)$ a family $X^\gamma \in \mathcal{U}(\mathfrak{G})$ so that

$$\text{(3 - 26)} \quad \sup_{X \in T} |f(X)| \le c_T \sum_\gamma \{ \int_K | X^\gamma * f(k) |^2 dk \}^{1/2}$$

(c_T a positive constant depending only on T), for all $\tilde{f} \in S_\lambda(\Gamma_o)$. We let $\psi \in (B_\lambda(\Gamma_o))_\infty$. From the comments in the beginning of section 3, we know there exists a sequence $\psi_n^\sim \in S_\lambda(\Gamma_o)$ so that $\|(V*\psi_n) - \tilde{\sigma}_\lambda(V)\psi\| \to 0$ as $n\to\infty$ for all $V \in \mathcal{U}(\mathfrak{G})$. Indeed $S_\lambda(\Gamma_o)$ is dense in $(B_\lambda(\Gamma_o))_\infty$ relative to the $(B_\lambda(\Gamma_o))_\infty$ topology, and we consider the seminorm $|\varphi|_{D,\{e\}} = \|D*\varphi\|$ $(\varphi \in C^\infty(O(Q), B_\lambda(\Gamma_o)), D \in \mathcal{U}(\mathfrak{G}))$ on the embedding of $(B_\lambda(\Gamma_o))_\infty \xhookrightarrow{\flat} C^\infty(O(Q), B_\lambda(\Gamma_o))$ given in the beginning of section 3. But note that for φ of the form $\beta^\flat$, $\beta \in (B_\lambda(\Gamma_o))_\infty$, we have $|\beta^\flat|_{D,\{e\}} = \|\tilde{\sigma}_\lambda(D)\beta\|$.

But we note from (3 - 26) that ψ_n is a Cauchy sequence in the Schwartz topology of $C_\lambda^\infty(O(Q))$. Thus there exists a unique function $\omega \in C_\lambda^\infty(O(Q))$ so that $V*\psi_n \to V*\omega$ uniformly on all compact subsets T of $O(Q)$ for all $V \in \mathcal{U}(\mathfrak{G})$. Thus we have $V*\omega(g) = \tilde{\sigma}_\lambda(V)\psi(g)$ for almost all $g \in O(Q)$ and all $V \in \mathcal{U}(\mathfrak{G})$.

Q.E.D.

Corollary 1 to Proposition 3.5 (1) If $\lambda_1 \neq \lambda_2$ and $\lambda_1 \neq \lambda_2^{-1}$, then $\mathrm{Hom}_{O(Q)}(\tilde{\sigma}_{\lambda_1}, \tilde{\sigma}_{\lambda_2}) = 0$. If $\lambda \neq (0,\pm 1)$ then $\mathrm{Hom}_{O(Q)}(\tilde{\sigma}_\lambda, \tilde{\sigma}_\lambda) = \mathbb{C}$.

(2) If $b = 1$ and $\lambda \notin \{(j,\epsilon) \mid j \leq 0 \text{ and } j \text{ integer}\}$, then $\dim \mathrm{Hom}_{O(Q)}(\tilde{\sigma}_\lambda, \tilde{\sigma}_{\lambda^{-1}}) \leq 1$.

If $b > 1$ and $\lambda \notin \{(j,\epsilon) \mid j \leq 0 \text{ and } j \text{ integer}\} \cup \{(\frac{1}{2}k - s, \epsilon(s)), s \geq 2, s \text{ integer}\}$ then $\dim \mathrm{Hom}_{O(Q)}(\tilde{\sigma}_\lambda, \tilde{\sigma}_{\lambda^{-1}}) \leq 1$.

Proof. We use the comments in the beginning of section 3 concerning intertwining maps and Corollaries 1 and 2 of Proposition 3.3.

Q.E.D.

We then consider a second application of Bruhat's theory. Again let $S = O(Q)$, $U_1 = G_t$, and $U_2 = P_a$, where G_t is the isotropy subgroup in

$O(Q)$ of the point $\tilde{v}_a + \frac{1}{2}tv_a$ with $t \neq 0$.

We first observe that G_t is unimodular for all $t \neq 0$.

Then recalling Proposition 2.4, we first let $y = e_{O(Q)}$. Then $U_e = G_t \cap P_a$. Since $M_a \subseteq G_t$, then for an element of the form $A_a(r)N_a(Y) \in G_t$, we must have $A_a(r)N_a(Y)(\tilde{v}_a + \frac{1}{2}tv_a) =$ $r^{-1}\tilde{v}_a + \frac{1}{2}(t - Q(Y,Y))rv_a + Y = \tilde{v}_a + \frac{1}{2}tv_a$; hence $Y = 0$ and $r = 1$. Thus $U_e = M_a$ for all $t \neq 0$. Since G_t is closed in $O(Q)$, we have that $\mathfrak{g}_t \cap \mathfrak{P}_a$ is the Lie algebra of $G_t \cap P_a$ ($\mathfrak{g}_t$, the Lie algebra of G_t). Then we note that $\mathfrak{S}_e = (\mathfrak{P}_a)_{\mathbb{C}} + (\mathfrak{g}_t)_{\mathbb{C}} = \mathfrak{S}_{\mathbb{C}}$ since $\dim(\mathfrak{g}_t)_{\mathbb{C}} + \dim(\mathfrak{P}_a)_{\mathbb{C}} -$ $\dim(\mathfrak{g}_t \cap \mathfrak{P}_a)_{\mathbb{C}} = \dim(\mathfrak{g}_t)_{\mathbb{C}} + \dim(\mathfrak{P}_a/\mathfrak{M}_a) = \dim \mathfrak{S}_{\mathbb{C}}$. Thus $t_e = 0$; hence we have $A_\ell(m) = I$ if $\ell = 0$ and $= 0$ if $\ell > 0$.

Then we assume $b > 1$ and let $y = c(a, a+1)$. Then $U_{c(a,a+1)} =$ $G_t \cap P_{a+1}$. Since $A_{a+1}(r) \subseteq G_t$, then for an element of the form $m_{a+1}N_{a+1}(Y) \in G_t$, we must have $m_{a+1}N_{a+1}(Y)(\tilde{v}_a + \frac{1}{2}tv_a) = m_{a+1}(\tilde{v}_a + \frac{1}{2}tv_a) +$ $(-Q(Y,\tilde{v}_a + \frac{1}{2}tv_a))v_{a+1} = \tilde{v}_a + \frac{1}{2}tv_a$; hence $Y \in \langle v_{a+1}, \tilde{v}_{a+1}, \tilde{v}_a + \frac{1}{2}tv_a\rangle^{\perp}$ in $\mathbb{R}^k$ and $m_{a+1} \in M_{a+1} \cap G_t$. Thus $G_t \cap P_{a+1}$ is the semidirect product $(M_{a+1} \cap G_t)\cdot A_{a+1}(r)\cdot N'_{a+1}$, where $N'_{a+1} = \{N_{a+1}(Y) \mid Y \in \langle v_{a+1}, \tilde{v}_{a+1}, \tilde{v}_a + \frac{1}{2}tv_a\rangle^{\perp}\}$. Then by the same reasoning as in the first application above, we have $\delta_{U_{c(a,a+1)}}(m_{a+1}A_{a+1}(r)N'_{a+1}(Z)) = \delta_{U_{c(a,a+1)}}(A_{a+1}(r)) = |r|^{3-k}$. Then $\mathfrak{g}_t \cap \mathfrak{P}_{a+1}$ is the Lie algebra of $G_t \cap P_{a+1}$, and we have $\dim(\mathfrak{g}_t)_{\mathbb{C}} + \dim(\mathfrak{P}_{a+1})_{\mathbb{C}} = \dim((\mathfrak{g}_t)_{\mathbb{C}} + (\mathfrak{P}_{a+1})_{\mathbb{C}}) + \dim((\mathfrak{g}_t)_{\mathbb{C}} \cap (\mathfrak{P}_{a+1})_{\mathbb{C}}) =$ $\dim \mathfrak{S}_{\mathbb{C}} - 1$. Thus $t_{c(a,a+1)}$ has dimension 1; since $\mathrm{Ad}(A_{a+1}(r))$ operates as r^{-1} on the $k-2$ dimensional space $\mathfrak{S}_{\mathbb{C}}/(\mathfrak{P}_{a+1})_{\mathbb{C}}$, it follows that $\Lambda_\ell(A_{a+1}(r)) = r^{-\ell}$ for all integers $\ell \geq 0$. Moreover since $t_{c(a,a+1)}$ is one dimensional and N'_{a+1} is unipotent, then $Ad(N'_{a+1})$ operates as the identity representation on $t_{c(a,a+1)}$. Thus we have $A_\ell(A_{a+1}(r)N'_{a+1}) =$

$[\delta_{P_a}(A_a(r))]^{1/2}\delta_{U_{c(a,a+1)}}(A_{a+1}(r)^{-1}) \wedge_\ell (A_{a+1}(r)) = |r|^{\frac{k}{2}-2} r^{-\ell}.$

Let $b = 1$ and $t > 0$. Let $y = \gamma_t$ (see Proposition 2.4). Then $G_t \cap \gamma_t^{-1}P_a\gamma_t = U_{\gamma_t}$ and $\gamma_t G_t \gamma_t^{-1}$ is the isotropy group of the point e_2. Then if $\gamma_t \times \gamma_t^{-1} \in \gamma_t G_t \gamma_t^{-1} \cap P_a$, we let $\gamma_t \times \gamma_t^{-1} = mA_a(r_x)N_a(Y_x)$. Then $mA_a(r_x)N_a(Y_x)(e_2) = m(e_2) - Q(Y_x,e_2)r_x \cdot v_a$. This gives that $Y_x \in \langle v_a,\tilde{v}_a,e_2\rangle^\perp$, r_x is arbitrary, and $m \in M_a \cap \gamma_t G_t \gamma_t^{-1}$. Thus $\gamma_t U_{\gamma_t}\gamma_t^{-1}$ is the semidirect product $(M_a \cap \gamma_t G_t \gamma_t^{-1})\cdot A_a(r)\cdot N_a^{\#}$, where $N_a^{\#} = \{N_a(Y) \mid Y \in \langle v_a,\tilde{v}_a,e_2\rangle^\perp\}$. As in the case above, we deduce that t_{γ_t} has dimension 1 and that $\gamma_t^{-1}A_a(r)\gamma_t$ operates as r^{-1} on t_{γ_t}; moreover $\gamma_t^{-1}N_a^{\#}\gamma_t$ is trivial on t_{γ_t}. Thus we have that

$A_\ell(\gamma_t^{-1}A_a(r)N_a(Y)\gamma_t) = [\delta_{P_a}(A_a(r))]^{1/2}\delta_{U_{\gamma_t}}(\gamma_t^{-1}A_a(r)^{-1}\gamma_t) \wedge_\ell (\gamma_t^{-1}A_a(r)\gamma_t) =$

$|r|^{\frac{k}{2}-2} r^{-\ell}.$

We then take the identity representation I_t of G_t and the differentiable representation $\beta \otimes \lambda$ of P_a constructed above.

Then if $y = e_{O(Q)}$ we have that $\psi_e(m) = \beta(m)$ on the space F_β. Thus $\dim I\,(I_t, \beta \otimes \lambda, e, \ell) \leq \dim \mathrm{Hom}_{M_a}(\beta, I)$ if $\ell = 0$ and $\dim I\,(I_t, \beta \otimes \lambda, e, \ell) = 0$ if $\ell > 0$.

Let $y = c(a, a+1)$ (and so $b > 1$). Then $\psi_{c(a,a+1)}(A_{a+1}(r)N'_{a+1}(Y)) = \lambda(r)I_\beta$, with I_β the identity operator on F_β. Thus $\dim I\,(I_t, \beta \otimes \lambda, c(a, a+1), s) = 0$ unless $\lambda = (\frac{1}{2}k - 2, 1) - (s, \epsilon(s))$.

Similarly if $y = \gamma_t$ (so that $b = 1$ and $t > 0$), we have $\psi_{\gamma_t}(\gamma_t^{-1} A_a(r) N_a^{\#}(Y)\gamma_t) = \lambda(r) I_\beta$ on F_β. Thus we deduce the same condition as above.

Proposition 3.6 We have

$$(3 - 27) \quad \dim I(\pi_{I_t}, \pi_{\beta\otimes\lambda}) \leq \dim (F_\beta^\vee)^{M_a}, \quad (t \neq 0)$$

($(F_\beta^\vee)^{M_a}$, the fixed point set of M_a in $F_\beta^\vee$) under the following conditions:

(A) if $b > 1$ then $\lambda \neq (\frac{1}{2}k - 2, 1) - (s, \varepsilon(s))$ with s a nonnegative integer;

(B) if $b = 1$ and (i) $t < 0$, then λ arbitrary, and (ii) $t > 0$, then $\lambda \neq (\frac{1}{2}k - 2, 1) - (s, \varepsilon(s))$ with s a nonnegative integer.

Q.E.D.

We now deduce from Propositions 3.6 and 3.2 a fairly interesting consequence. Let $\beta = I$ the identity representation of M_a. Take $C_c^\infty(\Gamma_t)$ $(t \neq 0)$ as the space of C^∞ functions of compact support on Γ_t (this space is canonically identified to the induced module $C_{I_t}^\infty(O(Q)) = \{f\colon O(Q) \to \mathbb{C} \mid f \in C^\infty(O(Q)),$ f has compact support mod G_t, and $f(g\, g_t) = f(g)$ for all $g_t \in G_t\}$).

Proposition 3.7 Under conditions (A) and (B) of Proposition 3.6 $(t \neq 0)$

$$(3 - 28) \quad \dim \operatorname{Hom}_{O(Q)}(C_c^\infty(\Gamma_t), S_{\lambda^{-1}}(\Gamma_o)) \leq 1$$

Q.E.D.

Remark 3.12 We note that by removing conditions (A) and (B) above, we have for any λ that $\dim \operatorname{Hom}_{O(Q)}(C_c^\infty(\Gamma_t), S_\lambda(\Gamma_o)) \leq 2$ $(t \neq 0)$.

We now consider a third application of Bruhat's theory. Let $b = 1$ above. We let $U_1 = P_a$ and $\beta \otimes \lambda$ the representation given before with

β a unitary finite-dimensional irreducible representation of M_a (recall M_a is compact). Let $U_2 = N_a$ and σ_2: $N_a(Y) \to \tau([Y,Z])$ a nontrivial unitary character on N_a $(Z \neq 0 \in \mathbb{R}^{k-2})$.

Then using the Bruhat decomposition of $P_a \backslash O(Q)/N_a$, we see by taking $y = d(a)$, that $U_{d(a)} = P_a \cap d(a)N_a d(a) = \{e\}$, and hence

$\dim I(\beta \otimes \lambda, \sigma_2, d(a), s) \leq \begin{cases} \dim F_\beta & \text{if } s = 0 \\ 0 & \text{otherwise} \end{cases}$ Then by taking $y = e_{O(Q)}$, we have $U_e = N_a$ and the representation of N_a on $S^\ell(t_e)$ $(t_e = \mathfrak{S}_{\mathbb{C}}/(\mathfrak{P}_a)_{\mathbb{C}}$ as above) is supertriangular relative to some basis of $S^\ell(t_e)$. But N_a is unimodular; thus $A_\ell(N_a(Y)) = \wedge_\ell(N_a(Y))$. Since $\sigma_2(N_a(Y)) = \tau([Y,Z])$ is nontrivial unitary, we have that $I(\beta \otimes \lambda, \sigma_2, e, s) = 0$ for all $s \geq 0$.

<u>Proposition 3.8</u> (<u>"Existence of Whittaker model weakly for continuous series of Lorentz group"</u>) $(b = 1)$

<u>For any β an irreducible unitary representation of M_a on a finite dimensional space F_β and any quasi character λ, the space</u>

$F_{\sigma_2} = \{T \in C^\infty_{\beta\otimes\lambda}(O(Q))' \mid \pi^c_{\beta\otimes\lambda}(N_a(Y))T = \tau([Y,Z])\, T \text{ for all } Y \in \mathbb{R}^{k-2}\}$

<u>has dimension at most</u> $\dim F_\beta$.

§ 4. REPRESENTATION THEORY OF METAPLECTIC GROUP

Let $\underline{G}$ be any connected covering group of $S\ell_2(\mathbb{R})$. Let $s\ell_2(\mathbb{R})$, the set of all 2×2 real matrices of trace zero, be the Lie algebra of the group $\underline{G}$ and exp: $s\ell_2(\mathbb{R}) \to \underline{G}$ be the usual exponential map. Let

$$K = \begin{bmatrix} 0 & 1 \\ -1 & 0 \end{bmatrix}, \ H = \begin{bmatrix} 1 & 0 \\ 0 & -1 \end{bmatrix}, \text{ and } E = \begin{bmatrix} 0 & 1 \\ 0 & 0 \end{bmatrix} \text{ be a basis of } s\ell_2(\mathbb{R}).$$

Let $\underline{\widetilde{K}}$, $\underline{A}$, and $\underline{N}$ be the associated one parameter groups $\exp_{\underline{G}}(tK)$, $\exp_{\underline{G}}(tH)$, and $\exp_{\underline{G}}(tE)$. We recall that $\underline{G}$ is diffeomorphic to $\underline{\widetilde{K}} \cdot \underline{A} \cdot \underline{N}$, the Iwasawa decomposition of $\underline{G}$. Let s: $\underline{G} \to S\ell_2(\mathbb{R})$ be the covering homomorphism. Let t_0 be the smallest positive number so that $s(\exp_{\underline{G}}(t_0 K)) = e_{S\ell_2(\mathbb{R})}$; then Kernel $(s) = \{\exp_{\underline{G}}(st_0K) \mid s \in \mathbb{Z}\}$.

Since the group $Z_2 = \left\{ \begin{bmatrix} 1 & 0 \\ 0 & 1 \end{bmatrix}, \begin{bmatrix} -1 & 0 \\ 0 & -1 \end{bmatrix} \right\}$ is the centralizer of

$\left\{ \begin{bmatrix} x & 0 \\ 0 & x^{-1} \end{bmatrix} \mid x > 0 \right\}$ in $s(\underline{\widetilde{K}})$, we have that $Ctz_{\underline{\widetilde{K}}}(A) = s^{-1}(Z_2)$; moreover $s^{-1}(Z_2)$ is a central extension of Z_2 by kernel(s). Let $\underline{M} = s^{-1}(Z_2)$; since Z_2 is the center of $S\ell_2(\mathbb{R})$, we have that $Ad_{\underline{G}}(\underline{M}) = e_{Ad(\underline{G})}$. Let $\underline{P} = \underline{MAN}$; we observe that the connected component of the identity in $\underline{P}$ is $\underline{AN}$. Let $Norm_{\underline{\widetilde{K}}}(M)$ be the normalizer of $\underline{M}$ in $\underline{\widetilde{K}}$. Then noting that $Norm_{\underline{\widetilde{K}}}(\underline{M})/\underline{M}$ has order two, we then choose w_o as an element of the nontrivial coset of $Norm_{\underline{\widetilde{K}}}(\underline{M})/\underline{M}$. We then lift the Bruhat decomposition of $S\ell_2(\mathbb{R})$ to $\underline{G}$ so that $\underline{G} = \underline{P} \cup \underline{N}w_o\underline{P}$ $(= \underline{P} \cup \underline{P}w_o\underline{P})$.

Let $\hat{\underline{M}}$ = set of unitary characters of $\underline{M}$. Since $s(\underline{A}) = \left\{ \begin{bmatrix} x & 0 \\ 0 & x^{-1} \end{bmatrix} \mid x > 0 \right\}$ and the map s restricted to $\underline{A}$ is an isomorphism to $s(\underline{A})$, we then can take $\hat{\underline{A}}$ as the set of quasi-characters on $\mathbb{R}_+^*$ $(=\{r \in \mathbb{R}^* \mid r > 0\})$.

We then can apply Bruhat's theory to $\underline{G}$. We let $U_1 = U_2 = \underline{P}$, and let $\sigma_1(man) = \sigma_{11}(m)\lambda_{11}(a)$, $\sigma_2(man) = \sigma_{12}(m)\lambda_{12}(a)$ for m, a, n $\in \underline{M}, \underline{A}$, and $\underline{N}$ and $\sigma_{11}, \sigma_{12} \in \hat{\underline{M}}$, $\lambda_{11}, \lambda_{12} \in \hat{\underline{A}}$. We then recall that the modular function $\delta_{\underline{P}}(m \exp_{\underline{G}}(xH)\exp_{\underline{G}}(yE)) = e^{-2x}$. Then a rho function for $\underline{G}/\underline{P}$ is $\rho(g) = \rho(k \exp_{\underline{G}}(xH)\exp_{\underline{G}}(yE)) = \delta_{\underline{P}}(\exp_{\underline{G}}(xH))$ (relative to the Iwasawa decomposition). Let $d\dot{\mu}_\rho$ be the quasi-invariant measure on $\underline{G}/\underline{P}$. Let $C^\infty_{\sigma_1}(\underline{G})$ and $C^\infty_{\sigma_2}(\underline{G})$ be the induced modules defined relative to σ_1 and σ_2. Then using the same arguments as in section 3, we deduce the following.

<u>Proposition 4.1</u> <u>Consider the following conditions</u>: (A) $\sigma_{11}\sigma_{12} \neq 1$ (B) $\lambda_{11} \neq \lambda_{12}$. <u>Then if</u> (A) <u>holds</u> $\dim I(\pi_{\sigma_1}, \pi_{\sigma_2}) = 0$. <u>If</u> (B) <u>holds then</u> $\dim I(\pi_{\sigma_1}, \pi_{\sigma_2}) = \dim I(\sigma_{11} \otimes \lambda_{11}, \sigma_{12} \otimes \lambda_{12})$

$$= \begin{cases} 1 & \text{if } \lambda_{11} = \lambda_{12}^{-1} \text{ and } \sigma_{11}\sigma_{12} = 1 \\ 0 & \text{otherwise} \end{cases}.$$

<u>Finally if</u> $\sigma_{11}\sigma_{12} = 1$ <u>and</u> $\lambda_{11} = \lambda_{12}$, <u>then</u> $\dim I(\pi_{\sigma_1}, \pi_{\sigma_2}) \leq 1$ if $\lambda_{11} \neq -k$, k integer ≥ 0 (<u>and if</u> $\sigma_{11}\sigma_{12} = 1$ <u>with</u> $\lambda_{11} = \lambda_{12} = -k$, <u>an integer as above, then</u> $\dim I(\pi_{\sigma_1}, \pi_{\sigma_2}) \leq 2$).

<u>Hence if</u> $\lambda = it$ <u>with</u> $t \neq 0$ <u>and</u> σ <u>arbitrary, then the unitarily induced representation</u> $\tilde{\pi}_{\sigma \otimes \lambda}$ <u>of</u> $\underline{G}$ <u>on</u> $E_{\sigma \otimes \lambda}$ <u>is irreducible</u>.

Q.E.D.

<u>Proposition 4.2</u> (<u>Existence of Whittaker model weakly for</u> $\underline{G}$)

<u>For any</u> $\sigma \in \hat{\underline{M}}$, $\lambda \in \hat{\underline{A}}$ <u>then the space</u>

$$(4-1) \quad \{T \in C^\infty_{\sigma\otimes\lambda}(\underline{G})' \mid \pi^c_{\sigma\otimes\lambda}(\exp(tE))T = \tau(t)T \text{ for all } t \in \mathbb{R}\}$$

<u>has dimension at most</u> 1.

Q.E.D.

Remark 4.1 From Proposition 4.1 and arguments similar to those in section 3, we see that the unique, nondegenerate $\underline{G}$ intertwining form between $C^\infty_{\sigma_1}(G)$ and $C^\infty_{\sigma_2}(G)$ $(\sigma_1 = \sigma_{11} \otimes \lambda$ and $\sigma_2 = \sigma_{11} \otimes (-\lambda))$ is given by

$$(4-2) \quad (f_1,f_2) = \oint_{\underline{G}/\underline{P}} f_1(x)f_2(x)\{\rho(x)\}^{-1}d\dot{\mu}_\rho(x)$$

$$= \int_{\underline{\tilde{K}}} f_1(k)f_2(k)d\underline{k} = c\int_{\mathbb{R}} f_1(n(y)w_o)f_2(n(y)w_o)dy,$$

where $d\underline{k}$ is a suitable normalized Haar measure on $\underline{\tilde{K}}$ and dy Lesbesque measure on $\mathbb{R}$ $(c \neq 0)$.

We then let $\underline{G} = \widetilde{S\ell}_2(\mathbb{R})$ be the two-fold covering of $S\ell_2(\mathbb{R})$ given in section 1. We let $k(\theta,\epsilon) = (\begin{bmatrix} \cos\theta & \sin\theta \\ -\sin\theta & \cos\theta \end{bmatrix}, \epsilon)$, $a(u) = (\begin{bmatrix} u & 0 \\ 0 & u^{-1} \end{bmatrix}, 1)$, and $n(s) = (\begin{bmatrix} 1 & s \\ 0 & 1 \end{bmatrix}, 1)$. Then in this case $\underline{\tilde{K}} = \{k(\theta,\epsilon) \mid -\pi < \theta \leq \pi\}$, $\underline{A} = \{a(u) \mid u > 0\}$ and $\underline{N} = \{n(s) \mid s \in \mathbb{R}\}$. From Iwasawa decomposition above for any $g, x \in \widetilde{S\ell}_2(\mathbb{R})$, there is a unique decomposition of $g^{-1}\cdot x = k_g(x)a_g(x)n_g(x)$ with $k_g(x) \in \underline{\tilde{K}}$, $a_g(x) \in \underline{A}$, and $n_g(x) \in \underline{N}$. Then the maps $(g,x) \to k_g(x), a_g(x)$, and $n_g(x)$ of $\widetilde{S\ell}_2(\mathbb{R}) \times \widetilde{S\ell}_2(\mathbb{R}) \to \underline{\tilde{K}}$, $\underline{A}$, and $\underline{N}$, respectively, are C^∞ maps.

We note here that $\underline{M} = \mathrm{Ctz}_{\underline{\tilde{K}}}(\underline{A}) = \mathrm{Center}\ (\widetilde{S\ell}_2(\mathbb{R})) = \{E, E^2, E^3, I\}$ with $E = (\begin{bmatrix} -1 & 0 \\ 0 & -1 \end{bmatrix}, -1)$. And $M' = \mathrm{Norm}_{\underline{\tilde{K}}}(\underline{A})$ is the cyclic group of order 8 with generator $w_o = (\begin{bmatrix} 0 & -1 \\ 1 & 0 \end{bmatrix}, -1)$, with the relation $w_o^2 = E$.

Then choosing w_o as the nontrivial element of M'/M, the Bruhat decomposition of $\widetilde{S\ell}_2(\mathbb{R})$ relative to $\underline{P} = \underline{M}\,\underline{A}\,\underline{N}$ is $\widetilde{S\ell}_2(\mathbb{R}) = \underline{N}w_o\underline{P} \cup \underline{P}$. We then note (as in the case of $S\ell_2(\mathbb{R})$ itself) the following computations:

(4 - 3) if $p \in P$ then $p = \left(\begin{bmatrix} x & y \\ 0 & x^{-1} \end{bmatrix}, \epsilon\right) = E^{\beta}\left(\begin{bmatrix} |x| & (-1)^{\beta}y \\ 0 & |x|^{-1} \end{bmatrix}, 1\right)$

(with $\beta = \beta(\mathrm{sgn}(x), \epsilon)$), where $\beta: \{\pm 1\} \times \{\pm 1\} \to \{0, 1, 2, 3\}$ is the function $\beta(1,1) = 0$, $\beta(1,-1) = 2$, $\beta(-1,1) = 3$, and $\beta(-1,-1) = 1$.

(4 - 4) if $g = \left(\begin{bmatrix} a & b \\ c & d \end{bmatrix}, \epsilon\right)$ with $c \neq 0$, we have relative to the

Bruhat decomposition $g = \left(\begin{bmatrix} 1 & \frac{a}{c} \\ 0 & 1 \end{bmatrix}, 1\right) w_o \left(\begin{bmatrix} c & d \\ 0 & c^{-1} \end{bmatrix}, -\epsilon\right)$.

Then using (4 - 3) and (4 - 4) we deduce the following:

(4 - 5) $w_o n(x) w_o = n(-\frac{1}{x}) w_o \varphi_x(E) a(|x|)$, where $\varphi_x(E) = \begin{cases} E^2 & \text{if } x > 0 \\ E^3 & \text{if } x < 0 \end{cases}$

and $x \neq 0$.

(4 - 6) $\left(\begin{bmatrix} s & 0 \\ 0 & s^{-1} \end{bmatrix}, 1\right) n(x) w_o = n(s^2 x) w_o \psi_s(E) a(|s|^{-1})$, where

$\psi_s(E) = \begin{cases} I & \text{if } s > 0 \\ E^3 & \text{if } s < 0 \end{cases}$.

(4 - 7) $a(q)k(\theta,\epsilon)a(u) = \left(\frac{1}{\phi(\theta,q)}\begin{bmatrix} q\cos\theta & q^{-1}\sin\theta \\ -q^{-1}\sin\theta & q\cos\theta \end{bmatrix}, \epsilon\right)$

$a(u\phi(\theta,q))n(s(\theta,u,q))$, with $\phi(\theta,q) = (q^2\cos^2\theta + q^{-2}\sin^2\theta)^{1/2}$

and $s(\theta,u,q) = \dfrac{(q^2 - q^{-2})\cos\theta\sin\theta}{u^2[\phi(\theta,q)]^2}$.

(4 - 8) $n(s)k(\theta,\epsilon)a(u) = \left(\frac{1}{\psi(\theta,s)}\begin{bmatrix} \cos\theta - s\sin\theta & \sin\theta \\ -\sin\theta & \cos\theta - s\sin\theta \end{bmatrix}, \epsilon\right)$

$a(u\psi(\theta,s))n(s'(\theta,u,s))$, with $\psi(\theta,s) =$ $[(\cos\theta - s\sin\theta)^2 + \sin^2\theta]^{1/2}$ and $s'(\theta,u,s) =$

$\dfrac{s\cos 2\theta - \frac{1}{2}s^2\sin 2\theta}{u^2[\psi(\theta,s)]^2}$.

(4 - 9) $k(\theta',\epsilon')k(\theta,\epsilon)a(u) = k(\theta + \theta', A(r(\theta'),r(\theta))\ \epsilon\ \epsilon')a(u)$,

with A the cocycle defined in section 1 and

$$r(\beta) = \begin{bmatrix} \cos\beta & \sin\beta \\ -\sin\beta & \cos\beta \end{bmatrix}, \ \beta \in \mathbb{R}.$$

At this point to avoid proliferation of symbols, we adopt the convention that $\widetilde{S\ell}_2 = \widetilde{S\ell}_2(\mathbb{R})$.

Let $u \in \mathbb{R}$ and $\tau_u\colon \underline{N} \to S^1$ be the unitary character on $\underline{N}$ given by $\tau_u(n(x)) = \tau(ux)$, $x \in \mathbb{R}$. Then as above we construct the complex line bundle $\widetilde{S\ell}_2 \times_u \mathbb{C} = (u)$ over the homogeneous space $\widetilde{S\ell}_2/\underline{N}$. Noting that both $\underline{N}$ and $\widetilde{S\ell}_2$ are unimodular, we let $S(u) = \{f\colon \widetilde{S\ell}_2 \to \mathbb{C} \mid f$ is measurable relative to Haar measure $d_{\widetilde{S\ell}_2}$ on $\widetilde{S\ell}_2$ and $f(gn(s)) = f(g)\tau_u(s)$ for all $g \in \widetilde{S\ell}_2$, $s \in \mathbb{R}\}$. Then $S(u)$ is exactly the set of measurable sections of the line bundle (u). Also we let $C^\infty(u)$ $(C_c^\infty(u))$ be the space of C^∞ sections (C_c^∞ sections of compact support) of (u).

We note that $\widetilde{S\ell}_2/\underline{N}$ admits an invariant measure $d\mu_o$, so that $C_c^\infty(u)$ has a positive definite Hermitian inner product which is $\widetilde{S\ell}_2$ invariant given by

$$(4 - 10) \quad (f \mid g) = \int_{\widetilde{S\ell}_2/\underline{N}} f(x)\overline{g(x)}d\mu_o(x) \quad \text{for } f, g \in C_c^\infty(u).$$

Since $\widetilde{S\ell}_2$ is semisimple, then by a well known formula of the decomposition of Haar measure $d_{\widetilde{S\ell}_2}(x) = |r|^2 dk d\underline{A}(r) d\underline{N}$, with dk, a Haar measure on $\widetilde{\underline{K}}$, $d\underline{A}(r) = \frac{dr}{|r|}$, a Haar measure on $\underline{A}$, and $d\underline{N}(s) = ds$, a Haar measure on $\underline{N}$. We then have in "local coordinates"

$$(4 - 11) \quad (f \mid g) = \int_{\underline{K} \times \mathbb{R}_+^*} f(ka(v))\overline{g(ka(v))}|v| dk dv,$$

for $f, g \in C_c^\infty(u)$.

We consider $L^2(u)$, the Hilbert space completion of $(\,|\,)$ on $C_c^\infty(u)$. The representation π_u of $\widetilde{S\ell}_2$ leaves $S(u)$, $C_c^\infty(u)$, $C^\infty(u)$, and $L^2(u)$ stable and is given in local coordinates by $\pi_u(g)(f)(ka(v)) = \overline{\tau_u(n_g(x))}f(k_g(x)a_g(x))$, with $x = k\cdot a(v)$.

Returning to the rotation subgroup $K_1 = \{r(\theta), -\pi \leq \theta < \pi\}$ of $S\ell_2(\mathbb{R})$, we restrict the multiplier A to K_1 (call it A_{K_1}). Then A_{K_1} is a Borel multiplier on K_1 and is locally trivial at (e_{K_1}, e_{K_1}). Indeed for nonzero θ, θ', $\theta + \theta' \in (-\frac{\pi}{2}, \frac{\pi}{2})$, we have $A_{K_1}(r(\theta), r(\theta')) = \left(\frac{-\sin\theta}{-\sin\theta'}\right)_H \left(\frac{-\sin\theta/\sin\theta'}{-\sin(\theta+\theta')}\right)_H = 1$. Thus equipping $\underline{\widetilde{K}}$ with the Weil topology, we see that for n sufficiently large (n integer), $T_n = \{k(\theta,1) \mid |\theta| < \frac{\pi}{n}\}$ is a neighborhood of $e_{\underline{\widetilde{K}}}$ in $\underline{\widetilde{K}}$ relative to the analytic structure on $\underline{\widetilde{K}}$. Let $C^\infty(\underline{\widetilde{K}})$ be the space of C^∞ functions on $\underline{\widetilde{K}}$ (relative to the unique analytic structure on $\underline{\widetilde{K}}$ which is compatible with the Weil topology). For $f \in C^\infty(\underline{\widetilde{K}})$ define $\psi_f(\theta) = \begin{cases} f(k(\theta+\pi, 1)), & -2\pi < \theta \leq 0 \\ f(k(\theta-\pi, -1)), & 0 < \theta \leq 2\pi \end{cases}$.

__Lemma 4.3__ *If* $f \in C^\infty(\underline{\widetilde{K}})$, *then* $\psi_f \in C^\infty((-2\pi, 2\pi))$ *and*

$$\lim_{\epsilon\to 0} \left(\frac{d}{d\theta}\right)^s \psi_f(-2\pi+\epsilon) = \lim_{\epsilon \to 0}\left(\frac{d}{d\theta}\right)^s \psi_f(2\pi - \epsilon) \quad (\epsilon > 0) \quad \textit{for}$$

all integers $s \geq 0$.

__Proof.__ Consider the set of $4n$ points $a_j^{\pm} = \{k(\frac{j}{n}\pi, \pm 1) \mid j = -(n-1), \ldots, 0, 1, \ldots, n\}$. Then we observe that $T_n \cdot a_j^{\pm}$ for all j form a differentiable covering of $\underline{\widetilde{K}}$. Indeed $T_n \cdot a_j^{\pm} = \{k(\theta, \pm 1) \mid (\frac{j-1}{n})\pi < \theta < (\frac{j+1}{n})\pi\}$ for $j = -(n-1), \ldots\ldots, (n-1)$; $T_n \cdot a_n^+ = \{k(\theta,1) \mid (\frac{n-1}{n})\pi < \theta \leq \pi\} \cup \{k(\theta,-1) \mid -\pi < \theta < -(\frac{n-1}{n})\pi\}$ and

$T_n \cdot a_n^- = \{k(\theta, -1) \mid (\frac{n-1}{n})\pi < \theta \leq \pi\} \cup \{k(\theta, 1) \mid -\pi < \theta < -(\frac{n-1}{n})\pi\}.$

We let $\beta_j^{\pm}: (-\frac{\pi}{n}, \frac{\pi}{n}) \to T_n \cdot a_j^{\pm}$ be the family of C^∞ functions defined by: $\beta_j^{\pm}(\theta) = k(\theta, \pm 1)$, $j = -(n-1), \ldots, (n-1)$;

$$\beta_n^+(\theta) = \begin{cases} k(\pi - \theta, 1), & -\frac{\pi}{n} < \theta \leq 0 \\ k(-\pi + \theta, -1), & 0 < \theta < \frac{\pi}{n}; \end{cases} \qquad \beta_n^-(\theta) = \begin{cases} k(\pi - \theta, -1), & -\frac{\pi}{n} < \theta \leq 0 \\ k(-\pi + \theta, 1), & 0 < \theta < \frac{\pi}{n} \end{cases}.$$

Thus $f \in C^\infty(\underline{\widetilde{K}})$ if and only if f restricted to $T_n \cdot a_j^{\pm}$ for all j determines a C^∞ map $\varphi_j^{\pm}: (-\frac{\pi}{n}, \frac{\pi}{n}) \xrightarrow{\beta_j^{\pm}} T_n \cdot a_j^{\pm} \xrightarrow{f} \mathbb{C}$. Then using the compatibility properties of the $\varphi_j^{\pm}$ on the various intersections, we observe that ψ_f has the desired properties.

Q.E.D.

<u>Remark 4.2</u> Let $C^\infty[-2\pi,2\pi] = \{f: [-2\pi,2\pi] \to \mathbb{C} \mid f \in C^\infty(-2\pi,2\pi)$ and for all $s \geq 0$ $\lim_{\epsilon\to 0}(\frac{d}{d\theta})^s f(-2\pi + \epsilon) = \lim_{\epsilon\to 0}(\frac{d}{d\theta})^s f(2\pi - \epsilon)$, $(\epsilon > 0)\}$. If $g \in C^\infty[-2\pi,2\pi]$, define $f_g(k(x,1)) = g(x - \pi)$, $-\pi < x \leq \pi$ and $f_g(k(x,-1)) = g(x + \pi)$, $-\pi < x \leq \pi$. By reversing the above argument in Lemma 4.3, $f_g \in C^\infty(\underline{\widetilde{K}})$ and hence we may identify $C^\infty(\underline{\widetilde{K}})$ to the space $C^\infty[-2\pi,2\pi]$ (carry the topology of $C^\infty(\underline{\widetilde{K}})$ to $C^\infty[-2\pi,2\pi]$).

From above we see easily that the space $C^\infty(u)$ of C^∞ sections of (u) can be identified to the space of C^∞ functions on $\underline{\widetilde{K}} \times \underline{A}$ $(C^\infty(\underline{\widetilde{K}} \times \underline{A}))$. But by using the same reasoning as above and defining for $f \in C^\infty(u) \cong C^\infty(\underline{\widetilde{K}} \times \underline{A})$,

$$\widetilde{\psi}_f(\theta,v) = \begin{cases} f(k(\theta + \pi,1)a(v)), & -2\pi < \theta \leq 0 \\ f(k(\theta - \pi,-1)a(v)), & 0 < \theta \leq 2\pi \end{cases},$$

we see that $C^\infty(u)$ can be identified (with the topology carried over) to the space $C^\infty([-2\pi,2\pi] \times \mathbb{R}_+^*) = \{f: [-2\pi,2\pi] \times \mathbb{R}^* \to \mathbb{C} \mid f \in C^\infty((-2\pi,2\pi) \times \mathbb{R}_+^*)$ and $f(\ ,x) \in C^\infty[-2\pi,2\pi]$ for all $x \in \mathbb{R}^*\}$.

Via this identification we are interested in determining the infinitesimal action $\pi_u(s\ell_2)$ on $C^\infty(u)$. We note at this point that for $|t| < \frac{\pi}{n}$, $\exp_{s\ell_2}(tK) = k(t,1)$. We let $F = \begin{bmatrix} 0 & 0 \\ 1 & 0 \end{bmatrix}$, $N_+ = \begin{bmatrix} 1 & -i \\ -i & -1 \end{bmatrix}$, $N_- = \begin{bmatrix} 1 & i \\ i & -1 \end{bmatrix}$ $(i = \sqrt{-1})$ be the elements of the complexification $(s\ell_2)_{\mathbb{C}}$ of $s\ell_2$; we recall the bracket relations $[K,N_+] = -2iN_+$; $[K,N_-] = 2iN_-$; $[H,E] = 2E$; $[H,F] = -2F$; $[E,F] = H$ and $[N_+,N_-] = 4iK$. We note that the Casimir element Ω of $\mathcal{U}(s\ell_2)$ is given by $H^2 + 2(EF + FE) = -K^2 + \frac{1}{2}(N_+N_- + N_-N_+)$.

<u>Lemma 4.4</u> <u>For</u> $f \in C^\infty([-2\pi,2\pi] \times \mathbb{R}_+^*)$ <u>we have</u>

$$(4-12)\quad \pi_u(K)f(\theta,y) = (-\frac{\partial}{\partial\theta})f(\theta,y) ,$$

$$(4-13)\quad \pi_u(H)f(\theta,y) = (\pi iu\frac{2\sin\theta}{y^2} + \sin 2\theta\frac{\partial}{\partial\theta} - y\cos 2\theta\frac{\partial}{\partial y})f(\theta,y)$$

$$(4-14)\quad \pi_u(E)f(\theta,y) = (\pi iu\frac{\cos 2\theta}{y^2} - \sin^2\theta\frac{\partial}{\partial\theta} + y\frac{\sin 2\theta}{2}\frac{\partial}{\partial y})\, f(\theta,y).$$

<u>Proof</u>. From above $f(\theta,y) = \tilde{\psi}_g(\theta,y)$ for $g \in C^\infty(\underline{\tilde{K}} \times \underline{A})$. Then for $Z \in s\ell_2$ we have $\pi_u(Z)(f)(\theta,y) = \frac{d}{dt}\tilde{\psi}_{\pi_u(\exp tZ)g}(\theta,y)\big|_{t=0}$. Then from (4 - 7) there exists a function θ': $\mathbb{R} \times \mathbb{R} \to \mathbb{R}$, so that $\cos\theta'(\theta,x) = \dfrac{e^{-x}\cos\theta}{\phi(\theta,e^{-x})}$ and $\sin\theta'(\theta,x) = \dfrac{e^{x}\sin\theta}{\phi(\theta,e^{-x})}$ (with $\theta,x \in \mathbb{R}$). Explicitly let $\theta'(\frac{k\pi}{2},x) = \frac{k\pi}{2}$ for k an odd integer and $\theta'(\theta,x) = \arctan(e^{2x}\tan\theta) + \mu(\theta)\pi$, where $\mu(\theta)$ is the unique integer so that $\theta = \theta_1 + \mu(\theta)\pi$ with $\theta_1 \in (-\frac{\pi}{2}, \frac{\pi}{2}]$. Then we see that θ' restricted to the open set $\{(\theta,z)|\ \theta \neq \frac{k}{2}\pi,\ k \text{ odd integer}\} \subseteq \mathbb{R} \times \mathbb{R}$

is differentiable. Moreover we have for any θ, $\phi(\theta \pm \pi, e^{-x}) = \phi(\theta,e^{-x})$ and $s(\theta \pm \pi,y,e^{-x}) = s(\theta,y,e^{-x})$. If $\theta \in (-2\pi,0]$, then $\theta'(\theta + \pi,x) \in (-2\pi,0]$. Thus for θ such that $2\pi < \theta < 0$ and $\theta \neq \frac{k}{2}\pi$ (k odd integer), $\widetilde{\psi}_{\pi_u(\exp xH)g}(\theta,y) - \widetilde{\psi}_g(\theta,y) =$

$\{e^{-\pi uis(\theta,y,e^{-x})}g(k(\theta'(\theta + \pi),1)a(y\phi(\theta,e^{-x})))\} -$

$g(k(\theta + \pi,1)a(y)) = \{e^{-\pi ius(\theta,y,e^{-x})}\widetilde{\psi}_g(\text{arc tan }(e^{2x} \tan \theta) +$

$\mu(\theta)\pi,\ y\phi(\theta,e^{-x}))\} - \widetilde{\psi}_g(\theta,y)$.

Then computing $\pi_u(H)$ we have that $(4 - 13)$ is valid for (θ,y) so that $-2\pi < \theta < 0$, $\theta \neq \frac{k}{2}\pi$ (k odd integer). Similar reasoning derives the same result valid for (θ,y) so that $0 < \theta < 2\pi$, $\theta \neq \frac{k}{2}\pi$ (k odd integer).

Then if $P\colon \mathbb{R}^3 \to \mathbb{C}$ is any polynomial function in 3 variables, we observe that a differential operator of the form $P(\cos k\theta, \sin \ell\theta, \frac{\partial}{\partial\theta})$ (k,ℓ integers) has the property that $P(g) \in C^\infty([-2\pi,2\pi])$ if $g \in C^\infty([-2\pi,2\pi])$.

Then the operator D_H defined by the expression on the right side of $(4 - 13)$ leaves stable $C^\infty([-2\pi,2\pi] \times \mathbb{R}_+^*)$, and since $D_H = \pi_u(H)$ on an open dense subset of $[-2\pi,2\pi] \times \mathbb{R}_+^*$, we see that $(4 - 13)$ is valid for all $(\theta,y) \in [-2\pi,2\pi] \times \mathbb{R}_+^*$.

Similar arguments work for $(4 - 12)$ and $(4 - 14)$.

Q.E.D.

<u>Remark 4.3</u> For $f \in C^\infty([-2\pi,2\pi] \times \mathbb{R}^*)$ we have

$$(4 - 15)\quad \pi_u(\Omega)f = \left(y^2 \frac{\partial^2}{\partial y^2} + 3y \frac{\partial}{\partial y} + \frac{4\pi iu}{y^2}\frac{\partial}{\partial\theta} - \frac{4\pi^2u^2}{y^4}\right)f \quad ,$$

$$(4 - 16)\quad \pi_u(N_-)f = e^{-2i\theta}\left(-y \frac{\partial}{\partial y} + i \frac{\partial}{\partial\theta} - \frac{2\pi u}{y^2}\right)f \quad ,$$

$$(4-17)\quad \pi_u(N_+)f = e^{2i\theta}\left(-y\frac{\partial}{\partial y} - i\frac{\partial}{\partial\theta} + \frac{2\pi u}{y^2}\right)f \quad .$$

To facilitate the algebraic study of the action of $\mathcal{U}(s\ell_2)$ on $C^\infty(u)$, we observe certain elementary facts.

We let

$$(4-18)\quad \begin{aligned} C_+^\infty(u) &= \{f \in C^\infty(u) \mid \text{for any } c > 0 \text{ and all } k \in \underline{\tilde{K}}, \\ &\int_{|y| \le c} |f(ka(y))|^2 y\,dy < \infty\} \\ C_-^\infty(u) &= \{f \in C^\infty(u) \mid \text{for any } c > 0 \text{ and all } k \in \underline{\tilde{K}}, \\ &\int_{|y| \ge c} |f(ka(y))|^2 y\,dy < \infty\}. \end{aligned}$$

<u>Lemma 4.5</u> <u>The spaces</u> $C_+^\infty(u)$ <u>and</u> $C_-^\infty(u)$ <u>are stable under</u> $\pi_u(\widetilde{S\ell_2})$. <u>The space</u> $C_+^\infty(u) \cap C_-^\infty(u)$ <u>is then stable under</u> $\pi_u(\widetilde{S\ell_2})$ <u>and</u> $C_+^\infty(u) \cap C_-^\infty(u) \subsetneq L^2(u)$.

<u>Proof</u>. From the polar decomposition of $\widetilde{S\ell_2}$, we have that $\widetilde{S\ell_2}$ is generated by $\underline{\tilde{K}}$ and $\underline{A}$. Then clearly $C_\pm^\infty(u)$ are stable under $\pi_u(\underline{\tilde{K}})$. For $a(e^x) \in \underline{A}$ and any $k \in \underline{\tilde{K}}$, we have from (4 - 7), and by the change of variables $y \to y\phi(\theta, e^x)$, (with $k = k(\theta, \epsilon)$)

$$(4-19)\quad \begin{aligned} &\int_{|y| \le c} |\pi_u(a(e^x))f(ka(y))|^2 |y|\,dy = \\ &\frac{1}{[\phi(\theta, e^{-x})]^2}\int_{|y'| \le c'} |f(k'a(y')|^2 |y'|\,dy' \quad , \end{aligned}$$

where k' depends only on x and θ and $c' > 0$ also depends only on x and θ. Thus $C_+^\infty(u)$ is stable under $\pi_u(\widetilde{S\ell_2})$ and we derive a similar result about $C_-^\infty(u)$.

Q.E.D.

The infinitesimal action of $\mathcal{U}(s\ell_2)$ on $C^\infty(u)$ is approached by first determining the spaces $(\lambda, m \in \mathbb{C})$

$$(4-20) \quad C^\infty(u)(\lambda) = \{f \in C^\infty(u) \mid \pi_u(\Omega)f = \lambda f\} \quad ,$$

$$(4-21) \quad C^\infty(u)(\lambda,m) = \{f \in C^\infty(u)(\lambda) \mid \pi_u(K)f = mf\} \quad .$$

If $f(\theta,y) \in C^\infty(u)(\lambda,m)$, then $\pi_u(\widetilde{K})f = -\frac{\partial}{\partial\theta} f(\theta,y) = mf(\theta,y)$ for all $y \in \mathbb{R}_+^*$. But $f(\ ,y) \in C^\infty([-2\pi,2\pi])$ for all y, and hence we have a separation of variables $f(\theta,y) = e^{-m\theta}\varphi_f(y)$, where $m = \frac{1}{2}ki$ (k integer) and $\varphi_f \in C^\infty(\mathbb{R}_+^*)$. But relative to the equation $\pi_u(\Omega)f = \lambda f$ we deduce that

$$(4-22) \quad (y^2 \frac{d^2}{dy^2} + 3y \frac{d}{dy} - (\frac{4\pi ium}{y^2} + \frac{4\pi^2 u^2}{y^4} + \lambda))\varphi_f = 0.$$

Thus we have $\dim C^\infty(u)(\lambda,m) = \begin{cases} 2 & \text{if } m = \frac{k}{2} i \text{ (k integer)} \\ 0 & \text{otherwise} \end{cases}$. If $y = x^{-\frac{1}{2}}$ we reduce $(4-22)$ to the classical "Whittaker equation"

$$(4-23) \quad (x^2 \frac{d^2}{dx^2} - (\pi\ imux + \pi^2u^2x^2 + \frac{\lambda}{4}))\{\ \} = 0.$$

Then using the classical theory of asymptotics of differential equations, we can deduce certain important properties of the solutions [5].

The indicial equation of $(4-23)$ at $x = 0$ is $X^2 - X - \frac{\lambda}{4} = 0$. The roots of the equation are $X = \frac{1}{2} \pm \frac{1}{2}\sqrt{1+\lambda}$, where for $\lambda \in \mathbb{C} - (-\infty,-1]$ we define $\sqrt{1+\lambda} = r^{\frac{1}{2}} e^{i\frac{1}{2}\theta}$ with $\lambda + 1 = r\,e^{i\theta}$ $(-\pi < \theta < \pi$ and $r > 0)$, and for $\lambda = x$ with $x \leq -1$, $\sqrt{1+\lambda} = |1+x|^{1/2}\sqrt{-1}$. We let $\beta(\lambda) = \frac{1}{2}\sqrt{1+\lambda}$ with the conventions above; β is a differentiable bijective map of $\mathbb{C} - (-\infty,-1]$ to $\{z \mid \mathrm{Re}(z) > 0\}$.

We recall that the fundamental system of solutions of (4 - 23) is given by:

(4 - 24) $\{x^{\frac{1}{2} \pm \beta} \varphi^{\pm}_{\beta,m,u}(x)\}$ with $\varphi^{\pm}_{\beta,m,u}(x)$ analytic at $x = 0$ and

$\varphi^{\pm}_{\beta,m,u}(0) \neq 0$ for $\beta \notin \{0, \frac{1}{2}, 1, \frac{3}{2}, \ldots\}$.

(4 - 25) $\{x^{\frac{1}{2} + \frac{k}{2}} \varphi^{+}_{\frac{k}{2},m,u}(x)\}$ and $\{\delta_k x^{\frac{1}{2} + \frac{k}{2}} \log(x)\varphi^{+}_{\frac{k}{2},m,u}(x) +$

$x^{\frac{1}{2} - \frac{k}{2}} \varphi^{-}_{\frac{k}{2},m,u}(x)\}$ with $\varphi^{\pm}_{\frac{k}{2},m,u,}(x)$ analytic at $x = 0$ and

$\varphi^{\pm}_{\frac{k}{2},m,u}(0) \neq 0$ ($\delta_k = 0$ if $u = 0$) for k integer > 0.

(4 - 26) $\{x^{\frac{1}{2}} \varphi^{+}_{0,m,u}(x)\}$ and $\{x^{\frac{1}{2}} \log(x)\varphi^{+}_{0,m,u}(x) + x^{\frac{1}{2}} \varphi^{-}_{0,m,u}(x)\}$

with $\varphi^{\pm}_{0,m,u}(x)$ analytic at $x = 0$, $\varphi^{+}_{0,m,u}(0) \neq 0$ and $\varphi^{-}_{0,m,u}(0) = 0$.

For the point $x = \infty$ of equation (4 - 23), we recall the theory of normal solutions and asymptotic expansions at $x = \infty$ (see [5] for discussion). We deduce that for $u < 0$, there exists a pair of linearly independent solutions of (4 - 23) of the form

$\{e^{\pi ux} x^{i \frac{1}{2} m} \phi_{\beta,m}(x),\ e^{-\pi ux} x^{-i \frac{1}{2} m} \tilde{\phi}_{\beta,m}(x)\}$, where $\phi_{\beta,m}$ and $\tilde{\phi}_{\beta,m}$ are

analytic functions in a sector $S_\epsilon = \{z \in \mathbb{C} |\ |z| \geq 1$ and $|\arg(z)| < \epsilon\}$ (for some $\frac{\pi}{2} > \epsilon > 0$) having the following asymptotic property; for any positive integer s, there exist constants $\ell_s \geq 1$ and $M^{\pm}_s > 0$ so that for $|z| > \ell_s$ and $z \in S_\epsilon$

$$(4-27)\qquad \begin{aligned} &|\phi_{\beta,m}(z) - (1 + \sum_{t=1}^{s-1} c_t z^{-t})| \le M_s^+ |z|^{-s} \quad , \\ &|\tilde{\phi}_{\beta,m}(z) - (1 + \sum_{t=1}^{s-1} C_t z^{-t})| \le M_s^- |z|^{-s} \quad , \end{aligned}$$

where $2utc_t = [(t - i\frac{1}{2}m)(t - 1 - i\frac{1}{2}m) - \frac{\lambda}{4}]c_{t-1}$ with $c_0 = 1$,

and $-2utC_t = [(t + i\frac{1}{2}m)(t - 1 + i\frac{1}{2}m) - \frac{\lambda}{4}]C_{t-1}$ with $C_0 = 1$. We observe that in any case $\{e^{\pi ux} x^{i\frac{1}{2}m} \sum_{v=0}^{\infty} c_v x^{-v}\}$ and $\{e^{-\pi ux} x^{-i\frac{1}{2}m} \sum_{v=0}^{\infty} C_v x^{-v}\}$ are formal solutions to (4 - 23). Moreover we know that $\phi'_{\beta,m}$, $\phi''_{\beta,m}$, $\tilde{\phi}'_{\beta,m}$, and $\tilde{\phi}''_{\beta,m}$ have asymptotic expansions in S_ε, similar to the above, which can be obtained by differentiating term by term.

Remark 4.4 We note at this point that $\phi_{\beta,m}$ $(\tilde{\phi}_{\beta,m})$ has a finite series expansion in case there exist γ (γ') an integer and s (s') a positive integer so that $\lambda = (\frac{\gamma^2}{4} - 1)$ and $\gamma = 4s - 2im - 2$ $(\lambda = (\frac{(\gamma')^2}{4} - 1)$ and $\gamma' = 4s' + 2im - 2)$.

The structure of $C^\infty(u)(\lambda,m)$ can be determined as follows.

For the remainder of this section m is of the form $\frac{1}{2}k\sqrt{-1}$ (k integer).

Proposition 4.6 (1) *Let* $u < 0$. *Then* $C_-^\infty(u)(\lambda,m)$ *is spanned by* $e^{-m\theta}y^{-(1+2\beta)}\varphi^+_{\beta,m,u}(y^{-2})$ *if* $\lambda \notin (-\infty, -1]$. *And* $C_-^\infty(u)(\lambda,m) = 0$ *if* $\lambda \in (-\infty, -1]$. *On the other hand* $C_+^\infty(u)(\lambda,m)$ *is spanned by* $e^{-m\theta}e^{\pi uy^{-2}}y^{-im}\phi_{\beta,m}(y^{-2})$ *for all* $\lambda \in \mathbb{C}$. *Moreover* $C_+^\infty(u)((\nu^2 + \nu - \frac{3}{4}), i(\nu - 2\ell - \frac{1}{2})) \subseteq L^2(u)$ *for* ℓ *any integer* ≥ 0 *and* ν *any integer* ≤ 0; *and* $C_+^\infty(u)((\nu^2 - 1), i(\nu - 1 - 2j)) \subseteq L^2(u)$ *for* j *any integer* ≥ 0 *and* ν *any integer* < 0.

(2) *Let* $u = 0$. *Then* $C_-^\infty(u)(\lambda,m)$ *is spanned by* $e^{-m\theta}y^{-(1+2\beta)}$ *if* $\lambda \notin (-\infty, -1]$, *and* $C_-^\infty(u)(\lambda,m) = 0$ *if* $\lambda \in (-\infty, -1]$. *Moreover* $C_+^\infty(u)(\lambda,m)$ *is spanned by* $e^{-m\theta}y^{2\beta-1}$ *if* $\lambda \notin (-\infty, -1]$ *and* $C_+^\infty(u)(\lambda,m) = 0$ *if* $\lambda \in (-\infty, -1]$.

Proof. First consider the local case $y = \infty$. We see that for $\varphi_f(y)$ of the form $y^{-(1+2\beta)}\varphi^+_{\beta,m,u}(y^{-2})$,

$$(4-28)\quad \int_{|y|\geq c_o} |\varphi_f(y)|^2 y\,dy \leq C \int_{|y|\geq c_o} |y|^{-4\mathrm{Re}(\beta)-1}dy < \infty$$

($C = \max_{|y|\geq c_o} |\varphi^+_{\beta,m,u}(y^{-2})|$) if $\mathrm{Re}\ (\beta) > 0$ (or $\lambda \notin (-\infty, -1]$). On the other hand since $\varphi^-_{\beta,m,u}(0) \neq 0$, we have for $\varphi_f(y) = y^{2\beta-1}\varphi^-_{\beta,m,u}(y^{-2})$ (where $\ell > 0$ so that $|\varphi^-_{\beta,m,u}(y^{-2})| \geq k_1$ for $|y| \geq \ell$)

$$(4-29)\quad \int_{|y|\geq k} |\varphi_f(y)|^2 y\,dy \geq k_1 \int_{|y|\geq k} |y|^{4\,\mathrm{Re}(\beta)-1}dy = \infty$$

if $\mathrm{Re}(\beta) > 0$ and $\beta \neq \frac{k}{2}$, k integer.

Then for the case $\varphi_f(y) = \delta_k y^{-(k+1)}\log y^{-2}\varphi^+_{\frac{k}{2},m,u}(y^{-2}) + y^{1+k}\varphi^-_{\frac{k}{2},m,u}(y^{-2})$, we easily observe that $y^{-(k+1)}\log y^{-2}\varphi^+_{\frac{k}{2},m,u}(y^{-2}) \in L^2([c,\infty), y\,dy)$. Thus $\varphi_f(y) \in L^2([c,\infty), y\,dy)$ if and only if $y^{k+1}\varphi^-_{\frac{k}{2},m,u}(y^{-2}) \in L^2([c,\infty), y\,dy)$. Then by a similar argument as above, we see that $\varphi_f(y) \notin L^2([c,\infty), y\,dy)$ for $\beta = \frac{1}{2}k$, k integer > 0.

If $\mathrm{Re}(\beta) = 0$ and $\beta \neq 0$ (or $\lambda \in (-\infty,-1)$), then we can write $\varphi_f(y) = A(\beta)y^{-(1+2\beta)} + B(\beta)y^{2\beta-1} + S(y,\beta)$ with $\int_{|y|\geq c} |S(y,\beta)|^2 y\,dy < \infty$

and where $A(\beta)$ and $B(\beta)$ are constants (we use the fact that $\varphi^{\pm}_{\beta,m,u}(x)$ are analytic at $x = 0$ and $\varphi^{\pm}_{\beta,m,u}(0) \neq 0$). Similarly if $\beta = 0$, we can write $\varphi_f(y) = Cy^{-1} + Dy^{-1} \log y^{-2} + S(y,0)$ with $S(y,0) \in L^2([c,\infty),ydy)$. Thus $\varphi_f(y)e^{-m\theta} \in C^{\infty}_{-}(u)(\lambda,m)$ if and only if $A(\beta)y^{-(1+2\beta)} + B(\beta)y^{2\beta-1}(Cy^{-1} + Dy^{-1} \log y^{-2}) \in L^2([c,\infty),ydy)$. Then it is an easy exercise to show that for $\beta = iq$ (q real), there exists no pair $(A(\beta),B(\beta))$ $((C,D))$ except $(0,0)$ so that $A(\beta)y^{-(1+2\beta)} + B(\beta)y^{2\beta-1}$ $(Cy^{-1} + Dy^{-1} \log y^{-2}) \in L^2([c,\infty),ydy)$.

We then consider the local case $y = 0$. First if $u < 0$ it is clear from the asymptotic properties of the solutions of (4 - 22) at $y = 0$, that $e^{\pi u y^{-2}} y^{-im} \phi_{\beta,m}(y^{-2})e^{-m\theta} \in C^{\infty}_{-}(u)(\lambda,m)$ for <u>all</u> λ and m, <u>and</u> $e^{-\pi u y^{-2}} y^{-im} \tilde{\phi}_{\beta,m}(y^{-2})e^{-m\theta} \notin C^{\infty}_{-}(u)(\lambda,m)$ for <u>all</u> λ and m. We observe that for a function $e^{-m\theta}e^{\pi u y^{-2}} y^{-im}\phi_{\beta,m}(y^{-2}) \in L^2(u)$, we must have $\int_0^1 |\phi_{\beta,m}(x)|^2 |x|^{-[\mathcal{I}(m)+2]}dx < \infty$, where $\mathcal{I}(m)$ = imaginary part of m.

Then using Remark 4.4 we derive the last part of (1).

Using arguments similar to those above, we derive the remaining results of the Proposition.

Q.E.D.

We fix $\lambda \in \mathbb{C}$ and m as above; normalize $\varphi^{\pm}_{\beta,m,u}$ so that $\varphi^{\pm}_{\beta,m,u}(0) = 1$.

We then define for all u

$$(4 - 30)\quad M^{+}_{u}(\lambda,m) = \{\text{the complex linear span of } f^{\lambda,+}_{m,k,u} = e^{-(m+2ki)\theta}y^{-(1+2\beta)}\varphi^{+}_{\beta,m+2ki,u}(y^{-2}) \text{ as } k \in \mathbb{Z}\}.$$

(4 - 31) $M_u^-(\lambda,m)$ = {the complex linear span of $f^{\lambda,-}_{m,k,u} = e^{-(m+2ki)\theta} B_{\beta,m+2ki,u}(y)$ as $k \in \mathbb{Z}$}, where $B_{\beta,m+2ki,u}(y) = y^{-1+2\beta} \varphi^-_{\beta,m+2ki,u}(y^{-2})$ if $\beta \neq \frac{v}{2}$, v integer > 0; $B_{\frac{v}{2},m+2ki,u}(y) = \delta_v y^{-(v+1)} (\log y^{-2}) \varphi^+_{\frac{v}{2},m+2ki,u}(y^{-2}) + y^{-1+v} \varphi^-_{\frac{v}{2},m+2ki,u}(y^{-2})$ for all v integer > 0; $B_{0,m+2ki,u}(y) = y^{-1} \log y^{-2} \varphi^+_{0,m+2ki,u}(y^{-2}) + y^{-1} \varphi^-_{0,m,u}(y^{-2})$.

(4 - 32) $W_u^+(\lambda,m)$ = {the complex linear span of $g^{\lambda,+}_{m,k,u} = e^{-(m+2ki)\theta} y^{-i(m+2ki)} e^{u\pi y^{-2}} \Phi_{\beta,m+2ki}(y^{-2})$ as $k \in \mathbb{Z}$} $(u < 0)$.

We then determine the $\mathcal{U}(sl_2)$ action on each of these spaces.

We observe that in the following cases the functions f and g are linearly independent over the ring of germs of analytic functions at $z = 0$: (1) $f(x) = x^{\gamma_1}, g(x) = x^{\gamma_2}$ where $\gamma_1 - \gamma_2 \notin \{0, 1, 2, \ldots\}$; (2) $f(x) = x^{\gamma_1} \log x$, $g(x) = x^{\gamma_2}$ where $\gamma_1 - \gamma_2$ is a nonnegative integer.

Lemma 4-7 (i) For all β and u

$$(4 - 33) \quad \pi_u(N_+)(f^{\lambda,+}_{m,k,u}) = (1 + 2\beta - 2k + im)\, f^{\lambda,+}_{m,k-1,u} \quad ,$$

$$(4 - 34) \quad \pi_u(N_-)(f^{\lambda,+}_{m,k,u}) = (1 + 2\beta + 2k - im)\, f^{\lambda,+}_{m,k+1,u} \quad .$$

(ii) For $u = 0$ and any β, or for $u < 0$ and $\beta \neq \frac{v}{2}$, v positive integer,

$$(4 - 35) \quad \pi_u(N_+)(f^{\lambda,-}_{m,k,u}) = (1 - 2\beta - 2k + im)\, f^{\lambda,-}_{m,k-1,u} + \sigma_1(u,\beta,k,m)\, f^{\lambda,+}_{m,k-1,u} \quad ,$$

$$(4-36)\quad \pi_u(N_-)(f^{\lambda,-}_{m,k,u}) = (1 - 2\beta + 2k - im)\, f^{\lambda,-}_{m,k+1,u} + \sigma_2(u,\beta,k,m)\, f^{\lambda,+}_{m,k+1,u} ,$$

<u>where</u> $\sigma_v(u,\beta,k,m) = 0$ <u>for all</u> $\beta \neq 0$; $\sigma_v(0,0,k,m) = 2$;

$$\sigma_v(u,0,k,m) = \begin{cases} 1-2k+im, & \text{if } v = 1 \text{ and } u < 0 \\ 1+2k-im, & \text{if } v = 2 \text{ and } u < 0 \end{cases} . \quad (\text{where } v = 1, 2).$$

(iii) <u>For all</u> β <u>and</u> $u < 0$

$$(4-37)\quad \pi_u(N_+)(g^{\lambda,+}_{m,k,u}) = 4\pi u g^{\lambda,+}_{m,k-1,u} ,$$

$$(4-38)\quad \pi_u(N_-)(g^{\lambda,+}_{m,k,u}) = \frac{1}{\pi u}\left(\frac{\lambda}{4} - (k - i\tfrac{m}{2})(k - i\tfrac{m}{2} + 1)\right)g^{\lambda,+}_{m,k+1,u} .$$

<u>Proof</u>. Via the substitution $y = x^{-\frac{1}{2}}$, we have that $\pi_u(N_+) = e^{2i\theta}(2x\frac{\partial}{\partial x} - i\frac{\partial}{\partial\theta} + 2\pi ux)$ and $\pi_u(N_-) = e^{-2i\theta}(2x\frac{\partial}{\partial x} + i\frac{\partial}{\partial\theta} - 2\pi ux)$.

Then for any φ of the form "f or g" above, we have that $\pi_u(\Omega)\pi_u(N_+)(\varphi) = \pi_u(N_+)\pi_u(\Omega)(\varphi) + \pi_u([\Omega,N_+])(\varphi) = \lambda\pi_u(N_+)(\varphi)$. Moreover $\pi_u(K)\pi_u(N_+)(\varphi) = \pi_u(N_+)\pi_u(K)(\varphi) + \pi_u([K,N_+])(\varphi) = (m - 2i)\pi_u(N_+)(\varphi)$.

Thus $\varphi \in C^\infty(u)(\lambda,m)$ implies that $\pi_u(N_+)(\varphi) \in C^\infty(u)(\lambda,m-2i)$.

Thus if $\varphi = f^{\lambda,+}_{m,k,u}$, we have $\pi_u(N_+)(\varphi) = e^{-(m + 2(k-1)i)\theta} y^{-(1 + 2\beta)}(1 + 2\beta + im - 2k)\varphi^{+}_{\beta,m+2ki,u} + x\psi_\beta$, where ψ_β is some analytic function at $x = 0$. But from the comment in the preceding paragraph, we can express

$$(4-39)\quad \pi_u(N_+)(\varphi) = Af^{\lambda,+}_{m,k-1,u} + Bf^{\lambda,-}_{m,k-1,u} .$$

Using the normalization of $\varphi^{\pm}_{\beta,m,u}$ and the comment preceding Lemma 4-7, we have the desired result for the case $\beta \notin \{0, \frac{1}{2}, 1, \ldots.\}$. If $\beta = \frac{v}{2}$, v integer > 0, we use the fact that $\pi_u(N_+)(\varphi)$ must belong to $C^\infty_-(u)(v^2 - 1, m+2(k-1)i)$ (recall $C^\infty_-(u)(v^2 - 1)$ is $\pi_u(\widetilde{S\ell}_2)$ stable). If $\beta = 0$ we deduce from $(4-39)$ a relation of the form

$B\{x \log x\}\varphi^{+}_{0,m+2ki,u}(x) + \ell(x) = 0$, with $\ell(x)$ analytic at $x = 0$. So if $B \neq 0$ we have a contradiction.

We repeat the same arguments for the case $\varphi = f^{\lambda,-}_{m,k,u}$. We note then that for $u = 0$, $\varphi^{\pm}_{\beta,m,0}(x) = 1$ $(\beta \neq 0)$ and $\varphi^{-}_{0,m,0}(x) = 0$; hence the computations can be done explicitly in this case.

Let $\varphi = g^{\lambda,+}_{m,k,u}$. Then $\pi_u(N_+)(\varphi) = e^{-(m + 2(k-1)i)\theta} e^{ux\pi} x^{i(\frac{m+2ki}{2})} V_{\beta,k,u}(x)$, where $V_{\beta,k,u}(x) = 4\pi ux\phi_{\beta,m+2ki} + 2x\phi'_{\beta,m+2ki} + (2m + 2ki)i\phi_{\beta,m+2ki}$. Further we must have $\pi_u(N_+)(\varphi) \in C^{\infty}_{+}(u)(\lambda, m + 2(k-1)i)$ (since $C^{\infty}_{+}(u)(\lambda)$ is stable by $\pi_u(\widetilde{S}\ell_2)$). Thus we can use the asymptotic expansion (see (4 - 27) of $V_{\beta,k,u}(x)$ at $x = \infty$ and derive that $\frac{1}{x} V_{\beta,k,u}(x)$ behaves at ∞ like $4\pi u(1 + \tilde{c}_1 x^{-1} + \ldots)$; hence we obtain the first part of (iii). By similar arguments (using explicitly the value of c_1 in (4 - 27)), we derive the second part of (iii).

Q.E.D.

We then define the following subspaces: (see Proposition 4.6 (i))

If ν any integer ≥ 0, let D_ν be the algebraic direct sum

$$(4 - 40) \quad \bigoplus_{j=0}^{\infty} C^{\infty}_{+}(u)((\nu^2 - \nu - \tfrac{3}{4}), -i(\nu + 2j + \tfrac{1}{2})).$$

If ν any integer ≥ 1, let H_ν be the algebraic direct sum

$$(4 - 41) \quad \bigoplus_{j=0}^{\infty} C^{\infty}_{+}(u)((\nu^2 - 1), -i(\nu + 2j + 1)).$$

<u>Remark 4.5</u> From Lemma 4.7 the spaces $M^{+}_{u}(\lambda,m), W^{+}_{u}(\lambda,m), M^{-}_{u}(\lambda,m)$ (for $\beta \notin \{0, \frac{1}{2}, 1, \frac{3}{2}, \ldots\}$ if $u \neq 0$ and for all $\beta \neq 0$ if $u = 0$), D_ν, and $H_{\nu'}$ (with ν integer ≥ 0, ν' integer ≥ 1) are $\pi_u(\mathcal{U}(sl_2))$ stable.

Moreover we have (i) for $u < 0$, $C_-^\infty(u)(\lambda) \supseteq M_u^+(\lambda,m)$ for all m and all $\lambda \notin (-\infty, -1]$; (ii) for $u < 0$, $C_+^\infty(u)(\lambda) \supseteq W_u^+(\lambda,m)$ for all m and all λ; (iii) $D_\nu \subseteq L^2(u)$ and $H_{\nu'} \subseteq L^2(u)$ for ν integer ≥ 0, ν' integer ≥ 1, and $u < 0$; (iv) for $u = 0$, $C_+^\infty(u)(\lambda) \supseteq M_o^+(\lambda,m)$ for all m and all $\lambda \notin (-\infty, -1]$; and (v) for $u = 0$, $C_-^\infty(u)(\lambda) \supseteq M_o^-(\lambda,m)$ for all m and all $\lambda \notin (-\infty, -1]$.

Since Ω is the Casimir element and hence central and symmetric (see section 3), we know that $\pi_u(\Omega)$ restricted to $C_c^\infty(u)$ is a self adjoint operator relative to (4 - 10) above, i.e., $(\pi_u(\Omega)f,g) = (f,\pi_u(\Omega)g)$ for all f and $g \in C_c^\infty(u)$. Moreover we recall that $\pi_u(\Omega)$ is an essentially self adjoint operator on $L^2(u)$. We also note from above that the representation π_u of $\widetilde{S\ell}_2$ on $C_c^\infty(u)$ is differentiable and that the contragredient representation π_u^c of $\widetilde{S\ell}_2$ on $C_c^\infty(u)'$ (= the space of distributions) is also differentiable (see section 3).

Then for any $\lambda \in \mathbb{C}$ we consider the generalized eigenspace $T_\lambda^u \subseteq C_c^\infty(u)'$, where $T_\lambda^u = \{T \in C_c^\infty(u)' \mid \pi_u(\Omega)T = \lambda T\}$. First it is easy to see that relative to the strong dual topology on $C_c^\infty(u)'$, T_λ^u is a closed subspace of $C_c^\infty(u)'$ invariant by $\pi_u^c(g)$ for all $g \in \widetilde{S\ell}_2$.

We recall that $C^\infty(u)$ can be embedded in $C_c^\infty(u)'$ by $g \in C^\infty(u)$ going to the linear functional $\langle T_g,\varphi\rangle = \int g(y)\varphi(y)d\mu_o(y)$ for all $\varphi \in C_c^\infty(u)$. Then we have that $\pi_u^c(\Omega)T_g = \lambda T_g$ if and only if $\pi_u(\Omega)(\overline{g}) = \overline{\lambda}\overline{g}$. We then note from (4 - 21) that $g \in C^\infty(u)(\lambda,m)$ if and only if $\overline{g} \in C^\infty(u)(\overline{\lambda},\overline{m})$. Thus we have for all λ, $T_\lambda^u \supseteq C^\infty(u)(\overline{\lambda},m)$ (relative to the embedding above) for any $m = \frac{1}{2}\nu i$, ν any integer.

Let $(T_\lambda^u)_K$ be the set of "K finite vectors" in T_λ^u (observe that the representation π_u^c of $\widetilde{S\ell}_2$ restricted to T_λ^u is differentiable). Then we know that $(T_\lambda^u)_K$ is dense in T_λ^u (relative to the strong topology induced on T_λ^u). As in the paragraph above, we note that $\pi_u^c(K)\cdot T_L = s\cdot T_L$ if and only if $\pi_u(K)(\overline{L}) = -\overline{s}\overline{L}$ (if $L \in C^\infty(u)$).

Let $(T_\lambda^u)_s = \{T \in (T_\lambda^u)_K \mid \pi_u^c(K)T = sT\}$.

<u>Lemma 4.8</u> $(T_\lambda^u)_s = 0$ <u>if</u> $s \neq \frac{v}{2}i$ (v integer). <u>Thus</u> $(T_\lambda^u)_m = C^\infty(u)(\lambda,\overline{m})$ <u>and</u> $(T_\lambda^u)_K$ <u>is the algebraic direct sum</u> $\bigoplus_{v\in\mathbb{Z}} C^\infty(u)(\lambda,\frac{v}{2}i)$.

<u>Proof</u>. The representation π_u of $\widetilde{S\ell}_2$ on $C_c^\infty(u)$ is differentiable, and $(C_c^\infty(u))_K$, the set of K finite vectors in $C_c^\infty(u)$, is dense in $C_c^\infty(u)$. Let $U \in (T_\lambda^u)_s$ so that $U \neq 0$; thus there exists a $\varphi \in C_c^\infty(u)$ so that $\pi_u(K)\varphi = \rho\varphi$, where $\rho = \frac{v}{2}i$, v integer, and where $\langle U,\varphi\rangle \neq 0$ (recall that $C_c^\infty(u)$ can be identified to the space $\{f \in C^\infty([-2\pi,2\pi] \times \mathbb{R}_+^*) \mid$ the function $x \to f(\theta,x)$ has compact support in $\mathbb{R}_+^*$ for all $\theta\}$. But this means that $s = \frac{1}{2}v_1 i$ for v_1 an integer.

Then we note that for $U \in (T_\lambda^u)_s$, $\{\pi_u^c(\Omega) + (\pi_u^c(K))^2\}(U) = (\lambda + s^2)U$. But the <u>highest</u> <u>order</u> term of the operator $\pi_u^c(\Omega + K^2)$ is $\frac{\partial^2}{\partial\theta^2} + y^2\frac{\partial^2}{\partial y^2}$; hence $\pi_u^c(\Omega + K^2)$ is <u>elliptic</u> and the distribution $U \in C_c^\infty(u)'$ is a C^∞ function. Thus $(T_\lambda^u)_m = C^\infty(u)(\lambda,\overline{m})$ from the comments above.

Q.E.D.

Let ζ be any fourth root of unity. Then define the closed subspace of T_λ^u, ${}_\zeta T_\lambda^u = \{T \in T_\lambda^u \mid \pi_u^c(E)T = \zeta T\}$. Then it is obvious that ${}_\zeta T_\lambda^u$ is invariant by $\pi_u(g)$ as $g \in \widetilde{S\ell}_2$. Also we have $T_\lambda^u = \oplus\, {}_\zeta T_\lambda^u$ as ζ ranges over the 4th roots of unity.

We note at this point that if $g \in C^\infty(u)(\lambda,\frac{1}{2}iv)$, then $\pi_u(E)g = e^{-\frac{1}{2}iv\pi} g$ (and hence $\pi_u^c(E)\cdot T_g = e^{-\frac{1}{2}iv\pi} T_g$).

For the remainder of this section, we assume $u \leq 0$.

Proposition 4.9. *The space* $({}_{\bar\zeta}T^u_\lambda)_K$ (*of* K *finite vectors in* ${}_{\bar\zeta}T^u_\lambda$ *and* $\bar\zeta$ = *complex conjugate of* ζ) *is the algebraic direct sum* $\bigoplus_{k\in\zeta} C^\infty(u)(\lambda,\frac{1}{2}ki)$, *where* $\zeta = \{k \in \mathbb{Z} \mid k \equiv s \bmod 4 \text{ with } \zeta = e^{\frac{1}{2}\pi is}\}$. *Moreover* $({}_{\bar\zeta}T^u_\lambda)_K = M^+_u(\lambda,\frac{1}{2}is) \oplus M^-_u(\lambda,\frac{is}{2})$. *For fixed* $s \in \{0, 1, 2, 3\}$ *let* $(* \text{-}s)$ *denote the following condition:*

$$(4 - 42) \quad \lambda \neq 4k^2 + 4k(\frac{s}{2} - 1) + (\frac{s^2}{4} - s) \quad \text{for } k \text{ any integer.}$$

If λ *satisfies* $(* \text{-}s)$, *then the* $\pi_u(\mathcal{U}(s\ell_2))$ *modules* $M^+_u(\lambda,\frac{is}{2})$ *and* $W^+_u(\lambda,\frac{is}{2})$ *are algebraically irreducible. And if* λ *satisfies* $(* \text{-}s)$ *and* $\lambda \neq k^2 - 1$ *with* k *integer* ≥ 0 $(\lambda \neq -1)$, *then* $M^-_u(\lambda,\frac{is}{2})$ *is an algebraically irreducible* $\pi_u(\mathcal{U}(s\ell_2))$ *module for* $u < 0$ $(u = 0)$. *The spaces* D_ν (ν *integer* ≥ 0) *and* $H_{\nu'}$ (ν' *integer* ≥ 1) *are the unique, nonzero* $\pi_u(\mathcal{U}(s\ell_2))$ *algebraically irreducible subspaces of* $W^+_u((\nu^2 - \nu - \frac{3}{4}), - i\,(\nu + \frac{1}{2}))$ *and* $W^+_u(((\nu')^2 - 1), - i\,(\nu' + 1))$ *respectively.*

Q.E.D.

Remark 4.6 If $m = \frac{is}{2}$ $(s \in \{0, 1, 2, 3\})$ and λ satisfies $(* \text{-}s)$, we define $A^u_{\lambda,s}(f^{\lambda,+}_{m,k,u}) = \beta_k(\lambda,s)f^{\lambda,-}_{m,k,u}$ and $A^u_{\lambda,s}(f^{\lambda,-}_{m,k,u}) = \sigma_k(\lambda,s)f^{\lambda,+}_{m,k,u}$, where

$$(4 - 43) \quad \begin{aligned} \beta_k(\lambda,s) &= \frac{\Gamma(\frac{1}{2} + \beta + \frac{s}{4})}{\Gamma(\frac{1}{2} - \beta + \frac{s}{4})} \; \frac{\Gamma(\frac{1}{2} - \beta + \frac{s}{4} + k)}{\Gamma(\frac{1}{2} + \beta + \frac{s}{4} + k)} \quad \text{and} \\ \sigma_k(\lambda,s) &= \frac{1}{\beta_k(\lambda,s)} \end{aligned}$$

(Γ, the classical gamma function). Thus $A^u_{\lambda,s}$ can be extended to a linear map on the space $({}_\zeta T^u_\lambda)_K$ for all λ satisfying $(*\text{-s})$, $(\zeta = e^{\pi is/2})$. It is then easy to see that $A^u_{\lambda,s}$ <u>defines a unique</u> (up to scalar multiple) $\mathcal{U}(s\ell_2)$ <u>intertwining isomorphism between</u> $M^+_u(\lambda,\frac{is}{2})$ <u>and</u> $M^-_u(\lambda,\frac{is}{2})$ in the following cases: (i) for $u < 0$ when λ satisfies $(*\text{-s})$ and $\lambda \neq k^2-1$ with k integer ≥ 0 and (ii) for $u = 0$ when $\lambda \neq -1$.

From (3 - 7) we recall that the map $f \to f^\sigma$ of $C^\infty_c(\widetilde{S}\ell_2)$ to $C^\infty_\sigma(\widetilde{S}\ell_2)$ is a continuous, linear, open, surjective $\widetilde{S}\ell_2$ intertwining map $(\sigma = \sigma_1 \otimes \lambda$, $\sigma_1 \in \hat{\underline{M}}$, $\lambda \in \hat{\underline{A}})$. Then it follows by inducing in stages that the linear map R_σ: $C^\infty_c(0) \to C^\infty_\sigma(\widetilde{S}\ell_2)$ given by

$$(4 - 44) \quad R_\sigma: \quad \varphi \to \varphi^\sigma(g) = \int_{\mathbb{R}^*_+ \times M} \varphi(gma(r))|r|\lambda(r)\sigma_1(m)\frac{dr}{|r|}\,dm$$

$(r \in (0,\infty)$, $m \in M$, and $g \in \widetilde{S}\ell_2)$ is a continuous, linear, open, surjective $\widetilde{S}\ell_2$ intertwining map.

Moreover recalling the Bruhat decomposition (4 - 4) of $\widetilde{S}\ell_2$, we define for $F \in C^\infty_\sigma(\widetilde{S}\ell_2), \varphi_F(x) = F(n(x)w_o)$ for $x \in \mathbb{R}$. Then $\varphi_F \in C^\infty(\mathbb{R})$. But then $\pi_\sigma(w_o^{-1})\, F \in C^\infty_\sigma(\widetilde{S}\ell_2)$, and hence $\varphi_{\pi_\sigma(w_o^{-1})F}(x) =$

$\varphi_F(-\frac{1}{x})\,|x|^{-(\lambda+1)}\sigma_1(\varphi_x(E))$. We let $\beta_s(x) = \begin{cases} \zeta^{\,s} & \text{if } x > 0 \\ \zeta^{\,2s} & \text{if } x < 0 \end{cases}$ with $\sigma_1(E) = \zeta^{\,s}$, where $\zeta = e^{\frac{\pi i}{2}}$. Then since $\varphi_{\pi_\sigma(w_o^{-1})F} \in C^\infty(\mathbb{R})$, we have that the function $h_F(x) = \varphi_F(-\frac{1}{x})\,|x|^{-(\lambda+1)}\beta_s(x)$ $(x \neq 0)$ extends to a C^∞ function on $\mathbb{R}$.

We define $C^\infty_{\lambda,s}(\mathbb{R}) = \{f \in C^\infty(\mathbb{R}) \mid x \to f(-\frac{1}{x})\,|x|^{-(\lambda+1)}\beta_s(x)$ $(x \neq 0)$ extends to a C^∞ function on $\mathbb{R}\}$. Define the linear form $Q_{\lambda,s}(\psi) = \lim_{x\to 0} \psi(-\frac{1}{x})\,|x|^{-(\lambda+1)}\beta_s(x)$ for $\psi \in C^\infty_{\lambda,s}(\mathbb{R})$.

<u>Lemma 4.10 The map</u> $F \to \varphi_F$ <u>defines a linear bijection between</u> $C^\infty_\sigma(\widetilde{S\ell}_2)$ <u>and</u> $C^\infty_{\lambda,s}(\mathbb{R})$.

<u>Proof</u>. The map is clearly injective since $F \in C^\infty_\sigma(\widetilde{S\ell}_2)$ is determined on the open cell of $\widetilde{S\ell}_2$. Let $\psi \in C^\infty_{\lambda,s}(\mathbb{R})$. Then define:

$$(4-45)\quad \begin{aligned} f_\psi(\left\{\begin{bmatrix} a & b \\ c & d \end{bmatrix}, \epsilon\right\}) &= \psi(\tfrac{a}{c})|c|^{-(\lambda+1)} \zeta^{s\beta(\mathrm{sgn}(c),\epsilon)} \quad \text{if } c \neq 0 \\ f_\psi(\left\{\begin{bmatrix} a & b \\ 0 & a^{-1} \end{bmatrix}, \epsilon\right\}) &= |a|^{-(\lambda+1)} \zeta^{s\beta(\mathrm{sgn}(a),\epsilon)} Q_{\lambda,s}(\psi). \end{aligned}$$

By definition it is clear that f_ψ is C^∞ on the open cell. Then let $p = \left\{\begin{bmatrix} u & s \\ 0 & u^{-1} \end{bmatrix}, \epsilon\right\}$. Choose ϵ_1 and $\epsilon_2 > 0$ and n sufficiently large so that relative to the Iwasawa decomposition of $\widetilde{S\ell}_2$, the set $B = \{k(\theta,1)a(r)n(ry) \mid |\theta| < \frac{\pi}{n},\ |r-1| < \epsilon_1 \text{ and } |y| < \epsilon_2\}$ is an open neighborhood of $e_{\widetilde{S\ell}_2}$ in $\widetilde{S\ell}_2$. Then $B\cdot p$ is a neighborhood of p; moreover $B\cdot p$ is diffeomorphic to the product $V = \{\theta \mid |\theta| < \frac{\pi}{n}\} \times \{r \mid |r-1| < \epsilon_1\} \times \{y \mid |y| < \epsilon_2\}$. Then on the set $B\cdot p$ we have for $\theta \neq 0$, $f_\psi(k(\theta,1)a(r)n(ry)p) = \psi(-\cot\theta)|ru|^{-(\lambda+1)}|\sin\theta|^{-(\lambda+1)}$
$\zeta^{s\beta(-\mathrm{sgn}(u\sin\theta),\, -\epsilon(\frac{\sin\theta}{-u})_H)}$ and for $\theta = 0$, $f_\psi(a(r)n(ry)p) =$
$Q_{\lambda,s}(\psi)|ru|^{-(\lambda+1)} \zeta^{s\beta(\mathrm{sgn}(u),\epsilon)}$. Then by hypothesis $\theta \to$
$\psi(-\cot\theta)|\sin\theta|^{-(\lambda+1)}|\cos\theta|^{\lambda+1}\beta_s(\cot\theta)$ $(\theta \neq 0)$ extends to a C^∞ function locally at $\theta = 0$; moreover we have

$$\zeta^{s\beta(-\mathrm{sgn}(u)\mathrm{sgn}(\sin\theta),\, -\epsilon(\frac{\sin\theta}{-u})_H)} = \beta_s(\cot\theta)\, \zeta^{s\beta(\mathrm{sgn}(u),\epsilon)}.$$

Thus the function $(\theta,r,y) \to f_\psi(k(\theta,1)a(r)n(ry)p)$ $(\theta \neq 0)$ extends to a C^∞ function on V. Then for all r and y so that $|r-1| < \epsilon_1$ and

$|y| < \epsilon_2$, $\lim_{\theta \to 0} f(k(\theta,1)a(r)n(ry)p) = \lim_{\theta \to 0} |ru|^{-(\lambda+1)}\psi(-\cot\theta)$

$|\sin\theta|^{-(\lambda+1)}\beta_s(\cot\theta)\,\zeta^{s\beta(\mathrm{sgn}(u),\epsilon)} = f_\psi(a(r)n(ry)p)$.

Thus $f_\psi \in C_\sigma^\infty(\widetilde{S\ell}_2)$ and $\varphi_{(f_\psi)}(x) = \psi(x)$ for all $x \in \mathbb{R}$. Hence the map $F \to \varphi_F$ is onto.

Q.E.D.

Remark 4.7 Using the definition of φ_F, we see that the representation π_σ carried over to $C_{\lambda,s}^\infty(\mathbb{R})$ is given by the formulae:

$$(4-46)\quad \varphi_{\pi_\sigma(n(x))F}(y) = \varphi_F(y-x) \quad ,$$

$$(4-47)\quad \varphi_{\pi_\sigma(w_o^{-1})F}(y) = \varphi_F(-\tfrac{1}{y})\,|y|^{-(\lambda+1)}\sigma_1(\varphi_y(E)) \quad ,$$

$$(4-48)\quad \varphi_{\pi_\sigma(a(r))F}(y) = \varphi_F(r^{-2}y)|r|^{-(\lambda+1)} \quad .$$

Remark 4.8 We note that $C_{\lambda,s}^\infty(\mathbb{R}) \subseteq L^p(\mathbb{R}, dx)$ for all λ so that $\frac{1}{p} - 1 < \mathrm{Re}(\lambda)$. Moreover by an easy argument, we note that $S(\mathbb{R}) \subseteq C_{\lambda,s}^\infty(\mathbb{R})$ for all λ.

Remark 4.9 By standard arguments similar to the case $S\ell_2(\mathbb{R})$, we have relative to any choice of Haar measure $d\underline{\tilde{k}}$ on $\underline{\tilde{K}}$, a constant $c_k \neq 0$ so that for all $F_1, F_2 \in C_\sigma^\infty(\widetilde{S\ell}_2)$

$$(4-49)\quad \int_{\underline{\tilde{K}}} F_1(k)\overline{F_2(k)}\,d\underline{\tilde{k}} = c_k \int_{\mathbb{R}} \varphi_{F_1}(x)\overline{\varphi_{F_2}(x)}(1+x^2)^{\mathrm{Re}(\lambda)}dx.$$

We then consider the nonempty Banach space $H_\lambda = \{\psi\colon \mathbb{R} \to \mathbb{C} \mid \psi$ measurable (identified modulo null Borel sets on $\mathbb{R}$) so that $\|\psi\|_\lambda^2 = \int_{\mathbb{R}} |\psi(x)|^2(1+x^2)^{\mathrm{Re}(\lambda)}dx < \infty\}$. It is clear that $C_{\lambda,s}^\infty(\mathbb{R})$ is dense in H_λ.

Lemma 4.11 The representation π_σ of $\widetilde{S\ell}_2$ on $C^\infty_{\lambda,s}(\mathbb{R})$ extends continuously to a representation $\hat{\pi}_\sigma$ of $\widetilde{S\ell}_2$ on H_λ so that

$$(4-50) \quad \|\hat{\pi}_\sigma(g)\psi\|_\lambda \leq e^{|\mathrm{Re}(\lambda_0)|x(g)}\|\psi\|_\lambda \quad \text{for} \quad \psi \in H_\lambda,$$

and where $g = \left\{\begin{bmatrix} a & b \\ c & d \end{bmatrix}, \epsilon\right\}$ and $x(g) = \text{arc cosh}\, \frac{1}{2}(a^2 + d^2 + b^2 + c^2)^{1/2}$.

Moreover the space $C^\infty_{\lambda,s}(\mathbb{R})$ coincides with the space $(H_\lambda)_\infty$ of C^∞ vectors of the representation $\hat{\pi}_\sigma$ of $\widetilde{S\ell}_2$ on H_λ.

Proof. For $g \in \widetilde{S\ell}_2$ we consider the polar decomposition $g = k_1 a(e^x)k_2$ with $k_i \in \widetilde{K}$. Then using Remark 4.9, we have that $\|\pi_\sigma(g)\psi\|_\lambda = \|\pi_\sigma(a(e^x))\psi\|_\lambda$ for $\psi \in C^\infty_\sigma(\widetilde{S\ell}_2)$. Then by (4 - 48) and change of variables, we easily deduce (4 - 50). We note that for g, the polar part

$$e^x = e^{\text{arc cosh}\,(\frac{1}{2}[a^2 + d^2 + b^2 + c^2]^{1/2})}.$$

Then using the same argument as in the proof of Proposition 3.5, we deduce the last statement of the Lemma.

Q.E.D.

Remark 4.10 The representation $\hat{\pi}_\sigma$ of $\widetilde{S\ell}_2$ on H_λ is unitary for $\lambda = it$, and from Proposition 4.1 it follows that for $t \neq 0$, $\hat{\pi}_\sigma$ is irreducible.

At this point we recall the notion of holomorphicity of a function defined by an integral (see [15], p. 4). If X is a locally compact topological space, $d\nu(x)$ some positive measure on X, and R any open subset of $\mathbb{C}$, then let $f(x,z)$ be a continuous function on $X \times R$. We assume that for any fixed $x \in X$, the function $z \to f(x,z)$ is holomorphic on R. If M is any relatively compact open set of R and if there exists a positive $L^1(X,d\nu)$ function ϕ_M on X so that $|f(x,z)| \leq \phi_M(x)$ for all $(x,z) \in X \times M$, then the integral

$$(4-51)\quad G(z) = \int_X f(x,z)d\nu(x)$$

is holomorphic in $z \in R$.

We define for $z > 0$ and $\lambda \in \mathbb{C}$ so that $\mathrm{Re}(\lambda) < 0$

$$(4-52)\quad E(z,\lambda) = \int_z^\infty x^{\lambda-1}e^{ix}dx$$

(absolutely convergent). Then by change of variable $x = x'z$, we deduce that $E(z,\lambda) = z^{-\lambda}\int_1^\infty x^{\lambda-1}e^{ixz}dx$. Thus we see that the map $(z,\lambda)\to E(z,\lambda)$ is continuous in $\mathbb{R}_+^* \times H$, where $H = \{z \in \mathbb{C} \mid \mathrm{Re}(z) < 0\}$ and for any fixed z, $\lambda\to E(z,\lambda)$ is analytic.

Lemma 4.12 *For $z > 0$ the map $\lambda\to E(z,\lambda)$ has a continuation to an analytic function $\lambda\to F(z,\lambda)$ on $\mathbb{C}$. Moreover the map $(z,\lambda)\to F(z,\lambda)$ is continuous on $\mathbb{R}_+^* \times \mathbb{C}$.*

Proof. By integration by parts we deduce that

$$(4-53)\quad \lambda E(z,\lambda) + iE(z,\lambda+1) = -z^\lambda e^{iz}$$

for λ so that $\mathrm{Re}(\lambda) < 0$. Then by iteration of this relation we deduce

$$(4-54)\quad E(z,\lambda+k) = z^\lambda e^{iz}\left(\sum_{j=0}^{k-1} P_j(\lambda)z^j\right) + d_k E(z,\lambda)\lambda(\lambda+1)\ldots(\lambda+k-1),$$

where P_j is some polynomial in λ of degree at most k and $d_k \neq 0$. Thus defining for $\mathrm{Re}(\lambda) > k$

$$(4-55)\quad F(z,\lambda) = z^{\lambda-k}e^{iz}\left(\sum_{j=0}^{k-1} P_j(\lambda-k)z^j\right) + d_k E(z,\lambda-k)(\lambda-k)\ldots(\lambda-1),$$

we see that $\lambda\to F(z,\lambda)$ is analytic for $\mathrm{Re}(\lambda) < k$ and for $\mathrm{Re}(\lambda) < 0$, $F(z,\lambda) = E(z,\lambda)$. Thus the statement of the Lemma follows.

Q.E.D.

Let $x \in \mathbb{C}$ and $u \in \mathbb{R}$ so that $u < 0$. Let $\mathcal{A}_s^x(B_u)$ be the space of all complex valued functions φ: $\mathbb{C} \times \widetilde{S\ell_2} \to \mathbb{C}$ so that

(4 - 56)
(i) $\varphi(\lambda, \ ) \in C^\infty_{\sigma_1 \otimes (x\lambda)}(\widetilde{S\ell_2}) \quad (\in C^\infty(u))$,

(ii) for any $\xi \in \mathcal{U}(s\ell_2)$ the map $(\lambda,g) \to \xi * \varphi(\lambda,g)$ ($\xi *$ means application in g variable) is continuous and analytic in λ.

We topologize $\mathcal{A}_s^x(B_u)$ by the following set of seminorms: let Ω_1, Ω_2 be compact subsets of $\widetilde{S\ell_2}$ and $\mathbb{C}$ respectively, and $\xi \in \mathcal{U}(s\ell_2)$

$$(4 - 57) \quad \nu_{\Omega_1,\Omega_2,\xi}(\varphi) = \sup_{\substack{g \in \Omega_1 \\ \lambda \in \Omega_2}} |\xi * \varphi(\lambda,g)| .$$

Then by standard arguments $\mathcal{A}_s^x(B_u)$ equipped with this family of seminorms is a Frechet space. Moreover we have the usual differentiable representation $\Sigma = \Sigma_s^x$ $(= \Sigma_u)$ of $\widetilde{S\ell_2}$ on $\mathcal{A}_s^x(B_u)$ given by $\Sigma(g)F(\lambda,g') = F(\lambda,g^{-1}g')$.

We define for $F \in \mathcal{A}_s^1$ and $\mathrm{Re}(\lambda) > 0$ $(F_\lambda(g) = F(\lambda,g))$

$$(4 - 58) \quad \psi_F(g,\lambda,u) = \gamma(u,\lambda) \int_{\mathbb{R}} \varphi_{\pi_\sigma}(g^{-1})F_\lambda(x)e^{-2\pi ixu}dx$$

with $\gamma(u,\lambda) = \begin{cases} 1 & \text{if } u \neq 0 \\ (\Gamma(\lambda))^{-1} & \text{if } u = 0 \end{cases}$ $\quad (\varphi_{\pi_\sigma}(g^{-1})F_\lambda \in L^1(\mathbb{R})$ with $\sigma = \sigma_1 \otimes \lambda$ for $\mathrm{Re}(\lambda) > 0$). We then observe that $\psi_{\Sigma_s^1(g')F}(g,\lambda,u) = \psi_F((g')^{-1}g,\lambda,u)$ and hence we deduce that for any $Z \in \mathcal{U}(s\ell_2)$ $\psi_{\Sigma_s^1(Z)F} = Z * \psi_F$ (where $*$ relative to first variable of ψ_F).

<u>Proposition 4.13</u> <u>For</u> $\mathrm{Re}(\lambda) > 0$

$$(4 - 59) \quad \psi_F(g,\lambda,u) = \gamma(u,\lambda) \int_{\mathbb{R}} \varphi_{\pi_\sigma}((gw_o)^{-1})F_\lambda(x)|x|^{\lambda-1}\gamma_s(x)e^{2\pi i \frac{u}{x}}dx$$

where $\gamma_s(x) = (\beta_s(x))^{-1} e^{-\frac{\pi i}{2}s}$ *and the latter integral is absolutely convergent. Moreover for fixed* g *and* u, *the function* $\lambda \to \psi_F(g,\lambda,u)$ *has a continuation to an analytic function* $\lambda \to A_F(g,\lambda,u)$ *for* $\lambda \in \mathbb{C}$. *Then* $A_F(\ ,\ ,u) \in B_u$ *if* $u \neq 0$ *and* $A_F(\ ,\ ,0) \in \mathcal{O}_s^{-1}$. *And if for* $\lambda \in \mathbb{C}$, *we have* $F(\lambda,\ ) \equiv 0$, *then* $A_F(\ ,\lambda,u) \equiv 0$. *Finally we have the following:*

(α) *Let* $u \neq 0$. *Then the map* $F \to A_F(\ ,\ ,u)$ *defines a continuous linear* $\widetilde{S\ell}_2$ *intertwining map between* Σ_s^1 and Σ_u.

(β) *Let* $u = 0$. *Then the map* $F \to A_F(\ ,\ ,0)$ *defines a continuous linear* $\widetilde{S\ell}_2$ *intertwining map between* Σ_s^1 *and* Σ_s^{-1}.

Proof. We let $\beta(g,\lambda,u,x) = \varphi_{\pi_\sigma((gw_o)^{-1})F_\lambda}(x)|x|^{\lambda-1}\gamma_s(x)e^{2\pi i\frac{u}{x}}$.

Then from above it is easy to see that $\varphi_{F_\lambda}(\frac{1}{x})|x|^{-2} \in L^1(\mathbb{R})$ for $\mathrm{Re}(\lambda) > 0$, and by application of Lesbeque dominated convergence, change of variables, and (4 - 47)

$$(4-60)\quad \psi_F(g,\lambda,u) = \gamma(u,\lambda) \lim_{R\to+\infty} \int_{\frac{1}{R}\le|x|\le R} \varphi_{\pi_\sigma(g^{-1})F_\lambda}(x)e^{-2\pi i x u}dx$$

$$= \gamma(u,\lambda) \lim_{R\to\infty} \int_{\frac{1}{R}\le|x|\le R} \varphi_{\pi_\sigma(g^{-1})F_\lambda}(\tfrac{1}{x})|x|^{-2}e^{-2\pi i\frac{u}{x}}dx =$$

$$= \gamma(u,\lambda) \int_{\mathbb{R}} \beta(g,\lambda,u,x)dx \quad .$$

Thus we have (4 - 59).

We then write

$$(4-61)\quad \gamma(u,\lambda) \int_{\mathbb{R}} \beta(g,\lambda,u,x)dx = C_1(g,\lambda,u) + C_2(g,\lambda,u) \quad ,$$

where $C_1(g,\lambda,u) = \gamma(u,\lambda) \int_{|x|\le 1} \beta(g,\lambda,u,x)dx$ and $C_2(g,\lambda,u) =$

$\gamma(u,\lambda) \int_{|x|>1} \beta(g,\lambda,u,x)dx$. Again by change of variables,

$C_2(g,\lambda,u) = \gamma(u,\lambda) \int_{|x|<1} \varphi_{\pi_\sigma(g^{-1})F_\lambda}(x)e^{-2\pi ixu}dx$. Then for a compact subset of $\widetilde{S\ell}_2$ of the form $S_1 = \Omega_1 \cdot X_1$, Ω_1 any compact subset of $\widetilde{S\ell}_2$ and $X_1 = \{n(x)w_o \mid |x| \leq 1\}$, and for $\hat{C}$ a compact subset of $\mathbb{C}$ and L a compact subset of $\mathbb{R}$, we have

$$(4-62) \quad |C_2(g,\lambda,u)| \leq \sup_{\substack{\lambda \in \hat{C} \\ u \in L}} |\gamma(u,\lambda)| \nu_{S_1,\hat{C}}(F_\lambda)$$

for $g \in \Omega_1$, $\lambda \in \hat{C}$, and $u \in L$. Moreover we have that $(g,\lambda,u) \to C_2(g,\lambda,u)$ is a continuous function of (g,λ,u), and by the criterion of holomorphicity above, $\lambda \to C_2(g,\lambda,u)$ is analytic for all $\lambda \in \mathbb{C}$.

We let $G_\lambda(g) = \pi_\sigma((gw_o)^{-1})F_\lambda$. We recall the Taylor expansion of $\varphi_{G_\lambda(g)}(x) = \sum_{k=0}^{T} \frac{1}{k!} \varphi_{G_\lambda(g)}^{(k)}(0)x^k + R_{G_\lambda(g)}(x)$, where $R_{G_\lambda(g)}$ is the remainder term put in the form $\frac{x^T}{T!} \int_0^1 (1-z)^T (\frac{d}{dz})^{T+1} (\varphi_{G_\lambda(g)})(zx)dz$.

Then we have that for $\mathrm{Re}(\lambda) > 0$

$$(4-63) \quad C_2(g,\lambda,u) = \gamma(u,\lambda) \sum_{k=0}^{T} \frac{1}{k!} \varphi_{G_\lambda(g)}^{(k)}(0) \{\alpha_k V(\lambda+k-1,\ u) + \beta_k V(\lambda+k-1,\ -u)\} + J_\lambda(g,x,u) \quad ,$$

where $V(\beta,u) = \int_0^1 x^\beta e^{2\pi iu/x}dx$ (defined for $\mathrm{Re}(\beta) > -1$), α_k and β_k certain constants and

$$(4-64) \quad J_\lambda(g,x,u) = \gamma(u,\lambda) \int_{|x| \leq 1} R_{G_\lambda(g)}(x)|x|^{\lambda-1}\gamma_s(x)e^{2\pi i\frac{u}{x}} dx.$$

We then deduce by differentiating (4 - 46) that $\varphi_{\pi_\sigma(E)F_\lambda} = \frac{d}{dx} \varphi_{F_\lambda}$.

Then we have that $\frac{d}{dx}\varphi_{G_\lambda}(g)^{(x)} = \varphi_{\pi_\sigma}((gw_o)^{-1})\pi_\sigma(\mathrm{Ad}(g)F)F_\lambda^{(x)}$. Moreover if $g = \left\{\begin{bmatrix} a & b \\ c & d\end{bmatrix}, \epsilon\right\}$ we see easily that $\mathrm{Ad}(g)F = -acH + a^2E + c^2F$. Then $(\mathrm{Ad}(g)F)^s = \sum_{\ell+m+n\leq s} \varphi_{\ell mn}(g)H^\ell E^m F^n$, where $\varphi_{\ell mn}$ is a polynomial function of (a, b, c, d); this means in particular that the maps $g\to\varphi_{\ell mn}(g)$ are continuous on $\tilde{S}\ell_2$.

Then we see by the change of variables argument again that for $\mathrm{Re}(\lambda) > -T$

$$(4 - 65)\quad J_\lambda(g,x,u) = \gamma(u,\lambda)\int_{|x|\geq 1} |x|^{-(T+\lambda+1)}\gamma_s\left(\frac{1}{x}\right)e^{2\pi ixu}$$
$$\left\{\int_0^1 (1-z)^T\left(\frac{d}{dz}\right)^{T+1}(\varphi_{G_\lambda}(g))\left(\frac{z}{x}\right)\,dz\right\}\,dx.$$

Thus we have a positive constant $C_1 > 0$ so that for $\mathrm{Re}(\lambda) > -T$ $(\lambda_o = \mathrm{Re}(\lambda))$

$$(4 - 66)\quad |J_\lambda(g,y,u)| \leq C_1\,|\gamma(u,\lambda)| \sup_{|s'|\leq 1} \left|\left(\frac{d}{ds}\right)^{T+1}\varphi_{G_\lambda}(g)(s')\right|$$
$$\left\{\int_1^\infty |x|^{-(T+\lambda_o+1)}dx\right\}$$
$$\leq C_1\frac{|\gamma(u,\lambda)|}{|\lambda_o + T|}\sum_{\ell+m+n\leq T+1} |\varphi_{\ell mn}(g)| \sup_{|s|\leq 1} |\varphi_{\pi_\sigma}((gw_o)^{-1})\pi_\sigma(H^\ell E^m F^n)F_\lambda^{(s)}|$$

Then we choose the compact subset $S_2 = \Omega_1\cdot\{w_o n(x)w_o \mid |x|\leq 1\}$ (Ω_1 any compact subset of $\tilde{S}\ell_2$), $\hat{D}$ any compact subset of $\{z \mid \mathrm{Re}(z) > -T\}$, and L as above; we have

$$(4 - 67)\quad |J_\lambda(g,y,u)| \leq C\left\{\sup_{\substack{\lambda\in\hat{D}\\ u\in L}}\left(\frac{|\gamma(u,\lambda)|}{|\lambda_o + T|}\right)\right\}\sum_{\ell+m+n\leq T+1} c_{\ell mn}\,\nu_{S_2,\hat{D},H^\ell E^m F^n}(F),$$

with $g \in \Omega_1, |y| \leq 1, \lambda \in \hat{D}, u \in L$, and $c_{\ell mn} = \sup_{g \in \Omega_1} |\varphi_{\ell mn}(g)|$. Moreover we deduce that $(\lambda,g,y,u)\to J_\lambda(g,y,u)$ is a continuous function of $\{z|\ \mathrm{Re}(z) > -T\} \times \widetilde{S\ell_2} \times \{y|\ |y| \leq 1\} \times \mathbb{R}$, and $\lambda\to J_\lambda(g,y,u)$ is analytic in λ for $\mathrm{Re}(\lambda) > -T$.

Then we note for $u > 0$, $V(\beta,u) = s(\beta)u^{\beta-1}F(2u,-(\beta+2))$, with $s(\beta)$ some nonvanishing analytic function on $\mathbb{C}$ (by change of variable). But by Lemma 4.12 $(u > 0)$, $F(2u, -(\beta+2))$ is continuous in (u,β) on $(\mathbb{R}_+^* \times \mathbb{C})$ and analytic in β. A similar result holds in case $u < 0$; if $u = 0$ then $V(\beta,u) = \frac{1}{\beta+1}$. Thus by similar reasoning as above, we deduce that the map $(\lambda,g,u)\to\gamma(u,\lambda)\varphi^{(k)}_{G_\lambda(g)}(0)V(\lambda + k-1, u)$ is continuous on $\mathbb{C} \times \widetilde{S\ell_2} \times \widetilde{\mathbb{R}}_+^*$ and analytic as a function of λ. Moreover we have that there exists $C' > 0$ so that

$$(4 - 68) \quad |\gamma(u,\lambda)\varphi^{(k)}_{G_\lambda(g)}(0)V(\lambda + k-1,u)| \leq C' \sup_{\substack{\lambda \in \hat{W} \\ u \in L}} |V(\lambda+k-1,u)\gamma(u,\lambda)|$$

$$\sum_{a+b+c\leq k} \tilde{c}_{abc}\ \nu_{S_2,\hat{W},H^aE^bF^c}(F_\lambda),$$

with S_2, L as above, $\hat{W}$ any compact subset of $\mathbb{C}$, and $\tilde{c}_{abc}$ appropriate constants.

Then collecting all that we have done, we see that from (4 - 61) the function $A_F(g,\lambda,u) = C_1(g,\lambda,u) + C_2(g,\lambda,u)$ has the property that $(g,\lambda,u)\to A_F(g,\lambda,u)$ is continuous on $\widetilde{S\ell_2} \times \mathbb{C} \times \mathbb{R}$. Moreover the map $\lambda\to A_F(g,\lambda,u)$ is analytic in λ and is the continuation of ψ_F. And if we take any $Z \in \mathcal{U}(s\ell_2)$ and repeat the same arguments above for $\Sigma_s^1(Z)F$, we deduce that $A_{\Sigma_s^1(Z)F} = Z * A_F$ has the same properties as A_F. Thus it is routine to check then that $A_F(\ ,\ ,u) \in B_u$ for $u \neq 0$, and $A_F(\ ,\ ,0) \in \mathcal{a}_s^{-1}$.

Then above we have also shown that the map $F \to A_F(\ , \ ,u)$ is a continuous $\widetilde{S\ell}_2$ intertwining map of $\mathcal{A}^1_s$ to $\mathcal{A}^{-1}_s(B_u)$ for $u = 0$ $(u \neq 0)$. Finally if $F(\lambda, \ ) \equiv 0$ for some $\lambda \in \mathbb{C}$, we see by the various expressions for C_1 and C_2 that $A_F(\ ,\lambda,u) \equiv 0$. Note, the argument used here is similar to that in [22].

Q.E.D.

We remark that for any $\lambda \in \mathbb{C}$ that the function $F_\lambda : \widetilde{S\ell}_2 \to \mathbb{C}$ given by the Iwasawa decomposition $F_\lambda(ka(r)n(y)) = \theta(k)|r|^{-(\lambda+1)}$ with $\theta \in C^\infty(K)$ and $\theta(km) = \theta(k)\sigma_1(m)$ (with $m \in M$) has the property that $F_\lambda(g) \in \mathcal{A}^1_s$.

Thus the map $\mathcal{A}^1_s \to C^\infty_{\sigma_1 \otimes \lambda}(\widetilde{S\ell}_2)$ given by $F \to F(\ ,\lambda)$ is surjective.

Thus by Proposition 4.13 we deduce the following.

<u>Corollary 1 to Proposition 4.13</u>

(1) *For* $u < 0$ *and all* $\lambda \in \mathbb{C}$, *the map* $F(\ ,\lambda) \xrightarrow{L_u(\sigma_1 \otimes \lambda)} A_F(\ ,\lambda,u)$ *defines a nonzero continuous* $\widetilde{S\ell}_2$ *intertwining map of* $C^\infty_{\sigma_1 \otimes \lambda}(\widetilde{S\ell}_2)$ *to* $C^\infty(u)$. *Moreover, up to a scalar multiple, this map is the unique* $\widetilde{S\ell}_2$ *intertwining map between* $C^\infty_{\sigma_1 \otimes \lambda}(\widetilde{S\ell}_2)$ *and* $C^\infty(u)$.

(2) *For* $u = 0$ *and all* $\lambda \in \mathbb{C}$, *the map* $F(\ ,\lambda) \xrightarrow{L_{\sigma_1 \otimes \lambda}} A_F(\ ,\lambda,0)$ *defines a continuous* $\widetilde{S\ell}_2$ *intertwining map of* $C^\infty_{\sigma_1 \otimes \lambda}(\widetilde{S\ell}_2)$ *to* $C^\infty_{\sigma_1 \otimes (-\lambda)}(\widetilde{S\ell}_2)$. *Moreover for* $\lambda \in \mathbb{C} - \{0, -1, -2, \ldots\}$, *this map is, up to scalar multiple, the unique intertwining* $\widetilde{S\ell}_2$ *map between* $C^\infty_{\sigma_1 \otimes \lambda}(\widetilde{S\ell}_2)$ *and* $C^\infty_{\sigma_1 \otimes (-\lambda)}(\widetilde{S\ell}_2)$.

<u>Proof</u>. We recall that $C^\infty_{\lambda,s}(\mathbb{R}) \supseteq C^\infty_c(\mathbb{R})$ for all $\lambda \in \mathbb{C}$. We let $\varphi \in C^\infty_c(\mathbb{R})$. Then it is possible by Lemma 4-10 to find for every $\lambda \in \mathbb{C}$ a unique function $F_\lambda \in C^\infty_{\sigma_1 \otimes \lambda}(\widetilde{S\ell}_2)$ and $F_\lambda(n(x)w_o) = \varphi(x)$ for all $x \in \mathbb{R}$. Thus

for all λ, $\varphi_{F_\lambda} = \varphi$. We note that for any $Z \in \mathcal{U}(s\ell_2)$, the function $Z*F_\lambda = \pi_\sigma(Z)F_\lambda$ has the property that support $(Z*F_\lambda) \subseteq$ support (F_λ). Thus support $(\varphi_{Z*F_\lambda}) \subseteq$ support (φ_{F_λ}) in $\mathbb{R}$. This implies that for all $Z \in \mathcal{U}(s\ell_2)$, $Q_{\lambda,s}(\varphi_{Z*F_\lambda}) \equiv 0$. Then by using the construction of (4 - 45) and arguments similar to those in the proof of Lemma 4.10, we deduce that the function $(\lambda,g) \xrightarrow{H} Z*F_\lambda(g)$ is continuous and analytic in λ. Then $H \in \mathcal{Q}_s^1$. We then have that for $u < 0$, $\psi_F(e_{\widetilde{S\ell}_2},\lambda,u) = \int_{\mathbb{R}} \varphi_{F_\lambda}(x)e^{-2\pi iux}dx$

is <u>defined</u> and analytic for all $\lambda \in \mathbb{C}$; hence the map $F(\ ,\lambda)\to A_F(\ ,\lambda,u)$ is a <u>nonzero</u> map for all $\lambda \in \mathbb{C}$ and $u < 0$.

On the other hand if $u = 0$, we observe by Proposition 4.13 that

$$(4 - 69)\quad \psi_F(w_0^{-1},\lambda,0) = \frac{c}{\Gamma(\lambda)} \int_0^\infty \delta_F(x)|x|^{\lambda-1}\, dx \quad \text{for } \operatorname{Re}(\lambda) > 0,$$

with $\delta_F(x) = \varphi_F(x) + e^{-\frac{1}{2}\pi is}\varphi_F(-x)$ (c a nonzero constant independent of φ_F and λ). Then we choose $F \in \mathcal{Q}_s^{+1}$ so that $F_\lambda(n(x)w) = \varphi(x)$ for $\varphi \in S(\mathbb{R})$ and all $x \in \mathbb{R}$. Hence $\delta_F \in S(\mathbb{R})$; moreover we know that for $f \in S(\mathbb{R})$ the function $V_f(\lambda,k) = \frac{1}{\Gamma(\lambda+k)} \int_0^\infty f^{(k)}(x)x^{\lambda+k-1}dx$ is analytic for $\operatorname{Re}(\lambda) > -k$

(k integer ≥ 1) and satisfies $V_f(\lambda,k) = \pm \frac{1}{\Gamma(\lambda)} \int_0^\infty f(x)x^{\lambda-1}dx$ for

$\operatorname{Re}(\lambda) > 0$. Thus for λ so that $\operatorname{Re}(\lambda) > -k$

$$(4 - 70)\quad A_F(w_0^{-1},\lambda,0) = c'[V_\varphi(\lambda,k) + e^{-\frac{1}{2}\pi is}V_{\varphi^*}(\lambda,k)]$$

where $\varphi^*(x) = \varphi(-x)$ and c' a nonzero constant. We know by classical arguments that for $\lambda \neq -m$ (m integer ≥ 0), it is possible to find $\varphi \in S(\mathbb{R})$ so that $V_\varphi(\lambda,k) \neq 0$ and $V_{\varphi^*}(\lambda,k) = 0$ (the zero of the function $\frac{1}{\Gamma(\lambda)}$ at $\lambda = -m$ is what makes it impossible to extend argument here to $\lambda = -m$). Thus $F_\lambda \to A_F(\ ,\lambda,0)$ is nonzero for all $\lambda \neq -m$ (m integer ≥ 0).

Uniqueness in both cases $(u = 0$ and $u < 0)$ follows from Propositions 4.1 and 4.2.

Q.E.D.

Remark 4.11. Let $F_{o,s} \in C^{\infty}_{\sigma_1 \otimes 0}(\widetilde{S\ell}_2)$ with $\sigma_1(E) = e^{\frac{1}{2}\pi is}$. Then from the proof of Proposition 4.13 it follows that the map

$$(4-71)\quad F_{o,s} \to S(F_{o,s}) = \int_0^1 (\delta_{F_{o,s}}(x) - \delta_{F_{o,s}}(0))\frac{dx}{x} + \int_1^{\infty} \delta_{F_{o,s}}(x)\frac{dx}{x}$$

(with $\delta_{F_{o,s}}$ as in (4 - 69)) is a continuous linear functional on $C_{\sigma_1 \otimes 0}(\widetilde{S\ell}_2)$. Indeed we use (4 - 61), (4 - 62), (4 - 64), and (4 - 67) and the expression for $\gamma_s(x)$. Moreover we note that

$$(4-72)\quad S(\pi_{\sigma}(Ea(r))F_{o,s}) = e^{-\frac{1}{2}\pi is}\, r\left\{\int_0^1 [\psi_{F_{o,s}}(r^2x) - \psi_{F_{o,s}}(0)]\frac{dx}{x}\right.$$

$$\left. + \int_1^{\infty} \psi_{F_{o,s}}(r^2x)\frac{dx}{x}\right\} .$$

By change of variable $x \to r^2x$, we deduce that $\{\ldots.\}$ in (4 - 72) is given by

$$(4-73)\quad \int_0^{r^2} [\psi_{F_{o,s}}(x) - \psi_{F_{o,s}}(0)]\frac{dx}{x} + \int_{r^2}^{\infty} \psi_{F_{o,s}}(x)\frac{dx}{x} =$$

$$\int_0^1 [\psi_{F_{o,s}}(x) - \psi_{F_{o,s}}(0)]\, dx + \int_1^{\infty} \psi_{F_{o,s}}(x)\frac{dx}{x} - \psi_{F_{o,s}}(0)\int_1^{r^2} \frac{dx}{x} .$$

Thus we have that for all $F_{o,s} \in C_{\sigma_1 \otimes 0}(\widetilde{S\ell}_2)$

$$(4-74)\quad S(\pi_{\sigma}(Ea(r))F_{o,s}) = \sigma(E)^{-1} rS(F_{o,s}) - 2r \log r[1+e^{-\frac{1}{2}\pi is}]\, F_{o,s}(e_{\widetilde{S\ell}_2}).$$

<u>Corollary 2 to Proposition 4.13</u>. <u>If</u> $s = 2$ (<u>so that</u> $\sigma_1(E) = e^{\pi i} = -1$), <u>then the representation</u> π_{σ} <u>of</u> $\widetilde{S\ell}_2$ <u>on</u> H_o ($C^{\infty}_{\sigma_1 \otimes 0}(\widetilde{S\ell}_2)$ resp.) <u>is the direct sum of</u> 2 <u>irreducible representations</u>. <u>Moreover</u> $\mathrm{Hom}_{\widetilde{S\ell}_2}(\pi_{\sigma}, \pi_{\sigma})$ <u>is a two dimensional space with basis</u> I <u>and</u> L, I <u>the identity operator and</u> L <u>the map given by</u>

$$(4-75)\quad L(F_{o,2})(g) = S(\pi_\sigma(g^{-1})F_{o,2})$$

for all $g \in \widetilde{S\ell}_2$, *with* $F_{o,2} \in C^\infty_{\sigma_1\otimes 0}(\widetilde{S\ell}_2)$.

Proof. We let $F \in \mathfrak{a}^1_s$. Then we note from $(4-69)$ and the arguments in Proposition 4.13 that

$$(4-76)\quad A_F(gw_o^{-1},0,0) = [1 + e^{-\frac{1}{2}\pi is}]\ \varphi_{\Sigma_s}(g^{-1})_F(0)$$

and

$$(4-77)\quad \frac{d}{dz} A_F(w_o^{-1},z,0)\Big|_{z=0} = \{1 + e^{-\frac{1}{2}\pi is}\}\ \frac{d}{dz} F(z,w_o)\Big|_{z=0} + S(F(0,\ )).$$

Moreover for any $G \in \mathfrak{a}^{+1}_s$ we have the functional equation:

$$(4-78)\quad \frac{d}{dz} G(z,gEa(r)n(y))\Big|_{z=\lambda_o} = \frac{d}{dz} G(z,g)\Big|_{z=\lambda_o}\ \sigma(E)^{-1}|r|^{-(\lambda_o+1)}$$
$$+ (\log r)\ G(\lambda_o,g)\sigma(E)^{-1}|r|^{-(\lambda_o-1)}.$$

Hence it follows for $s = 2$

$$(4-79)\quad \frac{d}{dz}A_F(\ ,z,0)\Big|_{z=0} \in C^\infty_{\sigma_1\otimes 0}(\widetilde{S\ell}_2).$$

But from $(4-77)$ we have that $\frac{d}{dz} A_F(w_o^{-1},\lambda,0) = S(F(0,\ ))$.
Hence if $s = 2$ the map $F_{0,2} \underset{L}{\rightsquigarrow} S(F_{0,2})$ defines an element of $\{T \in C_{\sigma_1\otimes 0}(\widetilde{S\ell}_2)' \mid T(\pi_\sigma(Ea(r)n(y)F_{0,2}) = r\sigma(E)T(F_{0,2})$ for all $F_{0,2}\}$. But Theorem 3.1 and Proposition 3.2 imply that the space $\{\ \}$ in the last sentence is isomorphic to $\mathrm{Hom}_{\widetilde{S\ell}_2}(\pi_\sigma,\pi_\sigma)$ (for $s = 2$). But from Proposition 4.1 we recall that $\mathrm{Hom}_{\widetilde{S\ell}_2}(\pi_\sigma,\pi_\sigma) \le 2$. Then if we show that

$S(F(0,\ ))$ is not a constant multiple of $F(0,\ )$ (i.e., the distributions S and $\delta_{\tilde{e}_{S\ell_2}}$ must be linearly independent), then $\dim \mathrm{Hom}_{\widetilde{S\ell}_2}(\pi_\sigma, \pi_\sigma) = 2$ (for $s = 2$).

We observe that via the inclusion of $C_c^\infty(\mathbb{R})$ in $C_{0,2}^\infty(\mathbb{R})$, S restricted to $C_c^\infty(\mathbb{R})$ defines a homogeneous distribution of degree 0. But clearly S does not vanish on $C_c^\infty(\mathbb{R})$ and hence, the set $\{n(x)w_o P\}$ intersects the support of S nontrivially. Thus the desired result follows.

Q.E.D.

By definition we see that $C_\sigma^\infty(\widetilde{S\ell}_2) \subseteq C^\infty(0)$ and that for $f \in C_\sigma^\infty(\widetilde{S\ell}_2)$, $\pi_o(\Omega)(f) = (\lambda^2 - 1)\ (f)$ and $\pi_o(E)f = (\sigma_1(E))^{-1} f\ (\sigma = \sigma_1 \otimes \lambda)$.

We then deduce from above that the transpose map R_σ^t (see (4 - 44)) maps the dual $C_\sigma^\infty(\widetilde{S\ell}_2)'$ continuously bijectively onto a closed subspace of $C_c^\infty(0)'$ (all topologies on dual space here are the strong topologies). Indeed recalling that $R_\sigma\colon C_c^\infty(0) \to C_\sigma^\infty(\widetilde{S\ell}_2)$ is an open surjective linear map, it follows that $R_\sigma^t(C_\sigma^\infty(\widetilde{S\ell}_2)')$ coincides with the space of all $T \in C_c^\infty(0)'$ so that $\langle T, f\rangle = 0$ for all $f \in \mathrm{Ker}(R_\sigma)$. Hence we have $R_\sigma^t(C_\sigma^\infty(\widetilde{S\ell}_2)') \subseteq {}_{\zeta_{\sigma_1}}T^0_{(\lambda^2-1)}$, where $\zeta_{\sigma_1} = \sigma_1(E)$.

We note that the homogeneous space $\underline{P}/\underline{N}$ admits an $\underline{P}$ invariant positive measure $ds(\xi)$ which is given in $M \times A$ coordinates by $ds(m,r) = dm\, \frac{dr}{|r|}$.

Then using standard decomposition formulae, we have for $f \in C_\sigma^\infty(\widetilde{S\ell}_2)$, $\varphi \in C_c^\infty(0)$ (see (4 - 10) and Remark 4.1).

$$(4 - 80)\quad \int_{\widetilde{S\ell}_2/\underline{N}} f(g)\varphi(g)d\mu_o(g) = \oint_{\widetilde{S\ell}_2/\underline{P}} \left\{ \int_{\underline{P}/\underline{N}} f(g\xi)\varphi(g\xi)ds(\xi) \right\} d\dot{\mu}_\rho(g)$$

$$= \oint_{\widetilde{S\ell}_2/\underline{P}} \left\{ \int_{M \times A} f(gma(r))\varphi(gma(r))\ dm\ \frac{dr}{|r|} \right\} d\dot{\mu}_\rho(g)$$

$$= \oint_{\widetilde{S}\ell_2/\underline{P}} f(g)R_{\sigma*}(\varphi)(g)d\dot{\mu}_\rho(g) \quad ,$$

where $\sigma^* = \sigma_1^{-1}\otimes(-\lambda)$. Thus we have (using the notation in the discussion preceding Lemma 4.8) for $f \in C_\sigma^\infty(\widetilde{S}\ell_2)$, $\varphi \in C_c^\infty(0)$,

$$(4 - 81) \quad \langle T_f,\varphi\rangle = (f , R_{\sigma*}(\varphi))$$

where $(,)$ is the intertwining form on $C_\sigma^\infty(\widetilde{S}\ell_2) \times C_{\sigma*}^\infty(\widetilde{S}\ell_2)$ given in Remark 4.1.

<u>Proposition 4.14</u> <u>For all</u> $\lambda \in \mathbb{C}$ <u>the space</u> $R_\sigma^t(C_\sigma^\infty(\widetilde{S}\ell_2)')$ <u>coincides exactly with the closure of</u> $M_0^{\rho(\lambda)}((\lambda^2-1),-\frac{is}{2})$ <u>in the strong dual topology on</u> $C_c^\infty(0)'$, <u>where</u> $\sigma_1(E) = e^{\pi i\frac{s}{2}}$ <u>and</u> $\rho(\lambda) = \begin{cases} + \begin{cases} \text{if } \operatorname{Re}(\lambda) < 0 \\ \text{if } \lambda = it \text{ with } t \leq 0 \end{cases} \\ - \begin{cases} \text{if } \operatorname{Re}(\lambda) > 0 \\ \text{if } \lambda = it \text{ with } t > 0 \end{cases} \end{cases}$.

<u>Proof</u>. We know that via the bilinear pairing $(,)$ above, $C_{\sigma*}^\infty(\widetilde{S}\ell_2)$ can be considered as a subspace of $C_\sigma^\infty(\widetilde{S}\ell_2)'$. Then with an abuse of notation, we have $R_\sigma^t(C_\sigma^\infty(\widetilde{S}\ell_2)') \supseteq R_\sigma^t(C_{\sigma*}^\infty(\widetilde{S}\ell_2))$. By the statement preceding the Proposition, $R_\sigma^t(C_{\sigma*}^\infty(\widetilde{S}\ell_2))$ coincides with the embedding of $C_{\sigma*}^\infty(\widetilde{S}\ell_2)$ in $C_c^\infty(0)'$ by the map $f\to T_f$ given above.

Then with the condition that $\operatorname{Re}(\beta) > 0$ or that $\operatorname{Re}(\beta) = 0$ with $\operatorname{Im}(\beta) \geq 0$ (for definition of β see paragraph preceding (4 - 24)), we derive that $M_0^{\rho(\lambda)}((\lambda^2 - 1),-\frac{is}{2})$, with $\rho(\lambda)$, s etc. given as above, is contained in $C_{\sigma*}^\infty(\widetilde{S}\ell_2)$; moreover $M_0^{\rho(\lambda)}((\lambda^2 -1),-\frac{is}{2})$ coincides exactly with the set of $\underline{K}$ finite vectors in $C_{\sigma*}^\infty(\widetilde{S}\ell_2)$.

We note that the space $C_\sigma^\infty(\widetilde{S}\ell_2)$ is a Montel space (i.e., a closed subspace of the Montel space $C^\infty(\widetilde{S}\ell_2)$), and hence is reflexive.

Since the pairing $(,)$ above is separately continuous and nondegenerate (and hence "hypocontinuous", see [23]), it follows that the embedding of

$C^\infty_{\sigma*}(\widetilde{S\ell}_2)$ in $C^\infty_\sigma(\widetilde{S\ell}_2)'$ is continuous with respect to the strong dual topology on $C^\infty_\sigma(\widetilde{S\ell}_2)'$. Moreover we see easily that $C^\infty_{\sigma*}(\widetilde{S\ell}_2)$ is dense in the strong topology of $C^\infty_\sigma(\widetilde{S\ell}_2)'$. Thus since R_σ is continuous, the strong closure of $R^t_\sigma(C^\infty_{\sigma*}(\widetilde{S\ell}_2)) = R^t_\sigma(C^\infty_\sigma(\widetilde{S\ell}_2)')$. However since the set of $\underline{K}$ finite vectors $M_0^{\rho(\lambda)}((\lambda^2 - 1), -\frac{is}{2})$ is dense in $C^\infty_{\sigma*}(\widetilde{S\ell}_2)$ (in the topology of $C^\infty_{\sigma*}(\widetilde{S\ell}_2)$), we have that the strong closure of $M_0^{\rho(\lambda)}((\lambda^2 - 1), -\frac{is}{2})$ in $C^\infty_c(0)'$ is $R^t_\sigma(C^\infty_\sigma(\widetilde{S\ell}_2)')$.

Q.E.D.

Remark 4.12. We let $\sigma = \sigma_1 \otimes \lambda$ be as above; define $\sigma^\vee = \sigma_1 \otimes - \lambda$ (with $\zeta = e^{\frac{1}{2}\pi is} = \sigma_1(E)$). From the fact that $({}_\zeta T^0{}_\lambda)_K = M_0^+(\lambda, -\frac{1}{2}is) \oplus M_0^-(\lambda, -\frac{1}{2}is)$ (see Proposition 4.9), we deduce that ${}_\zeta T^0_{(\lambda^2-1)} = R^t_\sigma(C^\infty_\sigma(\widetilde{S\ell}_2)') \oplus R^t_{\sigma^\vee}(C^\infty_{\sigma^\vee}(\widetilde{S\ell}_2)')$ for all $\lambda \neq 0$. Moreover from Proposition 4.14, the set of $\underline{K}$ finite vectors of $R^t_\sigma(C^\infty_\sigma(\widetilde{S\ell}_2)')$ is $M_0^{\rho(\lambda)}((\lambda^2-1), -\frac{1}{2}is)$ for all λ.

Recalling the definition of $\sigma^\vee$ in Remark 4.12, we consider the $\widetilde{S\ell}_2$ intertwining map L_σ defined between $C^\infty_\sigma(\widetilde{S\ell}_2)$ and $C^\infty_{\sigma^\vee}(\widetilde{S\ell}_2)$ in Corollary 1 to Proposition 4.13. We let L^t_σ be the transpose map: $C^\infty_{\sigma^\vee}(\widetilde{S\ell}_2)' \to C^\infty_\sigma(\widetilde{S\ell}_2)'$, and $\hat{L}^t_\sigma$ the corresponding induced map of $R^t_{\sigma^\vee}(C^\infty_{\sigma^\vee}(\widetilde{S\ell}_2)') \to R^t_\sigma(C^\infty_\sigma(\widetilde{S\ell}_2)')$ (i.e. $\hat{L}^t_\sigma = (R^t_\sigma)^{-1} L^t_\sigma (R^t_{\sigma^\vee})^{-1}$, with $(R^t_\sigma)^{-1}$ the inverse of the linear bijection R^t_σ: $C^\infty_\sigma(\widetilde{S\ell}_2)' \to R^t_\sigma(C^\infty_\sigma(\widetilde{S\ell}_2)')$). We then let

$$W_\sigma = \begin{cases} \hat{L}^t_\sigma \oplus \hat{L}^t_{\sigma^\vee} & \text{if } \lambda \neq 0 \\ \hat{L}^t_\sigma & \text{if } \lambda = 0 \end{cases} .$$

Corollary 1 to Proposition 4.14 *Let (A) denote the following condition:*

$$(4\text{-}82)\quad \lambda \neq \begin{cases} 2k + 1 & \text{if } s = 0 \\ 2k & \text{if } s = 2 \end{cases}, \text{ and } \lambda \neq \frac{\nu}{2} \text{ if } s = 1 \text{ or } s = 3$$

(k any integer and ν any odd integer).

Then $R^t_\sigma(C^\infty_\sigma(\widetilde{S\ell}_2)')$ is an algebraically irreducible $\widetilde{S\ell}_2$ module (i.e. the only closed $\widetilde{S\ell}_2$ invariant submodules (in the strong topology) in $R^t_\sigma(C^\infty_\sigma(\widetilde{S\ell}_2)')$ are $\{0\}$ and $R^t_\sigma(C^\infty_\sigma(\widetilde{S\ell}_2)')$). Moreover we have for λ satisfying (A) and $\lambda \neq 0$ that $\hat{L}^t_\sigma$ is a bijective $\widetilde{S\ell}_2$ intertwining map between $R^t_\sigma(C^\infty_\sigma(\widetilde{S\ell}_2)')$ and $R^t_{\sigma^\vee}(C^\infty_{\sigma^\vee}(\widetilde{S\ell}_2)')$. Explicitly for such λ we have that W_σ restricted to $({}_\zeta T^0_{(\lambda^2-1)})_K$ is a nonzero scalar multiple of $A^0_{\lambda^2-1,s}$.

Proof. We use Proposition 4.9 and Remark 4.12 to deduce the first statement. We then note that $W_\sigma(({}_\zeta T^0_{(\lambda^2-1)})_K) \subseteq ({}_\zeta T^0_{(\lambda^2-1)})_K$, since the intertwining operator $\hat{L}^t_\sigma$ preserves $\underline{K}$ finite vectors. However by irreducibility, $\hat{L}^t_\sigma$ must be a linear isomorphism of $M_0^{\rho(\lambda)}(\lambda^2-1,\frac{is}{2})$ to $M_0^{\rho(-\lambda)}(\lambda^2-1,\frac{is}{2})$ for λ satisfying (A) and $\lambda \neq 0$. Then from Remark 4.6 we deduce that W_σ restricted to $({}_\zeta T^0_{(\lambda^2-1)})_K$ is a nonzero multiple of $A^0_{\lambda^2-1,s}$. Then since $(A^0_{\lambda^2-1,s})^2 = I$, we deduce that $\hat{L}^t_\sigma\hat{L}^t_{\sigma^\vee}$ and $\hat{L}^t_{\sigma^\vee}\hat{L}^t_\sigma$ are nonscalar multiples of the identity operator on $R^t_\sigma(C^\infty_\sigma(\widetilde{S\ell}_2)')$ and $R^t_{\sigma^\vee}(C^\infty_{\sigma^\vee}(\widetilde{S\ell}_2)')$, respectively.

Q.E.D.

Remark 4.13 For λ satisfying (A) above and $\lambda \neq 0$, we have $L_{\sigma^\vee}L_\sigma$ and $L_\sigma L_{\sigma^\vee}$ are nonzero scalar multiples of the identity operator on $C^\infty_\sigma(\widetilde{S\ell}_2)$ and $C^\infty_{\sigma^\vee}(\widetilde{S\ell}_2)$, respectively.

Corollary 2 to Proposition 4.14 If λ satisfies (A), then the representations $\hat{\pi}_\sigma$ and π_σ of $\widetilde{S\ell}_2$ on the spaces H_λ and $C^\infty_\sigma(\widetilde{S\ell}_2)$, respectively, are algebraically irreducible (the only closed subspaces $\widetilde{S\ell}_2$ invariant in H_λ and $C^\infty_\sigma(\widetilde{S\ell}_2)$ are $\{0\}$ and H_λ, and $\{0\}$ and $C^\infty_\sigma(\widetilde{S\ell}_2)$, respectively).

Moreover if λ *satisfies* (A) *and* $\lambda \neq 0$, *then* L_σ *is an* $\widetilde{S\ell}_2$ *module isomorphism of* π_σ *to* $\pi_{\sigma^\vee}$.

Remark 4.14. We note that for $\mathrm{Re}(\lambda) > 0$, $L_u(C^\infty_{\sigma_1\otimes\lambda}(\widetilde{S\ell}_2)) \subseteq C^\infty_+(u)(\lambda^2-1)$. Indeed for $\varphi \in C_{\lambda,s}(\mathbb{R})$, we have $L_u(\sigma)(\varphi)(ka(r)) = (\pi_\sigma(k^{-1})\varphi)^\wedge(\frac{u}{r^2})r^{\lambda-1}$. And we know for any $k \in K$, $\pi_\sigma(k^{-1})(\varphi) \in L^1(\mathbb{R})$ so that $\pi_\sigma(k^{-1})(\varphi)^\wedge$ is a continuous function on $\mathbb{R}$ bounded at ∞. Thus $(\pi_\sigma(k^{-1})\varphi)^\wedge(\frac{u}{r^2})r^{\lambda-1} \in L^2([0,a],rdr)$ for $\mathrm{Re}(\lambda) > 0$.

Corollary 3 to Proposition 4.14. *The map* $L_u(\sigma)$ (*defined in Corollary* 1 *to Proposition* 4.13) *has the property that* $L_u(\sigma)(C^\infty_{\sigma_1\otimes\lambda}) \subseteq C^\infty_+(u)(\lambda^2-1)$ *for all* λ *so that* (i) $\mathrm{Re}(\lambda) \neq 0$ *and* (ii) λ *satisfies* (A). *Moreover for such* λ, $L_u(\sigma)$ *is a bijective intertwining map between the spaces* $M_0^{\rho(-\lambda)}(\lambda^2-1, \frac{1}{2}\text{ is})$ *and* $W_u^+(\lambda^2-1,\frac{1}{2}\text{ is})$.

Proof. We use Remark (4.14) to show the statement is valid for $\mathrm{Re}(\lambda) > 0$ and λ satisfying (A). If $\mathrm{Re}(\lambda) < 0$ and λ satisfies (A), then L_σ defines an $\widetilde{S\ell}_2$ isomorphism of $C^\infty_{\sigma_1\otimes\lambda}$ to $C^\infty_{\sigma_1\otimes-\lambda}$ (by Corollary 2 to Proposition 4.14). Hence the composition $L_u(\sigma_1\otimes(-\lambda)) \circ L_\sigma$ defines a nonzero $\widetilde{S\ell}_2$ intertwining map of $C^\infty_{\sigma_1\otimes\lambda}(\widetilde{S\ell}_2)$ to $C^\infty_+(u)(\lambda^2-1)$. But by Corollary 1 to Proposition 4.13, $L_u(\sigma_1\otimes(-\lambda)) \circ L_\sigma$ must be a nonzero scalar multiple of the map $L_u(\sigma_1\otimes\lambda)$.

Q.E.D.

We say a *closed* $\pi_u(\widetilde{S\ell}_2)$ invariant subspace $W \subseteq L^2(u)$ is "discretely irreducible" if $\pi_u(\widetilde{S\ell}_2)$ restricted to W gives an irreducible (unitary) representation.

<u>Proposition 4.15</u> <u>Every</u> $\widetilde{S\ell}_2$ <u>"discretely irreducible" subspace of</u> $L^2(u)$ <u>is</u> (i) $\{0\}$ <u>if</u> $u = 0$ <u>and</u> (ii) <u>either</u> $\{0\}$ <u>or of the form</u> $\overline{D}_\nu$ (ν <u>integer</u> ≥ 0) <u>or</u> $\overline{H}_{\nu'}$ (ν' <u>integer</u> ≥ 1), <u>where</u> $\overline{D}_\nu$ ($\overline{H}_{\nu'}$) <u>is the closure of</u> D_ν ($H_{\nu'}$) <u>in</u> $L^2(u)$, <u>if</u> $u < 0$.

<u>Proof</u>. Let W be such a subspace. By the criterion for differentiable vectors in $L^2(u)$ (see Remark 3.3), we have the set $(W_\infty)_K$ of K finite vectors in $W_\infty \subseteq C_+^\infty(u) \cap C_-^\infty(u)$. But W_∞ is also an eigenspace of $\pi_u(\Omega)$ with value (λ_1^2-1) ($\lambda_1 \in \mathbb{C}$ with $\mathrm{Re}(\lambda_1) \geq 0$).

The case $u = 0$ follows immediately from Proposition 4.6 since $C_+^\infty(0)(\lambda_1^2-1) \cap C_-^\infty(0)(\lambda_1^2-1) = \{0\}$.

Moreover if $u < 0$ and $\mathrm{Re}(\lambda_1) = 0$, then again by Proposition 4.6, $C_+^\infty(u)((\lambda_1^2-1)) \cap C_-^\infty(u)((\lambda_1^2-1)) = \{0\}$.

Thus we assume $\mathrm{Re}(\lambda_1) > 0$. Then if λ_1 satisfies (A), we have by Propositions 4.6 and 4.9 that W_K (= set of $\underline{K}$ finite vectors in W) = $M_u^+((\lambda_1^2-1),\frac{is}{2}) = W_u^+((\lambda_1^2-1),\frac{is}{2})$. But from Remark 4.5, $M_u^+((\lambda_1^2-1),\frac{is}{2}) \subseteq C_-^\infty(u)((\lambda_1^2-1))$.

From above, using arguments similar to those in the proof of Proposition 4.13, we know that the map $F \mapsto \psi_F(w_o,\lambda,0)$ of $C_\sigma^\infty(\widetilde{S\ell}_2) \to \mathbb{C}$ defines a continuous <u>nonzero</u> linear functional on $C_\sigma^\infty(\widetilde{S\ell}_2)$ for $\mathrm{Re}(\lambda) > 0$. Thus it follows that there exists a $\underline{K}$ finite function γ in $C_{\lambda,s}^\infty(\mathbb{R})$ so that $\check{\gamma}(0) \neq 0$. But from above if $F \in \mathcal{A}_{s_1}^1$ so that $F(\lambda,n(x)w_o) = \gamma(x)$ for all λ, then $\psi_{F_{\lambda_1}}(a(r),\lambda_1,u) = \check{\gamma}(\frac{u}{r^2})r^{\lambda_1-1}$. But then $\check{\gamma}$ is nonzero in a neighborhood of 0, and we deduce that $\check{\gamma}(\frac{u}{r^2})r^{\lambda_1-1} \notin L^2([a,\infty),rdr)$ for any $a > 0$.

Thus we have a contradiction to the above, and hence W_K is an irreducible subspace of $W_u^+((\lambda_1^2-1),\frac{is}{2})$ with λ_1 <u>not</u> satisfying (A).

Then using Proposition 4.9, W_K must be one of the spaces D_ν of $H_{\nu'}$.

Q.E.D.

Remark 4.15. We have deduced in Proposition 4.15 that for $\mathrm{Re}(\lambda) > 0$

(i) if λ satisfies (A), then $W_u^+(\lambda^2-1, \frac{1}{2} is) \cap M_u^+(\lambda^2-1, \frac{1}{2} is) = 0$ and

(ii) if λ does not satisfy (A), then $W_u^+(\lambda^2-1, \frac{1}{2} is) \cap M_u^+(\lambda^2-1, \frac{1}{2} is)$ coincides with one of the spaces D_ν or $H_{\nu'}$ (depending on λ and s).

§5. INTERTWINING DISTRIBUTIONS ATTACHED TO WEIL REPRESENTATION

At this point we recall some specific results concerning differentiable vectors. Let π be any continuous representation of S on a Banach space U with norm $\| \ \|$. Let $\{X_1,\ldots,X_\nu\}$ be a basis of the Lie algebra $\mathcal{S}$ of S. Let $\pi_\infty(X_i)$ denote the infinitesimal generator of the one parameter group $t \to \pi(\exp(t X_i))$.

Let U_∞ be the space of differentiable vectors of S relative to π; the space U_∞ can be characterized as

$$(5\text{-}1) \qquad U_\infty = \bigcap_{i=1}^{i=\nu} \ \bigcap_{s=1}^{\infty} D_{\{\pi_\infty(X_i)\}^s} \, ,$$

where $D_{\{\pi_\infty(X_i)\}^s}$ denotes the domain of $\pi_\infty(X_i)^s$ (see [20] for relevant definitions). Moreover we know that the topology on U_∞ given in §3 coincides with the Frechet topology on U_∞ given by the family of seminorms:

$$(5\text{-}2) \qquad \rho_n(\varphi) = \sum_{1 \le i_k \le \nu} \|\pi_\infty(X_{i_1} \ \ldots . \ X_{i_n})\varphi\|$$

as $n = 0, 1, \ldots.$ (with the convention $\rho_0(\varphi) = \|\varphi\|$).

Recalling the group $\beta(H_k)$ from §1, we know that from [14] the map $\beta(H_k) \times S(\mathbb{R}^k) \to S(\mathbb{R}^k)$ given by $(g,\varphi) \to g(\varphi)$ is a continuous map relative to the given topology on $\beta(H_k)$ in §1 and the usual Schwartz topology on $S(\mathbb{R}^k)$. Then from Theorem 1.2, we have that the representation $\pi_{\mathfrak{m}}$ of $\tilde{G}_2(Q) = \tilde{S\ell}_2(Q) \times O(Q)$ restricted to $S(\mathbb{R}^k)$ is continuous relative to the Schwartz topology on $S(\mathbb{R}^k)$.

From Theorem 1.2, we also know that the representation $\pi_{\mathfrak{m}}$ of $\tilde{G}_2(Q)$ on $L^2(\mathbb{R}^k)$ is unitary. Let F_Q be the space of differentiable vectors in $L^2(\mathbb{R}^k)$ relative to the representation $\pi_{\mathfrak{m}}$ of $\tilde{G}_2(Q)$. We then differentiate the representation $\pi_{\mathfrak{m}}$. We see that with $\mathfrak{S}$ the Lie algebra of $O(Q)$, and relative to the basis $\{e_i\}_{i=1}^k$ given in §2, that Image $\pi_{\mathfrak{m}}(\mathfrak{S})$ in $\operatorname{End}_{\mathbb{C}}(F_Q)$ is spanned by elements of the form

$$
\begin{aligned}
L_{ij} &= x_i\frac{\partial}{\partial x_j} - x_j\frac{\partial}{\partial x_i}, \quad 1 \le i < j \le a\,, \\
M_{st} &= x_s\frac{\partial}{\partial x_t} - x_t\frac{\partial}{\partial x_s}, \quad a+1 \le s < t \le k\,, \qquad (5\text{-}3)\\
N_{uv} &= x_u\frac{\partial}{\partial x_v} + x_v\frac{\partial}{\partial x_u}, \quad 1 \le u \le a \text{ and } a+1 \le v \le k\,.
\end{aligned}
$$

Moreover, we see by computation that for $\varphi \in F_Q$

$$\frac{d}{dt}\pi_{\mathfrak{m}}(a(e^t))(\varphi)(X)\Big|_{t=0} = (E_1+\tfrac{k}{2})(\varphi)(X)\,, \tag{5-4}$$

where E_1 is the Euler operator $\sum_{i=1}^k x_i \frac{\partial}{\partial x_i}$.

$$\frac{d}{ds}\pi_{\mathfrak{m}}(n(s))(\varphi)(X)\Big|_{s=0} = \pi i\, Q(X,X)(\varphi)(X)\,, \tag{5-5}$$

$$\frac{d}{du}\pi_{\mathfrak{m}}(\exp_{\widetilde{S\ell}_2}(uF))(\varphi)(X)\Big|_{u=0} = \frac{d}{ds}(\pi_{\mathfrak{m}}(w_o)\pi_{\mathfrak{m}}(n(s))\pi_{\mathfrak{m}}(w_o^{-1})) \tag{5-6}$$

$$(\varphi)(X)\Big|_{s=0} = -\pi i\pi_{\mathfrak{m}}(w_o)(Q\pi_{\mathfrak{m}}(w_o^{-1}))(\varphi)(X) = -\frac{1}{4\pi i}\,\partial(Q)(\varphi)(X)\,,$$

where $\partial(Q)$ is the differential operator on $\mathbb{R}^k$ which is obtained from Q by substituting $\frac{\partial}{\partial x_i}$ for x_i .

Let $\sigma_{\mathfrak{m}}$ denote the restriction of the representation $\pi_{\mathfrak{m}}$ of $\tilde{G}_2(Q)$ to $S(\mathbb{R}^k)$.

Lemma 5.1. *The representation $\sigma_{\mathfrak{m}}$ of $\tilde{G}_2(Q)$ on $S(\mathbb{R}^k)$ is differentiable (i.e. $F_Q \supseteq S(\mathbb{R}^k)$). The inclusion map $i : S(\mathbb{R}^k) \to F_Q$ defines a continuous linear $\tilde{G}_2(Q)$ intertwining map between $S(\mathbb{R}^k)$ and F_Q. Moreover the space $S(\mathbb{R}^k)$ is dense in F_Q with the topology on F_Q given above. The contragredient representation $\sigma_{\mathfrak{m}}^c$ on $T(\mathbb{R}^k)$ (the continuous dual space of $S(\mathbb{R}^k)$ provided with the strong dual topology) of $\tilde{G}_2(Q)$ is differentiable.*

Proof. From above we see that if $\{Z_j\}$ is a basis of the Lie algebra of $\tilde{G}_2(Q)$, then $(\pi_{\mathfrak{m}})_\infty(Z_j)$ is a polynomial coefficient differential operator;

hence $F_Q \supseteq S(\mathbb{R}^k)$. But then relative to the Schwartz topology on $S(\mathbb{R}^k)$, we know that $S(\mathbb{R}^k)$ is a Frechet and Montel space, so that by application of the closed graph Theorem, $\sigma_{\mathfrak{M}}$ is a differentiable representation. And since for any polynomial coefficient differential operator P on $\mathbb{R}^k$, $\|P \cdot \varphi\|_{L^2}$ is a continuous seminorm on $S(\mathbb{R}^k)$, we see that i is continuous above. Then noting that $S(\mathbb{R}^k)$ is dense in $L^2(\mathbb{R}^k)$ relative to the $\| \ \|_{L^2}$ topology, we have from §3 that $S(\mathbb{R}^k)$ is dense in F_Q relative to the topology on F_Q given above. Moreover by noting that $T(\mathbb{R}^k)$ with the strong dual topology is a complete Montel space, we derive the last part of the Lemma.

Q. E. D.

We let $P(\mathbb{R}^k)$ be the space of complex valued polynomials in $\mathbb{R}^k$, and $\partial(P)$, the space of complex constant coefficient differential operators on $\mathbb{R}^k$. There exists a bilinear nondegenerate pairing between $\partial(P)$ and $P(\mathbb{R}^k)$ given by $\langle R,f\rangle = R(f)(0)$, $R \in \partial(P)$, $f \in P$. We recall that P and $\partial(P)$ are graded algebras where $P = \sum_{t \geq 0} P_t$, $\partial(P) = \sum_{t \geq 0} \partial(P_t)$, P_t the space of homogeneous polynomials of degree t . If S is any Lie subgroup of $\mathrm{Aut}(\mathbb{R}^k)$, then the linear representation of S on $\mathbb{R}^k$ extends in the obvious way to a linear representation of S on $P(\mathbb{R}^k)$ and $\partial(P)$, so that one has $\langle g \cdot R , g f\rangle = \langle R,f\rangle$ for all $g \in S$.

Moreover we let $D(\mathbb{R}^k)$ be the space of polynomial coefficient differential operators on $\mathbb{R}^k$. Then we have algebraically that $D(\mathbb{R}^k) \simeq P(\mathbb{R}^k) \otimes_{\mathbb{C}} \partial(P)$. And again in the obvious way, the action of S extends to an action on $D(\mathbb{R}^k)$.

We let $D(\mathbb{R}^k)^S$ be the space of S invariant operators in $D(\mathbb{R}^k)$ ($P \in D^S$ if and only if $g \cdot P = P$ for all $g \in S$) .

Then we have the following characterization of the $O(Q)$ invariants.

<u>Lemma 5.2</u>. $D(\mathbb{R}^k)^{O(Q)}$ <u>is generated by</u> $\{I, Q, \partial(Q), E_1\}$. <u>Thus the ring of operators</u> $\mathbb{C}\left[\omega_{(s\ell_2)}\right]$ <u>of</u> $D(\mathbb{R}^k)$ <u>where</u> $\omega_{(s\ell_2)} = E_1^2 - Q\,\partial(Q) + (k-2)E_1 +$

$(\frac{k^2}{4} - k)\cdot I$ _is characterized as exactly the set_ $D_{\tilde{G}_2(Q)}$ _of all operators_ $\{L \in D(\mathbb{R}^k) \mid L \in D(\mathbb{R}^k)^{O(Q)}$ _and_ $[L,Q] = [L,\partial(Q)] = [L,E_1] = 0\}$.

Proof. We know that $D(\mathbb{R}^k)$ has a filtration defined by $D_\ell = \sum_{s+t\leq\ell} P_s\partial(P_t)$ $(D_\ell \subseteq D_{\ell+1}$ and $\cup D_\ell = D)$. Moreover D_ℓ is stable under the action of S .

First we let $S = \mathrm{Aut}(\mathbb{R}^k)$. By representation theory, we know that each P_j is an irreducible S module, and P_j and P_k are S inequivalent if $j \neq k$. Moreover $\partial(P_j)$ is equivalent to the contragredient representation of S on P_j (by using $\langle\ ,\ \rangle$ above). Then by Schur's Lemma we see that $D(\mathbb{R}^k)^{\mathrm{Aut}(\mathbb{R}^k)}$ is spanned by elements of the form

$$\ell_N = \sum_{m_1+m_2+\ldots+m_k=N} \frac{1}{m_1! \cdots m_k!} x_1^{m_1}\left(\frac{\partial}{\partial x_1}\right)^{m_1} \cdots\cdots (x_k)^{m_k}\left(\frac{\partial}{\partial x_k}\right)^{m_k} \tag{5-7}$$

$(N \geq 0)$. However one knows that $(x\frac{\partial}{\partial x})^L \equiv x^L(\frac{\partial}{\partial x})^L$ modulo terms of lower degree, and thus it follows that

$$\frac{1}{N!}\left(\sum_{i=1}^{k} x_i \frac{\partial}{\partial x_i}\right)^N \equiv \ell_N \text{ modulo } D_{2N-1} . \tag{5-8}$$

Then by using induction on the filtration, we see that $D(\mathbb{R}^k)^{\mathrm{Aut}(\mathbb{R}^k)}$ $= \mathbb{C}[E_1]$ (= the ring in $D(\mathbb{R}^k)$ generated by E_1) .

We then let $S = O(Q)$. By the separation of variables theorem, we know that P_t is $O(Q)$ equivalent to $H_t \oplus Q\, H_{t-2} \oplus \cdots\cdots \oplus Q^{[\frac{t}{2}]}H_{t-2[\frac{t}{2}]}$, where $H_m = \{f \in P_m \mid \partial(Q)\ f = 0\}$, the space of harmonic polynomials (an irreducible $O(Q)$ module) . We know that H_m is $O(Q)$ inequivalent to H_n with $m \neq n$. Thus by using the tensor decomposition of D and Schur's Lemma again, we see that $D(\mathbb{R}^k)^{O(Q)}$ is the linear span of elements of the form $Q^{j_1}R_\gamma\{\partial(Q)\}^{j_2}$, where $j_1, j_2 \geq 0$ and R_γ is the unique (up to scalar) $O(Q)$ invariant arising from Schur's Lemma determined by $H_\gamma \otimes \partial(H_\gamma)$, $\gamma > 0$.

It then suffices to show R_γ belongs to the ring generated by

I, Q, $\partial(Q)$, and E_1. We use induction on the filtration of D. First if $\gamma = 1$, $R_\gamma = E_1$. Then we know that the $\mathrm{Aut}(\mathbb{R}^k)$ invariant ℓ_γ in $D_{2\gamma}$ has a linear decomposition of the form $\ell_\gamma \equiv c_\gamma R_\gamma$ + terms of the form $Q^j R_{\gamma-2j}\{\partial(Q)\}^j$. Then $c_\gamma \neq 0$ since otherwise $\ell_\gamma(h) \equiv 0$ for all harmonic polynomials. But since $\ell_\gamma = R_1(E_1)$, R_1 a polynomial in one variable of degree γ, then $R_1(k) = 0$ for all integers $k \geq 0$, a contradiction.

For the operator $\omega_{(s\ell_2)}$ we take the Casimir element of the TDS determined by $X = E_1 + \frac{k}{2}$, $E^+ = \frac{1}{2} Q$, $E^- = -\frac{1}{2}\partial(Q)$ $(\omega_{(s\ell_2)} = X^2 - 2X + 4E^+E^-)$.

Q. E. D.

<u>Remark 5.1</u>. If $(\mathfrak{S} \oplus s\ell_2)_{\mathbb{C}}$ is the complexification of the Lie algebra of $\tilde{G}_2(Q)$, then the enveloping algebra $\mathcal{U}(\mathfrak{S} \oplus s\ell_2)$ has center $\mathcal{Z}_Q = \mathcal{Z}[(\mathfrak{S}_{\mathbb{C}})] \cdot \mathcal{Z}[(s\ell_2)_{\mathbb{C}}]$ where $\mathcal{Z}((\mathfrak{S})_{\mathbb{C}})$ and $\mathcal{Z}((s\ell_2)_{\mathbb{C}})$ are the centers of $\mathcal{U}(\mathfrak{S}_{\mathbb{C}})$ and $\mathcal{U}((s\ell_2)_{\mathbb{C}})$, respectively. Then Lemma 5.2 implies that

$$\sigma_{\mathfrak{m}}(\mathcal{Z}_Q) = \mathbb{C}\left[\omega_{(s\ell_2)}\right]\Big|_{S(\mathbb{R}^k)} .$$

<u>Remark 5.2</u>. We recall that the Casimir element $\omega_{0(Q)}$ of $\mathcal{U}(\mathfrak{S})$ is given by $\sum_{1 \leq i < j \leq a} L_{ij}^2 + \sum_{a+1 \leq s < t \leq k} M_{st}^2 - \sum_{\substack{1 \leq u \leq a \\ a+1 \leq v \leq k}} N_{uv}^2$. Then by easy computation we deduce that $\omega_{(s\ell_2)} + \omega_{0(Q)} = (\frac{k^2}{4} - k) \cdot I$ for all k.

We recall that the Fourier transform of any element $T \in T(\mathbb{R}^k)$ is defined by $\langle \hat{T}, \varphi \rangle = \langle T, \check{\varphi} \rangle$ for all $\varphi \in S(\mathbb{R}^k)$.

We also recall that for any open cone V in $\mathbb{R}^k$ ($X \in V$ implies that $t \cdot X \in V$ for all $t > 0$), we say an element $S \in C_c^\infty(V)'$ is homogeneous of degree λ, $\lambda \in \mathbb{C}$, if $\langle S, \varphi(t^{-1}X) \rangle = |t|^\lambda \langle S, \varphi(X) \rangle$ for all $t \in \mathbb{R}_+^*$ and $\varphi \in C_c(V)$.

Let $\phi_\lambda(V)$ be the subspace of $C_c^\infty(V)'$ of distributions homogeneous of degree λ. Then it is easy to see that $\phi_\lambda(V)$ is a closed subspace of $C_c^\infty(V)'$.

Moreover if $V = \mathbb{R}^k$, then we know that $\phi_\lambda(\mathbb{R}^k) \subseteq T(\mathbb{R}^k)$, i.e. every homogeneous distribution on $\mathbb{R}^k$ is <u>tempered</u>. Moreover if λ and

λ' satisfy $\lambda + \lambda' = -k$, then we have the relation $(\phi_\lambda(\mathbb{R}^k))^\wedge \subseteq \phi_{\lambda'}(\mathbb{R}^k)$.

We then define the subspace (λ_1 and λ_2 quasicharacters on $\mathbb{R}^*$ and $s \in \{0, 1, 2, 3\}$)

(5-9) $$\Pi_{\lambda_1,\lambda_2,s} = \{T \in T(\mathbb{R}^k) \mid \sigma_{\mathfrak{m}}^c(E\ a(r)\ \exp_{S\tilde{\ell}_2}(z\ F))\ T = e^{\pi i \frac{s}{2}} \lambda_1(r)\ T \text{ and } \sigma_{\mathfrak{m}}^c(mA_a(r)\ N_a(Y))T = \lambda_2(r)T\}.$$

Our problem is to determine the space $\Pi_{\lambda_1,\lambda_2,s}$.

We first observe that since $\sigma_{\mathfrak{m}}(w_o)$ is an automorphism of $S(\mathbb{R}^k)$, we have that $\sigma_{\mathfrak{m}}^c(w_o)\ (\Pi_{\lambda_1,\lambda_2,s}) = \Pi^{\lambda_1^{-1},\lambda_2,s} = \{T \in T(\mathbb{R}^k) \mid \sigma_{\mathfrak{m}}^c(E\ a(x)\ n(y))\ T = e^{\pi is/2}\lambda_1^{-1}(x)\ T \text{ and } \sigma_{\mathfrak{m}}^c(m\ A_a(r)\ N_a(Y))\ T = \lambda_2(r)\ T\}$.

Then by computation using Theorem 1.2, we see that $\langle\sigma_{\mathfrak{m}}^c(E)\ T, \varphi\rangle = \langle T, \sigma_{\mathfrak{m}}^c(E^3)\varphi\rangle = f_k(-1)\langle T, \varphi(-X)\rangle$. Thus for $T \neq 0$ in $\Pi^{\lambda_1^{-1},\lambda_2,s}$, we must have $e^{\pi i \frac{s}{2}} = \lambda_2(-1)\ f_k(-1)$. Thus if $s \not\equiv k \bmod 2$, $\Pi^{\lambda_1^{-1},\lambda_2,s} = \{0\}$.

<u>Lemma 5.3</u>. <u>If</u> $s \not\equiv k \bmod 2$, <u>then</u> $\Pi^{\lambda_1^{-1},\lambda_2,s} = 0$. <u>If</u> $s \equiv k \bmod 2$, <u>then</u>

$$\Pi^{\lambda_1^{-1},\lambda_2,s} = \left\{T \in \phi_{-(\lambda_1)_o+\frac{k}{2}}(\mathbb{R}^k) \,\middle|\, T \text{ is } \begin{cases}\text{even as } s \equiv \operatorname{sgn}(Q) \bmod 4\\ \text{odd as } s \equiv \operatorname{sgn}(Q) + 2 \bmod 4\end{cases},\right.$$

<u>support</u> $(T) \subseteq \Gamma_o$, <u>and</u> $\left.\sigma_{\mathfrak{m}}^c(m\ A_a(r)\ N_a(Y))\ T = \lambda_2(r)\ T\right\}$ ($\operatorname{sgn}(Q)$ = <u>signature of quadratic form</u> Q).

<u>Proof</u>. By computation $T \in \Pi^{\lambda_1^{-1},\lambda_2,s}$ and for $x > 0$, $\langle T, \sigma_{\mathfrak{m}}(a(x^{-1}))\varphi\rangle = |x|^{-k/2}\langle T, \varphi(x^{-1}\ )\rangle = \lambda_1^{-1}(x)\langle T, \varphi\rangle$. Moreover $\frac{d}{dy}\langle T, \sigma_{\mathfrak{m}}(n(y))\varphi\rangle|_{y=0} = \langle T, \pi i\ \ Q\cdot\varphi\rangle \equiv 0$ for all $\varphi \in S(\mathbb{R}^k)$. Thus $Q\cdot T = 0$ or support $(T) \subseteq \{X \mid Q(X,X) = 0\} = \Gamma_o$.

Q. E. D.

We then consider the Lie group $V = P_a \times E(r)$, where $E(r)$ is the group of dilations of $\mathbb{R}^k$ (i.e. $E(r)\ X = r\cdot X$, $r \in R^*$, $X \in \mathbb{R}^k$). Then we consider the linear action of V on $\mathbb{R}^k$. We let $\varphi_{\rho_1,\rho_2} : V \to \mathbb{C}^*$ be the character on V given by $\varphi_{\rho_1,\rho_2}(m\ A_a(r)\ N_a(Y)\ E(s)) = \rho_1(r)\rho_2(s)$, where ρ_1 and ρ_2 are quasi-characters on $\mathbb{R}^*$. We want to determine the space

$$(5\text{-}10)\qquad T_{\rho_1,\rho_2}(\mathbb{R}^k) = \{T \in C_c^\infty(\mathbb{R}^k)' \mid \langle T, \varphi(m\, A_a(r)\, N_a(Y)\, E(s)\, X)\rangle = \rho_1(r)\rho_2(s) \langle T, \varphi(X)\rangle \text{ for } X \in \mathbb{R}^k\} .$$

To do this we apply the results of §3.

Remark 5.3. In Lemma 5.3 we observe that if $s \equiv k \bmod 2$, then $\Pi^{\lambda_1^{-1},\lambda_2,s} = \left\{T \in T_{\lambda_2,((\lambda_1)_o - \frac{k}{2},\ \phi(s,Q))}(\mathbb{R}^k) \,\middle|\, \text{support }(T) \subseteq \Gamma_o\right\}$, where $\phi(s,Q) = 1$ if $s \equiv \text{sgn}(Q) \bmod 4$ and $\phi(s,Q) = -1$ if $s \equiv \text{sgn}(Q)+2 \bmod 4$.

We see that the orbits of V in $\mathbb{R}^k$ are the following (using Proposition 2.4).

$$(5\text{-}11)\qquad Q_+ = V(\tilde v_a + \tfrac12 v_a) \text{ with isotropy group } V^{(\tilde v_a + \frac12 v_a)} = M_a .$$

$$(5\text{-}12)\qquad Q_- = V(\tilde v_a - \tfrac12 v_a) \text{ with isotropy group } V^{(\tilde v_a - \frac12 v_a)} = M_a .$$

$$(5\text{-}13)\qquad T_+ = V\, c(a, a+1)\, (\tilde v_a + \tfrac12 v_a) \text{ with isotropy group } V^{(\tilde v_{a+1} + \frac12 v_{a+1})} = M_a^{(\tilde v_{a+1} + \frac12 v_{a+1})} \cdot A_a(r) \cdot \left\{N_a(Z) \,\middle|\, Z \in \langle v_a, \tilde v_a, \tilde v_{a+1} + \tfrac12 v_{a+1}\rangle^\perp\right\}$$

if $b > 1$, and

$$T_+ = V\, \gamma_t(\tilde v_a + \tfrac12 v_a) \text{ with isotropy group } V^{(e_2)} = M_a^{(e_2)} \cdot A_a(r) \cdot \left\{N_a(Z) \,\middle|\, Z \in \langle v_a, \tilde v_a, e_2\rangle^\perp\right\} \text{ if } b = 1 .$$

$$(5\text{-}14)\qquad T_- = V\, c(a, a+1)\, (\tilde v_a - \tfrac12 v_a) \text{ with isotropy group } V^{(\tilde v_{a+1} - \frac12 v_{a+1})} = M_a^{(\tilde v_{a+1} - \frac12 v_{a+1})} \cdot A_a(r) \cdot \left\{N_a(Z) \,\middle|\, Z \in \langle v_a, \tilde v_a, \tilde v_{a+1} - \tfrac12 v_{a+1}\rangle^\perp\right\}$$

if $b > 1$.

$$(5\text{-}15)\qquad L_0 = V(\tilde v_a) \text{ with isotropy group } V^{(\tilde v_a)} = M_a \cdot B(r), \text{ with } B(r) = \{A_a(r) \times E(r)\} .$$

(5-16) $T_0 = V(v_{a+1})$ with <u>isotropy group</u> $V^{(v_{a+1})} = \{M_a \cap \{M_{a+1} \cdot N_{a+1}\}\} \cdot A_a(r) \cdot \{N_a(Z) | Z \in \langle v_a, \tilde{v}_a, v_{a+1} \rangle^{\perp}\} \cdot C(r)$, with $C(r) = \{A_{a+1}(r^{-1}) \times E(r)\}$ if $b > 1$.

(5-17) $\ell_0 = V(v_a)$ with <u>isotropy group</u> $V^{(v_a)} = M_a \cdot N_a(Y) \cdot \tilde{B}(r)$, with $\tilde{B}(r) = \{A_a(r^{-1}) \times E(r)\}$.

(5-18) $\{0\} = V\{0\}$ with <u>isotropy group</u> $V^{\{0\}} = V$.

Then relative to the method used in §3, we have $\Omega_{Q_+} = Q_+$, $\Omega_{Q_-} = Q_-$, $\Omega_{T_+} = T_+ \cup Q_+ \cup Q_-$, $\Omega_{T_-} = T_- \cup Q_+ \cup Q_-$, $\Omega_{L_0} = L_0 \cup Q_+ \cup Q_-$, $\Omega_{T_0} = \mathbb{R}^k - \{\ell_0 \cup \{0\}\}$, $\Omega_{\ell_0} = \mathbb{R}^k - \{0\}$ and $\Omega_{\{0\}} = \mathbb{R}^k$.

We note at this point that if $\mathfrak{g}$ is the Lie algebra of any Lie subgroup G of $O(Q)$, then $\mathfrak{g} \cdot v$ spans the tangent space of the orbit $G \cdot v$ in $\mathbb{R}^k$, where we make the parallel identification of the tangent space of a point $v \in \mathbb{R}^k$ to the vector space $\mathbb{R}^k$ itself. Thus it follows by an easy argument that $\mathfrak{S} \cdot v$ is in fact the same as $\{Y \in \mathbb{R}^k | Q(Y,v) = 0\} = v^{\perp}$ ($\mathfrak{S}$ = Lie algebra of $O(Q)$).

We first consider Q_+ and observe that dim $Q_+ = k$. This implies that the representation "$\frac{\delta_S}{\delta_{S_p}} \tilde{\nu}_p^{\ell}$" is the identity representation ($\equiv I$) if $\ell = 0$ and 0 if $\ell > 0$. Thus dim $T_{\rho_1, \rho_2, Q_+}(\Omega_{Q_+}) \leq \dim \mathrm{Hom}_{M_a}(I,I) = 1$ for all ρ_1 and ρ_2. Similarly dim $T_{\rho_1, \rho_2, Q_-}(\Omega_{Q_-}) \leq 1$.

We then consider T_+ (with $b > 1$) and observe that $\mathscr{V} \cdot (\tilde{v}_{a+1} + \frac{1}{2} v_{a+1}) = \{s\, v_a + Y | Y \in \langle v_a, \tilde{v}_a \rangle^{\perp}$ and $s \in \mathbb{R}\}$, where $\mathscr{V}$ is the Lie algebra of V. Then the space $\mathbb{R}^k / \mathscr{V} \cdot (\tilde{v}_{a+1} + \frac{1}{2} v_{a+1})$ is one dimensional and the representation "$\frac{\delta_S}{\delta_{S_p}} \nu_p^{\ell}$" is in this case

$\phi_\ell(m\, A_a(r)\, N_a(Y)) = |r|^{-1} r^{-\ell}$, where $m \in M_a^{(\tilde{v}_{a+1} + \frac{1}{2} v_{a+1})}$ and $Y \in$

$\langle v_a, \tilde{v}_a, \tilde{v}_{a+1} + \frac{1}{2} v_{a+1}\rangle^{\perp}$. Thus $\dim T_{\rho_1,\rho_2,T_+}(\Omega_{T_+}) \le \sum_{\ell=0}^{\ell=\infty} \dim \mathrm{Hom}_{\mathbb{R}^*}$ $(\rho_1, -\underset{\sim}{\ell} - (1,1)) = \begin{cases} 0 & \text{if } \rho_1 \neq -\underset{\sim}{\ell} - (1,1) \ (\ell \text{ integer} \ge 0) \\ 1 & \text{otherwise} \end{cases}$. We derive a similar statement for the case of T_-.

We then assume $b = 1$ and again consider T_+. We have $\mathcal{V}\cdot e_2 = \{s\, v_a + Y \mid Y \in \langle v_a, \tilde{v}_a\rangle^{\perp}$ and $s \in \mathbb{R}\}$. Then we have the same result as in the case above.

We take L_o and see that $\mathcal{V}\cdot\tilde{v}_a = \{u\, \tilde{v}_a + Y \mid Y \in \langle v_a, \tilde{v}_a\rangle^{\perp}$ and $u \in \mathbb{R}\}$. Thus the space $\mathbb{R}^k/\mathcal{V}\cdot\tilde{v}_a$ is one dimensional, and we have the representation "$\frac{\delta_S}{\delta_{S_p}} \tilde{v}_p^{\ell}$" of $M_a\cdot B(r)$ is $\phi_\ell(m\, B(r)) = |r|^{2-k} r^{2\ell}$, and hence $\dim T_{\rho_1,\rho_2,L_o}(\Omega_{L_o}) \le \sum_{\ell=0}^{\ell=\infty} \dim \mathrm{Hom}_{\mathbb{R}^*}(\rho_1\rho_2, (2-k,1) + 2\underset{\sim}{\ell}) =$

$$\begin{cases} 0 & \text{if } \rho_1\rho_2 \neq (2-k,1) + 2\underset{\sim}{\ell} \ (\ell \text{ integer} \ge 0) \\ 1 & \text{otherwise} \end{cases}.$$

We then consider T_o $(b > 1)$ and see that $\mathcal{V}(v_{a+1}) = \{s\, v_a + Z \mid Z \in \langle v_a, \tilde{v}_a, v_{a+1}\rangle^{\perp}$ and $s \in \mathbb{R}\}$. Thus $\mathbb{R}^k/\mathcal{V}(v_{a+1})$ is a two dimensional space spanned by the elements $\{\tilde{v}_a, \tilde{v}_{a+1}\}$. Then the representation "$\frac{\delta_S}{\delta_{S_p}} \tilde{v}_p^{\ell}$" of $V^{(v_{a+1})} = L_a\cdot A_a(r)\ \{N_a(Z) \mid Z \in \langle v_a, \tilde{v}_a, v_{a+1}\rangle^{\perp}\}\cdot C(s)$ is

$|r|^{-1}|s|^{3-k}\ \Sigma_\ell(m\, A_a(r)\, N_a(Y)\, C(s))$, where Σ_ℓ is the representation of $V^{(v_{a+1})}$ on the quotient space $S^\ell(\mathbb{R}^k)/I_\ell$ with I_ℓ the intersection of the ideal in $S(\mathbb{R}^k)$ generated by v_a and $Z \in \langle v_a, \tilde{v}_a, v_{a+1}\rangle^{\perp}$ with $S^\ell(\mathbb{R}^k)$ (note $L_a = M_a \cap (M_{a+1}\cdot N_{a+1})$). But we note that the elements $\{v_a^\ell, \tilde{v}_a^{\ell-1}\tilde{v}_{a+1}, \ldots, \tilde{v}_{a+1}^\ell\}$ span $S^\ell(\mathbb{R}^k)/I_\ell$. Thus $A_a(r)\, C(s)\, (\tilde{v}_a^C\, \tilde{v}_{a+1}^D) = r^{-C} s^{C+2D} (\tilde{v}_a^C\, \tilde{v}_{a+1}^D)$. Then we note that for

$Z \in \langle v_a, \tilde{v}_a, v_{a+1}\rangle^{\perp}$, $(N_a(Z)\, \tilde{v}_a)^\ell \equiv (\tilde{v}_a - \frac{1}{2} Q(Z,Z)\, v_a + Z)^\ell \equiv (\tilde{v}_a)^\ell \bmod I_\ell$,

and $(N_a(Z)\cdot\tilde{v}_{a+1})^\ell \equiv (\tilde{v}_{a+1} - Q(Z,v_{a+1})\, v_a)^\ell \equiv (\tilde{v}_{a+1})^\ell \bmod I_\ell$. Moreover for $m \in L_a$ of the form $N_{a+1}(S)$, with $S \in \langle v_a, \tilde{v}_a, v_{a+1}, \tilde{v}_{a+1}\rangle^{\perp}$, we deduce $(m\, \tilde{v}_{a+1})^\ell \equiv (\tilde{v}_{a+1} - \frac{1}{2}Q(S,S)\, v_{a+1} + S)^\ell \equiv (\tilde{v}_{a+1})^\ell \bmod I_\ell$. Thus in any case if $x \in L_a\cdot\{N_a(Z) \mid Z \in \langle v_a, \tilde{v}_a, v_{a+1}\rangle^{\perp}\}$, then x operates as the identity on $S^\ell(\mathbb{R}^k)/I_\ell$. We let $\Omega = \{(-(1,1) - \underset{\sim}{t}, (3-k, 1) + \underset{\sim}{t'}) \mid t, t'$ integers ≥ 0 and $t + t' \equiv 0 \bmod 2\}$. Thus $\dim T_{\rho_1,\rho_2,T_o}(\Omega_{T_o}) \leq \sum_{\ell=0}^{\infty} \dim \mathrm{Hom}_{V^{(v_{a+1})}}(\rho_1(r)\rho_2(s), |r|^{-1}|s|^{3-k} \Sigma_\ell)$, and by the arguments above, we see that for all pairs $(\rho_1,\rho_2) \not\subset \Omega$, then $\dim T_{\rho_1,\rho_2,T_o}(\Omega_{T_o}) = 0$.

We then consider ℓ_o and see that $\mathcal{V}(v_a) = \{s\, v_a \mid s \in \mathbb{R}\}$. Thus $\mathbb{R}^k/\mathcal{V}\cdot v_a$ has dimension $k-1$ and is spanned by elements of the form $\{v_a, Z \mid Z \in \langle v_a, \tilde{v}_a\rangle^\perp\}$. Then the representation "$\frac{\delta_S}{\delta_{S_p}} \tilde{\nu}_p^\ell$" of $V^{(v_a)} = M_a N_a(Y)\, B(r)$ is $\psi_\ell(m\, N_a(Y)\, B(r))$, where ψ_ℓ is the representation of $V^{(v_a)}$ on the space $S^\ell(\mathbb{R}^k)/\tilde{I}_\ell$, where $\tilde{I}_\ell$ is the intersection of the ideal (in $S(\mathbb{R}^k)$) generated by v_a with $S^\ell(\mathbb{R}^k)$. Thus $\dim T_{\rho_1,\rho_2,\ell_o}(\Omega_{\ell_o}) \leq \sum_{\ell=0}^{\infty} \dim \mathrm{Hom}_{V^{(v_a)}}(\rho_1^{-1}\rho_2(r), \psi_\ell)$. Then we observe that by taking a basis of $S^\ell(\mathbb{R}^k)/\tilde{I}_\ell$ of the form $\{\tilde{v}_a{}^s P\}_{0 \leq s \leq \ell}$, where $P \in S^{\ell-s}(\langle v_a, \tilde{v}_a\rangle^\perp)$, we deduce $B(r)\, (\tilde{v}_a{}^s P) = r^{s+1}(\tilde{v}_a{}^s P)$. Moreover M_a leaves the span of elements $\{\tilde{v}_a{}^s P\}$ stable so that the M_a invariants in $S^\ell(\mathbb{R}^k)/\tilde{I}_\ell$ are given as follows: let P_1 be the restriction of Q to the space $\langle v_a, \tilde{v}_a\rangle^\perp$, then $\{\tilde{v}_a{}^\ell, \tilde{v}_a{}^{\ell-2} P_1, \ldots, \tilde{v}_a{}^{\ell - 2\left[\frac{\ell}{2}\right]} P^{\left[\frac{\ell}{2}\right]}_{\left[\frac{\ell}{2}\right]}\}$ span the M_a invariants. And in any case P_1 is $N_a(Y)$ invariant (mod $\tilde{I}_\ell$), and we observe that $(N_a(Y)\cdot \tilde{v}_a)^\ell \equiv (\tilde{v}_a - \frac{1}{2} Q(Y,Y)\, v_a + Y)^\ell \equiv (\tilde{v}_a + Y)^\ell \bmod \tilde{I}_\ell$; hence if ℓ is odd, there are no $M_a N_a(Y)$ invariants in $S^\ell(\mathbb{R}^k)/\tilde{I}_\ell$, and if ℓ is even, $(P_1)^{\left[\frac{\ell}{2}\right]}$ is (up to a scalar multiple) the only $M_a \cdot N_a(Y)$ invariant in $S^\ell(\mathbb{R}^k)/\tilde{I}_\ell$. Thus $\dim T_{\rho_1,\rho_2,\ell_o}(\Omega_{\ell_o}) \leq$

$$\begin{cases} 0 & \text{if } \rho_1^{-1}\rho_2 \neq (2j, 1),\ j \text{ integer} \geq 0 \\ 1 & \text{otherwise} \end{cases}.$$

Finally we take $\{0\}$ and then the representation "$\frac{\delta_S}{\delta_{S_p}} \tilde{\nu}_p^\ell$" of V is the standard representation t_ℓ of V on the space $S^\ell(\mathbb{R}^k)$. Then

$\dim T_{\rho_1,\rho_2,\{0\}}(\mathbb{R}^k) \leq \sum_{\ell=0}^{\infty} \dim \mathrm{Hom}_V(\rho_1(r)\rho_2(s), t_\ell)$. Then $t_\ell(E(s)) = s^\ell$, and by the classical theory of highest weight, we know that $\{v_a^\ell, v_a^{\ell-2}Q, \ldots, v_a^{\ell-2\left[\frac{\ell}{2}\right]}Q^{\left[\frac{\ell}{2}\right]}\}$ span the elements which are M_a and N_a invariant. Thus letting $\Omega' = \{(\underset{\sim}{v}, \underset{\sim}{s}) \mid s, v$ integers ≥ 0 with $v \leq s$ and $s - v \equiv 0 \mod 2\}$, we have if $(\rho_1,\rho_2) \notin \Omega'$, then $\dim T_{\rho_1,\rho_2,\{0\}}(\mathbb{R}^k) = 0$; otherwise $\dim T_{\rho_1,\rho_2,\{0\}}(\mathbb{R}^k) \leq 1$.

Theorem 5.4. We are given a pair of quasicharacters (λ_1,λ_2) and an integer $0 \leq s \leq 3$ subject to either (5-19) or (5-20):

(5-19)
(A) $(\lambda_1)_o \neq \ell_1 + \frac{k}{2}$ with ℓ_1 integer ≥ 0 .
(B) $(\lambda_2)_o \neq \ell_2$, ℓ_2 any integer ≤ -1 .
(C) $s \equiv k \mod 2$.

(5-20)
(A) $(\lambda_2)_o \neq \ell$, ℓ any integer.
(B) $s \equiv k \mod 2$.

then $\dim \Pi^{\lambda_1^{-1},\lambda_2,s} \leq 2$.

Moreover if $(\lambda_1)_o + (\lambda_2)_o \neq 2\ell + 2 - \frac{k}{2}$, ℓ integer ≥ 0 , then $\dim \Pi^{\lambda_1^{-1},\lambda_2,s} \leq 1$, and if $U \in \Pi^{\lambda_1^{-1},\lambda_2,s}$ with $U \neq 0$, then support (U) = $\{u\, v_a \mid u \in \mathbb{R}\}$.

And if $(\lambda_1)_o - (\lambda_2)_o \neq 2\ell' + \frac{k}{2}$, ℓ' integer ≥ 0 , then $\dim \Pi^{\lambda_1^{-1},\lambda_2,s} \leq 1$, and if $Y \in \Pi^{\lambda_1^{-1},\lambda_2,s}$ with $Y \neq 0$, then support (Y) = Γ_o .

Finally if both $(\lambda_1)_o + (\lambda_2)_o \neq 2\ell_1 + 2 - \frac{k}{2}$, ℓ_1 integer ≥ 0 , and $(\lambda_1)_o - (\lambda_2)_o \neq 2\ell_2 + \frac{k}{2}$, ℓ_2 integer ≥ 0 , then $\dim \Pi^{\lambda_1^{-1},\lambda_2,s} = 0$.

Proof. For any $Z \in \Pi^{\lambda_1^{-1},\lambda_2,s}$ we consider its restriction Z_{L_o} , Z_{T_o} , Z_{ℓ_o} , and $Z_{\{0\}}$ to the sets Ω_{L_o} , Ω_{T_o} , Ω_{ℓ_o} , and $\Omega_{\{0\}}$, respectively. It is easy to see that $Z_{L_o} \in T_{\lambda_2,((\lambda_1)_o - \frac{k}{2}, \phi(s,Q)),L_o}(\Omega_{L_o})$, and we can define a linear map

(5-21) $$\Pi^{\lambda_1^{-1},\lambda_2,s} \xrightarrow{E_1} T_{L_o}, \quad E_1(Z) = Z_{L_o}.$$

Then it follows that for $Z \in \text{Kernel}(E_1)$ that support $(Z) \cap \{Q_+ \cup Q_- \cup L_o\} = \emptyset$, and hence support $(Z_{T_o}) = T_o$ (we use here the fact that support (Z_{T_o}) is a closed V invariant subset of Ω_{T_o} which does not intersect L_o). Thus we have a map

(5-22) $$\text{Kernel}(E_1) \xrightarrow{E_2} T_{T_o}, \quad E_2(Z) = Z_{T_o}.$$

Similarly we get a sequence of maps

(5-23) $$\text{Kernel}(E_2) \xrightarrow{E_3} T_{\ell_o}, \quad E_3(Z) = Z_{\ell_o},$$

and

(5-24) $$\text{Kernel}(E_3) \xrightarrow{E_4} T_{\{0\}}, \quad E_4(Z) = Z_{\{0\}},$$

so that Kernel $(E_4) = 0$.

Thus $\dim \Pi^{\lambda_1^{-1},\lambda_2,s} = \dim (\Pi^{\lambda_1^{-1},\lambda_2,s}/\text{Ker } E_1) + \dim (\text{Ker } E_1/\text{Ker } E_2) + \dim (\text{Ker } E_2/\text{Ker } E_3) + \dim (\text{Ker } E_3) \leq \dim T_{L_o} + \dim T_{T_o} + \dim T_{\ell_o} + \dim T_o$. Then using what has been done above, we obtain the Theorem.

Q. E. D.

Let $A_a(r)$ and $N_a(Y)$ be as before. We consider the map

(5-25) $$\beta_a : \mathbb{R}^* \times \mathbb{R}^{k-2} \times \mathbb{R} \longrightarrow \mathbb{R}^k$$

given by $\beta_a(r,Y,t) = A_a(r)\, N_a(Y)\, (\tilde{v}_a + \frac{1}{2} t\, v_a) = \frac{1}{2} r\, (t - Q(Y,Y))\, v_a + r^{-1} \tilde{v}_a + Y$. Then β_a is an injective C^∞ differentiable map. Moreover Image $\beta_a = \{X \in \mathbb{R}^k \mid Q(X,v_a) \neq 0\} = T_a$ is an open dense subset of $\mathbb{R}^k$. We define the map $\gamma_a : T_a \longrightarrow \mathbb{R}^* \times \mathbb{R}^{k-2} \times \mathbb{R}$ given by $\gamma_a(X) = (1/Q(X,v_a),\ R_a(X),\ Q(X,X))$, where R_a is the orthogonal projection of X onto $\langle v_a, \tilde{v}_a\rangle^\perp$. Then γ_a is C^∞ differentiable and we easily see that

$\gamma_a \circ \beta_a$ = Identity on $\mathbb{R}^* \times \mathbb{R}^{k-2} \times \mathbb{R}$, and $\beta_a \circ \gamma_a$ = Identity on T_a.

Lemma 5.5. *Let f be any Borel measurable function on $\mathbb{R}^k$ so that $f \in L^1(\mathbb{R}^k, dX)$. Then $f \circ \beta_a \in L^1(A_a \times N_a \times \mathbb{R}, dA_a \times dN_a \times dt)$, where $dA_a(r) = \frac{dr}{|r|}$, $dN_a(Y) = dY$, and dt are left Haar measures on A_a, N_a, and $\mathbb{R}$, respectively. Moreover*

$$(5\text{-}26) \qquad \int_{\mathbb{R}^k} f(X)\, dX = \int_{A_a \times N_a \times \mathbb{R}} f(A_a(r)N_a(Y)(\tilde{v}_a + \tfrac{1}{2} t\, v_a))\, dA_a(r)\, dN_a(Y)\, dt$$

$$= \int_{\mathbb{R}^* \times \mathbb{R}^{k-2} \times \mathbb{R}} f(\beta_a(r,Y,t)) \frac{dr}{|r|}\, dY\, dt .$$

Proof. We write $X \in \mathbb{R}^k$ in coordinates as $X = x_a v_a + \tilde{x}_a \tilde{v}_a + Y$ with $Y \in \langle v_a, \tilde{v}_a \rangle^{\perp}$. By simple computation $dX = dx_a \wedge d\tilde{x}_a \wedge dY = \frac{1}{r} dr \wedge dt \wedge dY$, and hence the Jacobian of the mapping β_a is $\frac{1}{r}$. Then by general change of variables Theorem (see [7]), we have our result.

Q. E. D.

Remark 5.4. We recall that for any orbit $\Gamma_t - \{0\}$, there is a unique $k-1$ differential form ω_t given on $\Gamma_t - \{0\}$ which is $O(Q)$ invariant, and which is given on the set $T_a \cap \Gamma_t - \{0\}$ by $\omega_t = \frac{1}{\frac{\partial Q}{\partial x_a}} dx_a \wedge dY$. Then noting that $\Gamma_t - \{0\}$ is orientable, we consider the measure $d\mu_t$ defined on $\Gamma_t - \{0\}$ by ω_t. Then we know that $d\mu_t$ defines an $O(Q)$ invariant measure on $\Gamma_t - \{0\}$, and as in Lemma 5.5, we have the following: if φ is a Borel measurable function on $\Gamma_t - \{0\}$ so that $\varphi \in L^1(\Gamma_t - \{0\}, d\mu_t)$, then $\varphi \circ \beta_a(\ ,\ , t) \in L^1(A_a \times N_a, dA_a \times dN_a)$ and

$$(5\text{-}27) \qquad \int_{\Gamma_t - \{0\}} \varphi(Z)\, d\mu_t(Z) = \int_{A_a \times N_a} \varphi \circ \beta_a(r,Y,t)\, dA_a(r) \times dN_a(Y) .$$

Remark 5.5. We consider the map $Q : \mathbb{R}^k \longrightarrow \mathbb{R}$ and apply the theory of disintegration of measures relative to the measure dX on $\mathbb{R}^k$ (see [16],

p. 462). We recall that if ω_o is any finite Borel measure on $\mathbb{R}^k$ equivalent to the measure dX ($\omega_o = g(X)\,dX$ where $\int |g(X)|\,dX < \infty$) , then there exists a family of $0(Q)$ invariant positive measures $d\lambda_t$ concentrated on Γ_t for all $t \in \mathbb{R}$ so that for any $f_1 \in L^1(\mathbb{R}^k, dX)$

$$\text{(5-28)} \qquad \int_{\mathbb{R}^k} f_1(X)\,dX = \int_{\mathbb{R}} \left\{ \int_{\Gamma_t} f_1(X)\,d\lambda_t(X) \right\} d\tilde{\omega}_o(t) ,$$

where $\tilde{\omega}_o$ is the image measure of ω_o under Q (i.e. $\tilde{\omega}_o[(a,b)] = \omega_o\{X \mid Q(X,X) \in (a,b)\}$) . We then note that by the uniqueness of $0(Q)$ invariant measure on each orbit $\Gamma_t - \{0\}$, we have $d\lambda_t = c(t)\,d\mu_t$ $(t \neq 0)$ with $c(t) \neq 0$. Moreover $f \in L^1(\mathbb{R}^k)$ implies that $f\,|_{\Gamma_t - \{0\}} \in L^1(\Gamma_t - \{0\}, d\mu_t)$ for almost all $t \in \mathbb{R}$. It then follows from Lemma 5.5 and Remark 5.4 that for $f \in L^1(\mathbb{R}^k, dX)$

$$\text{(5-29)} \qquad \int_{\mathbb{R}^k} f(X)\,dX = \int_{\mathbb{R}} \left\{ \int_{\Gamma_t - \{0\}} f(X)\,d\mu_t(X) \right\} dt .$$

We recall that every element $g \in 0(Q)$ can be written in the form $g = k_1 \cdot a \cdot k_2$ with $k_i \in K$ and $a \in A$. We know that a is defined uniquely up to Weyl group conjugacy so that if $a = \prod_{s=a}^{k-1} A_s(r_s)$, then we define $A(g) = \max(|r_s|, |r_s|^{-1})$, with $s = a,\ldots,k-1$.

Let $\varphi : \mathbb{R} \longrightarrow \mathbb{C}$ be any function on $\mathbb{R}$ so that $\varphi \in C^\infty(\mathbb{R})$, and for all $t \in \mathbb{R}$ the function $F_t(X) = \varphi(t\,Q(X,X))$ is a <u>slowly increasing function</u> at ∞, i.e. for every t, $F_t \in \mathcal{O}_{\mathfrak{m}}$ (see [23], p. 275) or for every $f \in S(\mathbb{R}^k)$, the multiplication map $f \longrightarrow F_t \cdot f$ is a continuous map of $S(\mathbb{R}^k)$ to $S(\mathbb{R}^k)$.

At this point we adopt the convention that $f_g(X) = \pi_{\mathfrak{m}}(g^{-1})\,f(X)$ for all $X \in \mathbb{R}^k$.

For an arbitrary measurable f on $\mathbb{R}^k$, we consider the abstract projection maps:

$$\theta_f^\chi(g,t,\varphi) = \int_{A_a \times N_a \times \mathbb{R}} f_g(A_a(r)N_a(Y)(\tilde{v}_a + \tfrac{s}{2} v_a))\chi(r)\varphi(s\,t)\, dA_a(r)\, dN_a(Y)\, ds\ , \tag{5-30}$$

$$R_f^\chi(g,t) = \int_{A_a \times N_a} f_g(A_a(r)N_a(Y)(\tilde{v}_a + \tfrac{t}{2} v_a))\chi(r)\, dA_a(r)\, dN_a\ . \tag{5-31}$$

<u>Proposition 5.6</u>. *If* $\mathrm{Re}(\chi_o) < 0$ *and* $f \in S(\mathbb{R}^k)$, *then* $\theta_f^\chi(g,t,\varphi)$ *and* $R_f^\chi(g,t)$ *are absolutely convergent integrals. If* $f \in C_c^\infty(\mathbb{R}^k)$, *then we have the estimate : there exist* C_1 *and* $C_2 > 0$ *so that*

$$|\theta_f^\chi(g,t,\varphi)| \le C_1\, L(f,g,t)\, \|f\|_\infty \frac{(R(f)\ A(g))^{k - \mathrm{Re}(\chi_o)}}{|\mathrm{Re}\ (\chi_o)|}\ , \tag{5-32}$$

and

$$|R_f^\chi(g,t)| \le C_2\, L(f,g,t)\, \|f\|_\infty \frac{(R(f)\ A(g))^{k - 2 - \mathrm{Re}(\chi_o)}}{|\mathrm{Re}\ (\chi_o)|}\ , \tag{5-33}$$

where $R(f)$ = *smallest* $x > 0$ *so that* $\mathrm{support}(f) \subseteq B_x$, *with* B_x *a ball of radius* x *about* 0 *in* $\mathbb{R}^k$ *and* $L(f,g,t) = \max_{|z| \le |t|(A(g)R(f))^2} |\varphi(z)|$.

<u>Proof</u>. Consider the polynomial $P_{k_1,k_2,k_3}(X) = (1 + \tilde{x}_a^2)^{k_1} \cdot (1 + Q(X,X)^2)^{k_2} \cdot (1 + \|Y\|^2)^{k_3}$ in $\mathbb{R}^k$ and the corresponding seminorm on $S(\mathbb{R}^k)$ given by $\nu_{P_{k_1,k_2,k_3}}(f) = \sup_{X \in \mathbb{R}^k} |P_{k_1,k_2,k_3}(X)\, f(X)|$. Then we have for $f \in S(\mathbb{R}^k)$

$$\int_{A_a \times N_a \times \mathbb{R}} |f_g(A_a(r)\, N_a(Y)\, (\tilde{v}_a + \tfrac{s}{2} v_a))|\, |r|^{\mathrm{Re}(\chi_o)} |\varphi(st)| \frac{dr}{|r|}\, dY\, ds \tag{5-34}$$

$$\le \nu_{P_{k_1,k_2,k_3}}(F_t \cdot f_g) \left\{ \int_{-\infty}^{+\infty} |r|^{\mathrm{Re}(\chi_o)-1} (1 + r^{-2})^{-k_1}\, dr \right\} .$$

$$\left\{ \int_{\mathbb{R}} (1+s^2)^{-k_2}\, ds \right\} \left\{ \int_{\mathbb{R}^{k-2}} (1+\|Y\|^2)^{-k_3}\, dY \right\}.$$

Then we note that there exist constants C_1, C_2, and C_3 so that

$$\int_{\mathbb{R}} \left(\frac{r^2}{1+r^2}\right)^{k_1} |r|^{\mathrm{Re}(\chi_0)-1}\, dr \le C_1 \int_0^1 |r|^{2k_1+\mathrm{Re}(\chi_0)-1}\, dr +$$

$$C_2 \int_1^\infty |r|^{\mathrm{Re}(\chi_0)-1}\, dr \le C_3 \left\{ \frac{1}{|2k_1+\mathrm{Re}(\chi_0)|} + \frac{1}{|\mathrm{Re}(\chi_0)|} \right\}.$$ Then for

$-2k_1 < \mathrm{Re}(\chi_0) < 0$ and choosing $k_2 > 0$ and $k_3 > \frac{1}{2}(k-2)$, we have that there exists a positive constant D (independent of f) so that

$$(5\text{-}35) \qquad |\theta_f^\chi(g,t,\varphi)| \le D \max\left(\frac{1}{|\mathrm{Re}(\chi_0)|}, \frac{1}{|\mathrm{Re}(\chi_0)+2k_1|}\right) \nu_{P_{k_1,k_2,k_3}}(F_t f_g).$$

We then observe by the same reasoning that for the same conditions on χ, that there exists $D' > 0$ so that

$$(5\text{-}36) \qquad |R_f^\chi(g,t)| \le D' \max\left(\frac{1}{|\mathrm{Re}(\chi_0)|}, \frac{1}{|\mathrm{Re}(\chi_0)+2k_1|}\right) \nu_{P_{k_1,k_2,k_3}}(F_t f_g).$$

Let $f \in C_c^\infty(\mathbb{R}^k)$. Then we observe that $\gamma_a(\text{support }(f) \cap T_a) \subseteq \{(r,Y,t) \mid |r| \ge (R(f))^{-1}, \|R_a(\text{supp }(f))\| \le R(f),\ |t| \le (R(f))^2\}$. Then we have for $\mathrm{Re}(\chi_0) < 0$

$$(5\text{-}37) \qquad |\theta_f^\chi(g,t,\varphi)| \le \left\{ \int_{\mathrm{supp}(f_g)} \left| f_g\left(A_a(r)\, N_a(Y)\left(\tilde{v}_a + \frac{s}{2} v_a\right)\right)\right| |r|^{\mathrm{Re}(\chi_0)} \right.$$

$$\left. |\varphi(st)| \frac{dr}{|r|}\, dY\, ds \right\} \le L(f,g,t) \max_{x \in \mathbb{R}^k} |f_g| \left\{ \int_{|r| \ge (R(f_g))^{-1}} |r|^{\mathrm{Re}(\chi_0)-1}\, dr \right\}$$

$$\cdot \left\{ \int_{\|Y\| \le R(f_g)} dY \right\} \cdot \left\{ \int_{|s| \le (R(f_g))^2} ds \right\}.$$

Let $\|f\|_\infty = \max_{x \in \mathbb{R}^k} |f|$ for $f \in S(\mathbb{R}^k)$. Then for $f \in C_c^\infty(\mathbb{R}^k)$ and $\mathrm{Re}(\chi_0) < 0$, there exists $A > 0$ so that

$$(5\text{-}38) \qquad |\theta_f^{\chi}(g,t,\varphi)| \leq A \frac{1}{|\mathrm{Re}(\chi_0)|} \{R(f_g)\}^{k - \mathrm{Re}(\chi_0)} .$$

However using the polar decomposition $g = k_1 a k_2$ above, it is easy to see that since $k_i \in K$ preserves Euclidean distance, we have $R(f_g) \leq A(g) R(f)$.

Q. E. D.

Corollary 1 to Proposition 5.6. *For* $\mathrm{Re}(\chi_0) < 0$ *the maps* $S(\mathbb{R}^k) \longrightarrow \mathbb{C}$ *given by* $f \longrightarrow \theta_f^{\chi}(e_{0(Q)}, t, \varphi)$ *and* $f \longrightarrow R_f^{\chi}(e_{0(Q)}, t)$ *are continuous linear functions of* $S(\mathbb{R}^k)$ *(hence tempered distributions).*

Proof. We use the estimates (5-35) and (5-36) above and then recall that the representation $\pi_{\mathfrak{m}}$ of $G_2(Q)$ on $S(\mathbb{R}^k)$ is continuous.

Q.E.D.

At this point we make a note on the use of notation. The addition of two quasicharacters $\lambda_1 = ((\lambda_1)_0, \varepsilon)$ and $\lambda_2 = ((\lambda_2)_0, \varepsilon')$ is given by $\lambda_1 + \lambda_2 = ((\lambda_1)_0 + (\lambda_2)_0, \varepsilon\varepsilon')$; moreover λ_1^{-1}, the inverse of λ_1, will sometimes be denoted by $-\lambda_1 = (-(\lambda_1)_0, \varepsilon)$.

For $\chi = (\chi_0, \varepsilon)$ a quasicharacter and fixed $Y_0 \neq 0$, we define the function

$$(5\text{-}39) \qquad [Q(X,Y_0)]^{\chi} = \begin{cases} 0 & \text{if } Q(X,Y_0) = 0 \\ e^{\chi_0 \log|Q(X,Y_0)|} (\mathrm{sgn}(Q(X,Y_0)))^{\varepsilon} & \text{if } Q(X,Y_0) \neq 0 \end{cases}$$

Then $[Q(X,Y_0)]^{\chi}$ is a C^{∞} differentiable function for all X so that $Q(X,Y_0) \neq 0$, and is C^s for all X if $\mathrm{Re}(\chi_0) > s$ (i.e. $x \longrightarrow |x|^{\lambda}$ is C^s at $x = 0$ if $\mathrm{Re}(\lambda) > s$). If $t \neq 0$, then $[Q(X,tY_0)]^{\chi} = \chi(t) [Q(X,Y_0)]^{\chi}$ for all $X \in \mathbb{R}^k$ (homogeneity property). Moreover if $\mathrm{Re}(\chi_0) > 2$, we observe that $\partial(Q) [Q(X,Y_0)]^{\chi} = Q(Y_0,Y_0) [Q(X,Y_0)]^{\chi-(2,1)}$.

Corollary 2 to Proposition 5.6. *If* $\mathrm{Re}(\chi_0) < 0$ *and* $f \in S(\mathbb{R}^k)$, *we have*

$$(5\text{-}40) \qquad \theta_f^{\chi}(g,t,\varphi) = \int_{\mathbb{R}^k} f(X) [Q(X, gv_a)]^{-\chi} \varphi(t\, Q(X,X))\, dX ,$$

where the integral in (5-40) is absolutely convergent for $\mathrm{Re}(\chi_0) < 1$, and

(5-41) $$R_f^{\chi}(g,t) = \int_{\Gamma_t - \{0\}} f(X)\,[Q(X, gv_a)]^{-\chi}\, d\mu_t(X),$$

where the integral in (5-41) is absolutely convergent for $\mathrm{Re}\,(\chi_0) < 0$. Moreover the maps defined by $f \longrightarrow \theta_f^{\chi}(g,t,\varphi)$ by (5-40) and $f \longrightarrow R_f^{\chi}(g,s)$ by (5-41) are continuous linear functions on $S(\mathbb{R}^k)$ for all χ so that $\mathrm{Re}(\chi_0) < 1$ and $\mathrm{Re}(\chi_0) < 0$, respectively.

Proof. To apply Lemma 5.5 we must show that $\psi_g(X) = f_g(X)\,[Q(X,v_a)]^{-\chi}\varphi(t\,Q(X,X)) \in L_1(\mathbb{R}^k, dX)$. Then let $P_k(X) = (1 + \tilde{x}_a^{\,2})^k$ and $T_k(X) = (1 + x_a^{\,2} + \|Y\|^2)^k$, and consider the seminorm $\nu_{T_k P_k}(f) = \sup_{x \in \mathbb{R}^k} |T_k\, P_k\, f|$. Then for $f \in S(\mathbb{R}^k)$

(5-42) $$\int_{\mathbb{R}^k} |\psi_g(X)|\, dX \leq \nu_{T_k P_k}(F_t\, f_g) \left\{ \int_{\mathbb{R}} |\tilde{x}_a|^{-\mathrm{Re}(\chi_0)} (1 + \tilde{x}_a^{\,2})^{-k} d\tilde{x}_a \right\}$$

$$\cdot \left\{ \int_{\mathbb{R}^{k-1}} |T_k(X)|^{-1}\, dx_a\, dY \right\}.$$

However by choosing k sufficiently large, the product above is finite for $(1 - 2k) < \mathrm{Re}(\chi_0) < 1$. To deduce (5-40), we then apply Lemma 5.5, make the linear change of variable $X \longrightarrow g\,X$, and observe that $\det(g) = \pm 1$.

Similarly we recall that $T_a \cap (\Gamma_s - \{0\})$ is an open dense set in $\Gamma_s - \{0\}$ for all s; it then follows that on $T_a \cap (\Gamma_s - \{0\})$, $\omega_t = \frac{1}{\frac{\partial Q}{\partial x_a}}\, d\tilde{x}_a \wedge dY = \frac{1}{r}\, dr \wedge dY$. Hence we have for $\zeta_g(X) = f_g(X)[Q(X, v_a)]^{-\chi}$ that

(5-43) $$\int_{T_a \cap (\Gamma_t - \{0\})} |\zeta_g(X)|\, d\mu_t(X) \leq \int |f_g(\beta_a(r,Y,t))|\, |r|^{\mathrm{Re}(\chi_0) - 1}\, dr\, dY \leq$$

$$\nu_{S_{k_1} P_{k_1}}(f_g) \left\{ \int_{-\infty}^{+\infty} |r|^{\mathrm{Re}(\chi_0) - 1} (1 + r^{-2})^{-k_1} dr \right\} \cdot \left\{ \int_{\mathbb{R}^{k-2}} (1 + \|Y\|^2)^{-k_1} dY \right\},$$

where $S_{k_1} = (1 + \|Y\|^2)^{k_1}$ and $\nu_{S_{k_1} P_{k_1}}$, the similar seminorm on $S(\mathbb{R}^k)$ chosen as above. Then by choosing k_1 sufficiently large, we have convergence of the above integrals for $-2k_1 < \mathrm{Re}(\chi_0) < 0$.

Q. E. D.

Remark 5.5. The proof of Corollary 2 to Proposition 5.6 shows easily that for $(1 - 2k_1) < \mathrm{Re}(\chi_0) < 1$, the expression $|f_g(xv_a + r\tilde{v}_a + Z)\chi(r)\,\varphi(t\,(2xr + Q(Z,Z)))|$ can be bounded by a function $\eta(r,x,Y)$ independent of χ so that $\eta \in L^1(\mathbb{R}^k)$. *Thus for* $\mathrm{Re}(\chi_0) < 0$, *the map given (for fixed* g *and* t) *by* (see (5-40))

$$\chi_0 \longrightarrow \theta_f^{\chi}(g,t,\varphi) \tag{5-44}$$

is analytic. Similarly we see that for $\mathrm{Re}(\chi_0) < 0$ *and fixed* g *and* t, *the map given by* (see (5-41))

$$\chi_0 \longrightarrow R_f^{\chi}(g,t) \tag{5-45}$$

is analytic. Moreover by the Fubini Theorem, we have valid for $\mathrm{Re}(\chi_0) < 0$

$$\theta_f^{\chi}(g,t,\varphi) = \int_{-\infty}^{+\infty} \chi^{-1}(r)\, h_{f_g}(r,\varphi)\, dr \,, \tag{5-46}$$

where $h_{f_g}(r,\varphi) = \int_{\mathbb{R}^{k-1}} f_g(xv_a + r\tilde{v}_a + Y)\,\varphi(t\,(2xr + Q(Y,Y)))\,dx\,dY$,

and for $\mathrm{Re}(\chi_0) < 0$ *(using* (5-30)),

$$\theta_f^{\chi}(g,t,\varphi) = \int_{-\infty}^{+\infty} \varphi(st)\, R_f^{\chi}(g,s)\, ds \,. \tag{5-47}$$

Corollary 3 to Proposition 5.6. *The map* $\chi_0 \longrightarrow \theta_f^{\chi}(g,t,\varphi)$ *given by* (5-44) *has an analytic continuation to a meromorphic function* $\chi_0 \longrightarrow \Xi_f^{\chi}(g,t,\varphi)$ *on* $\mathbb{C}$, *with possible simple poles in the set* $\{n \in \mathbb{Z} \mid n \geq 1\}$. *Moreover if* $\varepsilon = 1$, *the possible poles occur at* $\{2k+1,\ k \geq 0\}$ *with*

(5-48) $$\underset{\chi_0=2k+1}{\text{residue}}\ \Xi_f^{\chi}(g,t,\varphi) = \frac{1}{(2k)!}\left(\frac{d}{dr}\right)^{2k} h_{f_g}(0,\varphi) ,$$

and if $\varepsilon = -1$, *the possible poles occur at* $\{2k \mid k \geq 1\}$ *with*

(5-49) $$\underset{\chi_0 = 2k}{\text{residue}}\ \Xi_f^{\chi}(g,t,\varphi) = \frac{1}{(2k-1)!}\left(\frac{d}{dr}\right)^{2k-1} h_{f_g}(0,\varphi) .$$

Proof. By hypothesis on φ and by a standard argument, we see that $r \longmapsto h_{f_g}(r,\varphi)$ is a Schwartz function in r . Hence (5-46) is the Tate Mellin integral, which has an analytic continuation in χ_0 . By the usual argument involving Taylor expansion of h_{f_g} in r , we derive the residues.

Q. E. D.

Remark 5.6. From Remark 5.5 and Corollary 3 to Proposition 5.6, we deduce that

(5-50) $$\Xi_f^{\chi}(g,t,\varphi) = \int_{\mathbb{R}^k} \chi^{-1}(r)\, h_{f_g}(r,\varphi)\, dr$$

is valid for all χ_0 so that $\text{Re}(\chi_0) < 1$.

We define for $f \in S(\mathbb{R}^k)$ the integral, with $\xi \in \mathbb{R}^k$,

(5-51) $$\psi_f(\xi,t,s,\varphi) = \int_{\mathbb{R}^K} f(X)\, e^{-s\, Q(X,\xi)^2}\, \varphi(t\, Q(X,X))\, dX .$$

It is clear that the integral is absolutely convergent.

We then recall the formula:

(5-52) $$|a|^{-s} = \frac{1}{\Gamma(s)} \int_0^{\infty} e^{-ax}\, x^{s-1}\, dx \quad \text{for} \quad \text{Re}(s) > 0 .$$

Proposition 5.7. *We have the formulae*:

(A) *For* $1 > \text{Re}(\chi_0) > 0$ *and* $\varepsilon = 1$

(5-53) $$\Xi_f^{\chi}(g,t,\varphi) = \frac{1}{\Gamma\left[\frac{\chi_0}{2}\right]} \int_0^{\infty} s^{\frac{\chi_0}{2}-1}\, \psi_f(gv_a, t, s,\varphi)\, ds .$$

(B) *For* $0 > \text{Re}(\chi_0) > -1$ *and* $\varepsilon = -1$ *and* $f_1 = Q(X, gv_a)\, f$

(5-54) $$\Xi_f^{\chi}(g,t,\varphi) = \frac{1}{\Gamma\left[\frac{1+\chi_0}{2}\right]} \int_0^{\infty} s^{\frac{\chi_0}{2}-\frac{1}{2}} \psi_{f_1}(gv_a, t, s, \varphi)\, ds .$$

Proof. We let $u(X,\xi) = Q(X,\xi)^2$. Then

(5-55) $$[Q(X,\xi)]^{-\chi} = \begin{cases} (u(X,\xi))^{-\chi_0/2} & \text{if } \varepsilon = 1 \\ Q(X,\xi)\ (u(X,\xi))^{-(1+\chi_0)/2} & \text{if } \varepsilon = -1 \end{cases} .$$

Then let $\xi \in \Gamma_0 - \{0\}$ and choose $\eta \in \Gamma_0 - \{0\}$ so that $<\xi,\eta>$ is a hyperbolic plane. Then consider the seminorm on $S(\mathbb{R}^k)$ defined by $\nu_k(f) = \sup|(1+\eta^2)^k(1+\|Z\|^2)^k f(\eta,Z)|$, with $\eta \oplus (Z) = \mathbb{R}^k$. Then we have

(5-56) $$\int_{\mathbb{R}^k \times \mathbb{R}_+^*} |f(X)\, F_t(X)|\, e^{-s\, Q(X,\xi)^2} |s|^{Re(\lambda)}\, ds\, dX =$$

$$\int_{\mathbb{R}_+^* \times \mathbb{R} \times \mathbb{R}^{k-2}} |f(\eta,Z)\, F_t(\eta,Z)|\, e^{-s\eta^2} |s|^{Re(\lambda)}\, ds\, d\eta\, dZ \leq$$

$$\nu_k(F_t \cdot f) \int_{\mathbb{R}_+^* \times \mathbb{R} \times \mathbb{R}^{k-2}} (1+\|Z\|^2)^{-k} (1+\eta^2)^{-k} s^{Re(\lambda)} e^{-s\eta^2}\, ds\, d\eta\, dZ .$$

We then observe that the integral

(5-57) $$\int_{\mathbb{R}_+^* \times \mathbb{R}} (1+\eta^2)^{-k}\, \eta^{-2Re(\lambda)-2} |u|^{Re(\lambda)}\, e^{-u}\, du\, d\eta$$

is absolutely convergent for $-\frac{1}{2} > Re(\lambda) > -1$ and k sufficiently large. Thus letting $u = s\,\eta^2$, we get by change of variables in (5-56) that

$$\int_{\mathbb{R}_+^* \times \mathbb{R}} (1+\eta^2)^{-k} |s|^{Re(\lambda)} e^{-s\eta^2}\, ds\, d\eta$$ is absolutely convergent. Then taking $(u(X,\xi))^{-\chi_0/2}$ above and assuming that $\lambda = \frac{\chi_0}{2} - 1$, we see that for $1 > Re(\chi_0) > 0$, the substitution in (5-40) for $[u(X,gv_a)]^{-\chi_0/2}$ can be made,

and the resulting two integrals permuted.

Then for (B) we repeat the same argument and assume here that $f_1(X) = Q(X,\xi)\, f(X)$.

Q. E. D.

We recall at this point the Tate-Mellin integral defined for $h \in S(\mathbb{R})$ and a quasicharacter χ by

$$(5\text{-}58) \qquad I_\chi(h) = \int_{\mathbb{R}} h(x)\, \chi(x) \frac{dx}{|x|} \quad \text{defined for } \operatorname{Re}(\chi_0) > 0 .$$

Then for $0 < \chi_0 < 1$, we have the functional equation:

$$(5\text{-}59) \qquad I_\chi(h) = \gamma(\chi)\, I_{\chi^{-1}+(1,1)}(\hat{h}) ,$$

where $\gamma(\chi) = \pi^{-\chi_0 + \frac{1}{2}} \dfrac{\Gamma(\chi_0/2)}{\Gamma((1-\chi_0)/2)}$ if $\varepsilon(\chi) = 1$, $\gamma(\chi) = -\sqrt{-1}\, \pi^{-\chi_0+\frac{1}{2}} \left[\dfrac{\Gamma((1+\chi_0)/2)}{\Gamma\left[1 - \frac{\chi_0}{2}\right]}\right]$ if $\varepsilon(\chi) \neq 1$.

<u>Proposition 5.8</u>. <u>If</u> $f \in S(\mathbb{R}^k)$ <u>and</u> $\operatorname{Re}(\chi_0) > 0$, <u>then</u>

$$(5\text{-}60) \qquad \Xi_f^\chi(g,t,\varphi) = \gamma(\chi^{-1}+(1,1)) \int_{-\infty}^{+\infty} \chi(r)\, \tilde{h}_f(-r\, M_Q(g\, v_a),\, t) \frac{dr}{|r|} ,$$

<u>where</u> $\tilde{h}_f(Z,t) = \int_{\mathbb{R}^k} f(X)\, \varphi(t\, Q(X,X))\, e^{2\pi i [X,Z]}\, dX$ <u>(the Fourier transform of</u> $f\, \varphi(t\, Q(X,X))$) <u>and</u> M_Q , <u>the symmetric matrix associated to</u> Q <u>so that</u> $Q(X,Y) = [X, M_Q(Y)]$ <u>for all</u> $X, Y, \in \mathbb{R}^k$. <u>Then the map defined by</u> $f \longrightarrow \dfrac{1}{\Gamma(1-\chi_0)} \Xi_f^\chi(g,t,\varphi)$, <u>with</u> $\Xi_f^\chi(g,t,\varphi)$ <u>given by</u> (5-60) <u>is a continuous linear functional on</u> $S(\mathbb{R}^k)$ <u>for all</u> χ <u>so that</u> $\operatorname{Re}(\chi_0) > 0$.

<u>Proof</u>. From (5-50) we know that for $\operatorname{Re}(\chi_0) < 1$, $\Xi_f^\chi(g,t,\varphi) = I_{\chi^{-1}+(1,1)}(h_{f_g}(\ ,\varphi))$, and for $0 < \chi_0 < 1$, we have by (5-59) that $\Xi_f(g,t,\varphi) = \gamma(\chi^{-1}+(1,1))\, I_\chi(\hat{h}_{f_g}(\ ,\varphi))$, where

$$(5\text{-}61)\qquad \hat{h}_{f_g}(u,\varphi) = \int_{-\infty}^{+\infty} e^{2\pi i u r}\, h_{f_g}(r,\varphi)\,dr = \left\{ \int_{\mathbb{R}\times\mathbb{R}\times\mathbb{R}^{k-2}} f_g(xv_a + rv_a + Y) \right.$$

$$\left. \varphi\,(t(2xr + Q(Y,Y)))\, e^{2\pi i u r}\, dx\, dr\, dY \right\} =$$

$$\int_{\mathbb{R}^k} f_g(X)\, \varphi\,(t\, Q(X,X))\, e^{2\pi i u x Q(X,v_a)}\, dX$$

(order of integration can be changed since $f \in S(\mathbb{R}^k)$).

Then we observe that $\tilde{h}_f(\ ,t)$ is a Schwartz function on $\mathbb{R}^k$, and hence $\int_{-\infty}^{+\infty} \chi(r)\, \tilde{h}_f(-r\, M_Q(g\, v_a),t)\, \frac{dr}{|r|}$ is an analytic function of χ_0 for $\mathrm{Re}(\chi_0) > 0$. Moreover we then observe directly that $M_Q(v_a) = \tilde{v}_a$. Then using an argument similar to that in the proof of Corollary 2 of Proposition 5.6 applied to $\tilde{h}_f(\ ,t)$, we see that

$$(5\text{-}62)\qquad \int_{-\infty}^{+\infty} |r|^{\mathrm{Re}(\chi_0)-1}\, |\tilde{h}_{f_g}(-r\tilde{v}_a, t)|\, dr \le$$

$$\nu_{P_k}(\tilde{h}_{f_g}(\ ,t)) \int_{-\infty}^{+\infty} (1 + r^2)^{-k}\, |r|^{\mathrm{Re}(\chi_0)-1}\, dr .$$

But since $\tilde{h}_f(\ ,t)$ is the Fourier transform of $f\cdot F_t$, we have that there exists a continuous seminorm ν' on $S(\mathbb{R}^k)$ so that $\nu_{P_k}(h_{f_g}(\ ,t)) \le \nu'(f_g\, F_t)$.

Finally we observe from the definition of $\gamma(\chi^{-1} + |1|)$ above, that the possible simple poles of $\gamma(\chi^{-1} + |1|)$ occur at $\chi_0 = 1, 2, 3,\ldots.$

Q. E. D.

<u>Remark 5.7.</u> It is easy to see that $M_Q \in O(Q)$, and with the notation of §2, we see by an easy exercise that $M_Q = \prod_{s=a}^{k-1} d(s)$.

<u>Remark 5.8.</u> In terms of the metaplectic representation: $f \in S(\mathbb{R}^k)$ and $\mathrm{Re}(\chi_0) > 0$,

(5-63) $$\Xi_f^{\chi}(g,t,\varphi) = (-1)^k \frac{\gamma(\chi^{-1} + (1,1))}{c_k \begin{bmatrix} 0 & -1 \\ 1 & 0 \end{bmatrix}} \left\{ \int_{-\infty}^{+\infty} \pi_{\mathfrak{m}}((M_Q\ g)^{-1}) \right.$$

$$\left. \pi_{\mathfrak{m}}(\left(\begin{bmatrix} r & 0 \\ 0 & r^{-1} \end{bmatrix}, 1\right)) \pi_{\mathfrak{m}}(w_0)(f\cdot F_t)(v_a)\chi(r)|r|^{-\frac{k}{2}}(f_k(r))^{-1} \frac{dr}{|r|} \right\}.$$

And if $\varphi(x) = \tau(x) = e^{\pi i x}$, we have for $f \in S(\mathbb{R}^k)$ and $\mathrm{Re}(\chi_0) > 0$,

(5-64) $$\Xi_f^{\chi}(g,t,\tau) = (-1)^k \frac{\gamma(\chi^{-1} + (1,1))}{c_k \begin{bmatrix} 0 & -1 \\ 1 & 0 \end{bmatrix}} \left\{ \int_{-\infty}^{+\infty} \pi_{\mathfrak{m}}((M_Q\ g)^{-1})\pi_{\mathfrak{m}}(w_o) \right.$$

$$\left. \pi_{\mathfrak{m}}(\left(\begin{bmatrix} r^{-1} & 0 \\ 0 & r \end{bmatrix}, 1\right)) \pi_{\mathfrak{m}}(n(t))(f)(v_a)\chi(r)|r|^{-\frac{k}{2}} (f_k(r))^{-1} \frac{dr}{|r|} \right\}.$$

We let $f^*(X) = \pi_{\mathfrak{m}}(w_o^{-1})\, f(X)$. Then we obtain the functional equation of the projection map.

We let $\Theta_f^{\chi}(g,t,\tau) = \frac{1}{\Gamma(1-\chi_o)} \Xi_{f^*}^{\chi}(g,-t,\tau)$ for all $\chi_o \in \mathbb{C}$.

<u>Theorem 5.9.</u> <u>If</u> $f \in S(\mathbb{R}^k)$ <u>and</u> $\chi_o \in \mathbb{C}$ <u>and</u> $t \neq 0$, <u>then</u>

(5-65) $$\Theta_{f^*}^{\chi}(g,t,\tau) = \delta_k \psi_k(t)\ \chi(t)\ |t|^{-k/2}\ \Theta_f^{\chi}(g,-\tfrac{1}{t},\tau)$$

<u>where</u> $\delta_k = (-1)^k$ <u>and</u>

(5-66) $$\psi_k(t) = \begin{cases} \left(\frac{t}{-1}\right)_H^{\delta(Q)} & \text{, if } k \text{ is even} \\ \begin{cases} 1 & \text{, if } t > 0 \\ e^{\pi i\, \mathrm{sgn}(Q)/2} & \text{, if } t < 0 \end{cases} & \text{, if } k \text{ is odd.} \end{cases}$$

<u>Proof.</u> Using the Bruhat decomposition (4-4) of $\tilde{S\ell}_2$, we deduce that

$$(5\text{-}67)\qquad w_o\left(\begin{bmatrix} r^{-1} & 0 \\ 0 & r \end{bmatrix}, 1\right) n(t)\, w_o^{-1} = \left(\begin{bmatrix} r & 0 \\ -\frac{t}{r} & r^{-1} \end{bmatrix}, \left(\frac{r}{-t}\right)_H\right) =$$

$$n\left(-\frac{r^2}{t}\right) w_o\left(\begin{bmatrix} -\frac{t}{r} & 0 \\ 0 & -\frac{r}{t} \end{bmatrix}, -\left(\frac{r}{-t}\right)_H\right) n\left(-\frac{1}{t}\right).$$

Then noting that $M_Q\, g\, v_a \in \Gamma_o - \{0\}$ and from (5-64), we have for $\mathrm{Re}(\chi_o) > 0$, $\chi_o \neq$ integer,

$$(5\text{-}68)\qquad \Xi^{\chi}_{f^*}(g,t,\tau) = \frac{\gamma(\chi^{-1} + (1,1))}{c_k\begin{bmatrix} 0 & -1 \\ 1 & 0 \end{bmatrix}} \left\{ \int_{-\infty}^{+\infty} \pi_{\mathfrak{m}}((M_Q\, g)^{-1}) \right.$$

$$\pi_{\mathfrak{m}}(w_o)\pi_{\mathfrak{m}}\left(\left(\begin{bmatrix} -\frac{t}{r} & 0 \\ 0 & -\frac{r}{t} \end{bmatrix}, 1\right)\right)\pi_{\mathfrak{m}}\left(n\left(-\frac{1}{t}\right)\right) (f)\ (v_a)\ \chi(r)\ |r|^{-k/2}$$

$$\left. (f_k(r))^{-1} \left\{\left(\frac{r}{-t}\right)_H\right\}^k \frac{dr}{|r|} \right\}.$$

We then let $r = -ut$ (the integrand as a function of r is integrable for $\mathrm{Re}(\chi_o) > 0$), and the integral of (5-68) is

$$(5\text{-}69)\qquad \chi(-t)\,|t|^{-\frac{k}{2}} \left\{ \int_{-\infty}^{+\infty} \pi_{\mathfrak{m}}((M_Q\, g)^{-1})\pi_{\mathfrak{m}}(w_o)\pi_{\mathfrak{m}}\left(\begin{bmatrix} u^{-1} & 0 \\ 0 & u \end{bmatrix}, 1\right) \right.$$

$$\left. \pi_{\mathfrak{m}}\left(n\left(-\frac{1}{t}\right)\right) (f)(v_a)\chi(u)\,|u|^{-\frac{k}{2}}(f_k(-ut))^{-1} \left\{\left(\frac{-ut}{-t}\right)_H\right\}^k \frac{du}{|u|} \right\}.$$

Then we deduce that $\{f_k(ut)\}^{-1}\left\{\left(\frac{ut}{t}\right)_H\right\}^k = \psi_k(t)\ \{f_k(u)\}^{-1}$, where ψ_k is as above. Then using the analyticity of Θ^{χ}_f in χ_o, we get the desired result.

Q. E. D.

We recall from Lemma 5.1 that the representation $\sigma_{\mathfrak{m}}$ of $\tilde{G}_2(Q)$ is differentiable on the space $S(\mathbb{R}^k)$. Then since $f \longrightarrow \Theta^{\chi}_f(g,t,\tau)$ is a tempered distribution for all quasicharacters χ $(f \in S(\mathbb{R}^k))$, it follows that $\Theta^{\chi}_f(\ ,t,\tau) \in C^{\infty}(O(Q))$ and $\Theta^{\chi}_f(g,\ ,\tau) \in C^{\infty}(\mathbb{R})$.

We note at this point certain generalities from [3] and §3. Let σ_1

(resp. σ_2) be a differentiable representation of a closed subgroup A_1 (resp. A_2) of a Lie group G_1 (resp. G_2) in a Frechet space E_1 (resp. E_2). Then the representation $\pi_{\sigma_1} \hat{\otimes} \pi_{\sigma_2}$ of $G_1 \times G_2$ in $C^\infty_{\sigma_1}(G_1, E_1) \hat{\otimes} C^\infty_{\sigma_2}(G_2, E_2)$ (given by $\pi_{\sigma_1} \hat{\otimes} \pi_{\sigma_2} = \pi_{\sigma_1}(g_1) \otimes \pi_{\sigma_2}(g_2)$) is equivalent to the differentiable induced representation $\pi_{\sigma_1 \hat{\otimes} \sigma_2}$ of $G_1 \times G_2$ on $C_{\sigma_1 \hat{\otimes} \sigma_2}(G_1 \times G_2, E_1 \hat{\otimes} E_2)$.

We recall the representation $\pi_{\sigma_1 \otimes \lambda}$ of $S\tilde{\ell}_2$ on $C^\infty_{\sigma_1 \otimes \lambda}(S\tilde{\ell}_2)$ discussed in §4. In the remaining discussion, we will also denote by $\pi_{\sigma_1 \otimes \lambda}$ the representation of $S\tilde{\ell}_2$ on the space $C^\infty_{\lambda,s}(\mathbb{R})$ (by Lemma 4-10 and Remark 4-7) with the topology of $C^\infty_{\sigma_1 \otimes \lambda}(S\tilde{\ell}_2)$ carried over to $C^\infty_{\lambda,s}(\mathbb{R})$.

We define ($\tilde{\varepsilon} = 0$ if $\varepsilon = 1$ and $\tilde{\varepsilon} = 1$ if $\varepsilon = -1$)

$$\tilde{\phi}(k,Q,\varepsilon) = \begin{cases} 0 \text{ if } k \text{ even and } \delta(Q) + \tilde{\varepsilon} \equiv 0 \mod 2 \\ 2 \text{ if } k \text{ even and } \delta(Q) + \tilde{\varepsilon} \equiv 1 \mod 2 \\ 1 \text{ if } k \text{ odd and } \operatorname{sgn}(Q) + 2\tilde{\varepsilon} \equiv 1 \mod 4 \\ 3 \text{ if } k \text{ odd and } \operatorname{sgn}(Q) + 2\tilde{\varepsilon} \equiv 3 \mod 4 . \end{cases} \tag{5-70}$$

Corollary 1 to Theorem 5.9. *For all quasicharacters χ and $f \in S(\mathbb{R}^k)$, $\theta^\chi_f(\ ,t,\tau) \in C^\infty_{\chi+(\frac{2-k}{2},1)}(O(Q))$ and $\theta^\chi_f(g,\ ,\tau) \in C^\infty_{(\frac{k}{2} - \chi_o - 1, \tilde{\phi}(k,Q,\varepsilon(\chi)))}(\mathbb{R})$. Moreover the map $f \longrightarrow \theta_f(\ ,\ ,\tau)$ is a continuous nonzero $\tilde{G}_2(Q)$ intertwining map between the representation $\sigma_{\mathfrak{m}}$ of $\tilde{G}_2(Q)$ on $S(\mathbb{R}^k)$ and the representation V_χ of $\tilde{G}_2(Q)$ on $C^\infty_{\chi+(\frac{2-k}{2},1)}(O(Q)) \hat{\otimes} C^\infty_{(\frac{k}{2} - \chi_o - 1, \tilde{\phi}(k,Q,\varepsilon(\chi)))}(\mathbb{R})$, where $V_\chi(g,G) = \pi_{\chi+(\frac{2-k}{2},1)}(g) \hat{\otimes} \pi_{\sigma_1 \otimes (\frac{k}{2} - \chi_o - 1)}(G)$, $g \in O(Q)$, $G \in S\tilde{\ell}_2$, and $\sigma_1(E) = e^{\frac{\pi i}{2}\tilde{\phi}(k,Q,\varepsilon(\chi))}$. Finally if $\chi_o \notin \mathbb{Z}$ and $\chi_o \neq \frac{k}{2} - s$, s integer ≥ 1, then the map $f \longrightarrow \theta^\chi_f(\ ,\ ,\tau)$ is the unique $\tilde{G}_2(Q)$ intertwining map (up to scalar multiple) with the property above.*

Proof. By the functional equation in Theorem 5.9, we deduce that $\theta^\chi_f(g,\ ,\tau) \in C^\infty_{(\frac{k}{2} - \chi_o - 1, \tilde{\phi}(k,Q,\varepsilon(\chi)))}(\mathbb{R})$ (see (4-47)). Moreover by definition of θ^χ_f in (5-30), we have $\theta^\chi_f(g\, m\, A_a(r)\, N_a(Y),\ t,\ \tau) = \chi^{-1}(r)\, \theta^\chi_f(g,t,\tau)$ for

all χ. It is clear by the various integral representations of $\Theta_f^{\chi}(\ ,\ ,\tau)$ above, that for a fixed χ, there exists $f \in S(\mathbb{R}^k)$ so that $\Theta_f^{\chi}(\ ,\ ,\tau) \neq 0$. At this point we recall the functional $Q_{\lambda,s}$ on the space $C^{\infty}_{\lambda,s}(\mathbb{R})$ defined in the paragraph preceding Lemma 4.10. It is a simple exercise to observe that $Q_{\lambda,s}$ is continuous on $C^{\infty}_{\lambda,s}(\mathbb{R})$ and $Q_{\lambda,s}(\pi_{\sigma}(n(x))\, f) = Q_{\lambda,s}(f)$ for all $x \in \mathbb{R}$, $f \in C^{\infty}_{\lambda,s}(\mathbb{R})$. Thus it is easy to see that the distribution $(f \longrightarrow \frac{1}{\Gamma(1-\chi_o)} \Xi_f^{\chi}(e_{O(Q)},0,\tau))$ belongs to $\Pi_{(\frac{1}{2}-k_o,\varepsilon),\chi^{-1},\phi(k,Q,\varepsilon(\chi))}$.
And from Theorem 5.4 we deduce the uniqueness property of Θ_f^{χ}.

Q. E. D.

Corollary 2 to Theorem 5.9. *For any quasicharacter* $\chi = (\chi_o,\varepsilon)$, $g \in O(Q)$, *and* $G \in \widetilde{s\ell}_2$, *we define* $H_f(\chi_o,\varepsilon,G,g) = \Theta^{\chi_1}_{\pi_{\mathfrak{m}}(G^{-1})f}(g,0,\tau)$ *for* $f \in S(\mathbb{R}^k)$, *where* $\chi_1 = (-\chi_o + \frac{k-2}{2}, \varepsilon(\chi))$. *Then* $H_f(\ ,\ ,\ ,g) \in \mathcal{A}^1_{\tilde{\phi}(k,Q,\varepsilon)}$ *(see* (4-59)).

Proof. From (5-64), $H_f(-\chi_o + \frac{k}{2} - 1,\varepsilon,G,e_{O(Q)})$ can be expressed for $\mathrm{Re}(\chi_o) > 0$ as the linear combination $\alpha_1\Omega_f^o(\chi_o,G,+) + \beta_1\Omega_f^o(\chi_o,G,-)$, where $\alpha_1, \beta_1 \in \mathbb{C}$, and
$$\Omega_f^o(\chi_o,G,\pm) = \ell_1(\chi)\left\{\int_{\mathbb{R}_+} \pi_{\mathfrak{m}}(G^{-1})\,(f)\,(\pm r\, v_a)\,|r|^{\chi_o}\frac{dr}{|r|}\right\}$$
with $\ell_1(\chi) = \dfrac{\gamma(\chi^{-1}+(1,1))}{\Gamma(1-\chi_o)}$. But then we note that the function

$$\Omega_f^k(\chi_o,G,\pm) = \frac{\ell_1(\chi)}{\chi_o\ldots(\chi_o+k-1)}\left\{\int_0^{\infty}\left(\frac{\partial}{\partial x_a}\right)^k \pi_{\mathfrak{m}}(G^{-1})\,(f)\right.$$
$$\left.(\pm\, r\, v_a)\,|r|^{\chi_o+k}\frac{dr}{|r|}\right\} \tag{5-71}$$

is analytic for $\mathrm{Re}(\chi_o) > -k$ and satisfies $\Omega_f^{(k)}(\chi_o,G,\pm) = \pm\,\Omega_f^o(\chi_o,G,\pm)$ for $\mathrm{Re}(\chi_o) > 0$. Then we note that there exists a continuous seminorm ν_1 on $S(\mathbb{R}^k)$ so that $\left|\left(\frac{\partial}{\partial x_a}\right)^k \pi_{\mathfrak{m}}(G^{-1})\,(f)\,(\pm\, r\, v_a)\right| \le (1+r^2)^{-\ell}\nu_1(\pi_{\mathfrak{m}}(G^{-1})f)$ for any integer $\ell \ge 0$. Thus it follows that the map $(\lambda,G) \longrightarrow \Omega_f(\lambda,G,\pm)$ is continuous on $\{\lambda \in \mathbb{C} \mid \mathrm{Re}(\lambda) > -k\} \times \widetilde{s\ell}_2$. If $Z \in \mathcal{U}(s\ell_2)$, then

$Z * H_f(-\chi_o + \frac{k}{2} - 1, \varepsilon, G, e_{0(Q)}) = \Theta^{\chi}_{\pi_{\mathfrak{m}}(G^{-1})\pi_{\mathfrak{m}}(Z)f}(e_{0(Q)}, 0, \tau)$, and using the same arguments as above, we deduce that $(\chi_o, G) \longrightarrow Z * H_f(\chi_o, \varepsilon, G, e_{0(Q)})$ is continuous on $\mathbb{C} \times \widetilde{S\ell}_2$.

Q. E. D.

Then as in (4-62), we define for $f \in S(\mathbb{R}^k)$ and χ so that $\mathrm{Re}(\chi_o) < \frac{k}{2} - 1$ $(f^* = \pi_{\mathfrak{m}}(w_o^{-1})f)$

$$\psi^{\chi}_f(g) = \frac{1}{\Gamma(\frac{k}{2} - \chi_o - 1)} \int_{\mathbb{R}} \Theta^{\chi}_{f*}(g,t,\tau)\, dt . \tag{5-72}$$

Then from Corollary 2 of Theorem 5.9 and Proposition 4.13, we deduce that the function $\chi_o \longrightarrow \psi^{\chi}_f(g)$ $(\mathrm{Re}(\chi_o) < \frac{k}{2} - 1)$ has a continuation to an analytic function $\chi_o \longrightarrow \sigma_f(\chi, g)$ on all of $\chi_o \in \mathbb{C}$. Moreover using Corollary 1 of Proposition 4.13 again for $g = e_{0(Q)}$, the map $f \longrightarrow \sigma_f(\chi, e_{0(Q)})$ defines a continuous linear functional on $S(\mathbb{R}^k)$ for all $\chi_o \in \mathbb{C}$.

Q. E. D.

<u>Corollary 3 to Theorem 5.9</u>. <u>The distribution</u> $f \longrightarrow \sigma_f(\chi, e_{0(Q)}) \in \Pi_{(\chi_o - \frac{k}{2} + 2, \varepsilon), \chi^{-1}, \tilde{\phi}(k, Q, \varepsilon(\chi))}$. <u>Then for all</u> χ <u>so that</u> $\mathrm{Re}(\chi_o) < 0$ <u>and</u> $f \in S(\mathbb{R}^k)$

$$\sigma_f(\chi, g) = \psi^{\chi}_f(g) = \frac{2}{\Gamma(1 - \chi_o)\Gamma(\frac{k}{2} - \chi_o - 1)} R^{\chi}_{f*}(g, 0) . \tag{5-73}$$

<u>Proof</u>. We have

$$\psi^{\chi}_{\pi_{\mathfrak{m}}(a(r))f}(e_{0(Q)}) = \frac{1}{\Gamma(\frac{k}{2} - \chi_o - 1)} \int_{\mathbb{R}} \Theta^{\chi}_{\pi_{\mathfrak{m}}(a(r^{-1}))f*}(g,t,\tau)\, dt \tag{5-74}$$

$$= |r|^{\frac{k}{2} - \chi_o - 2} \psi^{\chi}_f(e_{0(Q)})$$

(use (4-48) and change of variable $u = r^2 t$) . Similarly $\psi^{\chi}_{\pi_{\mathfrak{m}}(E)f}(e_{0(Q)}) = \{\sigma_1(E)\}^{-1}\psi^{\chi}_f(e_{0(Q)})$ (since $E \in \mathrm{Center}\,(\widetilde{S\ell}_2)$) . Thus $\sigma_f(\chi, e_{0(Q)}) \in$

$\Pi_{(\chi_o - \frac{k}{2} + 2,\varepsilon),\ \chi^{-1},\tilde{\phi}(k,Q,\varepsilon(\chi))}$.

Then using the definition of R_f^{χ} in (5-31), we see that

$R^{\chi}_{\pi_{\mathfrak{m}}(w_o^{-1})\pi_{\mathfrak{m}}(E)\pi_{\mathfrak{m}}(a(r))\pi_{\mathfrak{m}}(\exp_{\tilde{s\ell}_2}(sF))f}(e_{0(Q)}, 0) =$

$(-1)^k f_k(-1)\chi(-1)|r|^{-\chi_o + \frac{k}{2} - 2} R^{\chi}_{\pi_{\mathfrak{m}}(w_o^{-1})f}(e_{0(Q)}, 0)$. However we have that $(-1)^k f_k(-1)\chi(-1) = e^{-\pi i\tilde{\phi}(k,Q,\varepsilon(\chi))/2}$. Thus for $\mathrm{Re}(\chi_o) < 0$, the distribution $(f \longrightarrow R^{\chi}_{\pi_{\mathfrak{m}}(w_o^{-1})f}(e_{0(Q)}, 0)) \in$

$\Pi_{(\chi_o - \frac{k}{2} + 2,\varepsilon),\ \chi^{-1},\phi(k,Q,\varepsilon))}$. Then by Theorem 5.4, it follows that for all χ_o so that $\mathrm{Re}(\chi_o) < 0$, there exists a function $c(\chi)$ so that $\sigma_f(\chi, e_{0(Q)}) = c(\chi)\, R^{\chi}_{f*}(e_{0(Q)}, 0)$ for all $f \in S(\mathbb{R}^k)$.

Then we let $g \in C_c^{\infty}(\mathbb{R}^k)$ so that support $(g) \subseteq$ Image $\beta_a = \{X \in \mathbb{R}^k |\ Q(X,v_a) \neq 0\}$, and so that on Image β_a, $g \circ \beta_a(r,Y,t) = \varphi_1(r)\varphi_2(Y)\varphi_3(t)$ with $\varphi_1 \in C^{\infty}(\mathbb{R}^*)$ so that support $(\varphi_1) \subseteq [a,b]$ with $a, b > 0$; $\varphi_2 \in C^{\infty}(\mathbb{R}^{k-2})$ so that support (φ_2) is compact; and $\varphi_3 \in C^{\infty}(\mathbb{R})$ so that support $(\varphi_3) \subseteq [-a', a']$ with $a' > 0$. Then by computation, using the definition of R_f^{χ} in (5-31), we have $R_g^{\chi}(e_{0(Q)},t) = \varphi_3(t)\ I_{\chi}(\varphi_1)\ M(\varphi_2)$ with $I_{\chi}(\varphi)$ given in (5-50) (analytic for all χ_o by support property of φ) , and $M(\varphi) = \int_{\mathbb{R}^{k-2}} \varphi(Y)\, dY$.

But noting that φ_3 has compact support and by using (5-47) and the Fourier inversion formula $\Theta_g^{\chi}(e_{0(Q)},t,\tau) = \hat{\varphi}_3(\frac{t}{2})\ I_{\chi}(\varphi_1)\ M(\varphi_2)$ for all χ_o so that $\mathrm{Re}(\chi_o) < 0$. Then we have $\psi^{\chi}_{\pi_{\mathfrak{m}}(w_o)\, g}(e_{0(Q)}) =$

$$\frac{1}{\Gamma(\frac{k}{2} - \chi_o - 1)} \int_{\mathbb{R}} \Theta_g^{\chi}(e_{0(Q)}, t,\tau)\, dt = \frac{2}{\Gamma(1 - \chi_o)\Gamma(\frac{k}{2} - \chi_o - 1)} (\varphi_3(0)$$

$I_{\chi}(\varphi_1)\ M(\varphi_2))$ for $\mathrm{Re}(\chi_o) < 0$. Then choose φ_1, φ_2, and φ_3 so that $I_{\chi}(\varphi_1) \neq 0$, $M(\varphi_2) \neq 0$, and $\varphi_3(0) \neq 0$ for a fixed χ_o so that $\mathrm{Re}(\chi_o) < 0$.

Thus $c(\chi) = \dfrac{2}{\Gamma(1-\chi_o)\Gamma(\frac{k}{2}-\chi_o-1)}$.

Q. E. D.

Remark 5.9. From Corollary 3 to Theorem 5.9, we see that the function

$f \longrightarrow \dfrac{2}{\Gamma(1-\chi_o)\Gamma(\frac{k}{2}-\chi_o-1)} R_f^{\chi}(g,0)$ (defined for $\mathrm{Re}(\chi_o) < 0$) has a continuation to an analytic function on all of $\mathbb{C}$ given by $\sigma_{\pi_{\mathfrak{m}}(w_o)f}(\chi,g)$.

Then we let for all $t \in \mathbb{R}$ and all $f \in S(\mathbb{R}^k)$

$$\beta_f(\chi,g,t) = \sigma_{\pi_{\mathfrak{m}}(n(-t))\ f}(\chi,g) \quad . \tag{5-75}$$

Corollary 4 to Theorem 5.9. For all quasicharacters χ and $f \in S(\mathbb{R}^k)$, $\beta_f(\chi,\ ,t) \in C^{\infty}_{\chi+(\frac{2-k}{2},1)}(O(Q))$ and $\beta_f(\chi,g,\ ) \in C^{\infty}_{(1+\chi_o-\frac{k}{2},\tilde{\phi}(k,Q,\varepsilon(\chi)))}(\mathbb{R})$. Moreover the map $f \longrightarrow \beta_f(\chi,\ ,\ )$ is a continuous $\tilde{G}_2(Q)$ intertwining map between the representation $\sigma_{\mathfrak{m}}$ of $G_2(Q)$ on $S(\mathbb{R}^k)$ and the representation W_χ of $\tilde{G}_2(Q)$ on $C^{\infty}_{\chi+(\frac{2-k}{2},1)}(O(Q)) \hat{\otimes} C^{\infty}_{(1+\chi_o-\frac{k}{2},\tilde{\phi}(k,Q,\varepsilon(\chi)))}(\mathbb{R})$, where $W_\chi(g,G) = \pi_{\chi+(\frac{2-k}{2},1)}(g) \hat{\otimes} \pi_{\sigma_1 \otimes (1+\chi_o-\frac{k}{2})}(G)$, $g \in O(Q)$, $G \in \tilde{S\ell}_2$, and $\sigma_1(E) = e^{\frac{\pi i}{2}\tilde{\phi}(k,Q,\varepsilon(\chi))}$. Finally if $\chi_o \neq \ell$ integer ≥ 1 and $\chi_o \neq \frac{k}{2}+j$, j integer ≥ -1 , then the map $f \longrightarrow \beta_f(\chi,\ ,\ )$ is the unique nonzero $\tilde{G}_2(Q)$ intertwining map (up to scalar multiple) with the property above.

Proof. By definition of ψ_f^{χ} and using Corollary 1 to Proposition 4.13, we have $\beta_f(\chi,g,\ ) \in C^{\infty}_{(1+\chi_o-\frac{k}{2},\tilde{\phi}(k,Q,\varepsilon(\chi)))}(\mathbb{R})$. By choosing $f_1 = g$ as in the proof of the last Corollary, we have $\sigma_f(\chi,e_{O(Q)}) =$

$\dfrac{2}{\Gamma(1-\chi_o)\Gamma(\frac{k}{2}-\chi_o-1)} I_\chi(\varphi_1)\ M(\varphi_2)\varphi_3(0)$. Then for a fixed $\chi_o \in \mathbb{C}$ satisfying the hypotheses in the Corollary above, it is possible to choose φ_1, φ_2, and φ_3 so that $\sigma_f(\chi, e_{O(Q)}) \neq 0$.

The uniqueness follows again by Theorem 5.4.

Q. E. D.

From (5-60) we recall that for $f \in S(\mathbb{R}^k)$ and $\mathrm{Re}(\chi_o) > 0$

$$(5\text{-}76) \qquad \Theta^{\chi}_{\pi_{\mathfrak{m}}(w_o)f}(N_a(Y)\,d(a),\,t,\,\tau) = \frac{\gamma((1,1)-\chi_o)}{\Gamma(1-\chi_o)} \left\{ \int_{\mathbb{R}} \tilde{h}_f(-r\,M_Q N_a(Y)\,\tilde{v}_a,t)\, \chi(r)\,\frac{dr}{|r|} \right\} .$$

Noting that $\tilde{h}_f(\ ,t) \in S(\mathbb{R}^k)$ and $\tilde{h}_f(-r\,M_Q\,N_a(Y)\,\tilde{v}_a,t) =$ $\tilde{h}_f(-M_Q\,A_a(r^{-1})\,N_a(rY)\,\tilde{v}_a,t)$, we then deduce using the same argument as in the proof of Corollary 2 to Proposition 5.6, that $\tilde{h}_f(-M_Q\,A_a(r^{-1})\,N_a(rY)\,\tilde{v}_a,t)\,\chi(r) \in L^1(\mathbb{R}^* \times \mathbb{R}^{k-2}, \frac{dr}{|r|} \times dY)$ for $\mathrm{Re}(\chi_o) > k-2$ (making change of variable $r \longrightarrow r^{-1}$ and $Z = rY$) . Then it follows that for χ so that $\mathrm{Re}(\chi_o) > k-2$, the integral

$$(5\text{-}77) \qquad \delta_f(\chi,g,t) = L(\chi) \int_{\mathbb{R}^{k-2}} \Theta^{\chi}_{\pi_{\mathfrak{m}}(w_o)f}(g\,N_a(Y)\,d(a),\,t,\,\tau)\,dY$$

is absolutely convergent where $L(\chi) = \dfrac{\Gamma(1-\chi_o)}{\gamma(\chi^{-1}+(1,1))}\,\dfrac{2}{\Gamma(\chi_o-k+3)}\,\dfrac{1}{\Gamma(\chi_o-\frac{k}{2}+1)}$.

Moreover from the argument of the proof of Corollary 2 to Proposition 5.6, we have that there exists a seminorm ν_1 on $S(\mathbb{R}^k)$ so that

$$(5\text{-}78) \qquad |\delta_f(\chi,g,t)| \le m_\chi\,\nu_1(f_g) ,$$

where m_χ is a positive constant depending only on χ $(\mathrm{Re}(\chi_o) > k-2)$. Thus the map $f \longrightarrow \delta_f(\chi,g,t)$ defines a continuous linear functional on $S(\mathbb{R}^k)$ for $\mathrm{Re}(\chi_o) > k-2$.

<u>Corollary 5 to Theorem 5.9.</u> "<u>Functional Equation of Intertwining Operator and its Analytic Continuation</u>" . <u>For</u> $f \in S(\mathbb{R}^k)$ <u>and</u> $\mathrm{Re}(\chi_o) > k-2$, <u>we have the functional equation</u>

$$(5\text{-}79) \qquad \delta_f(\chi,g,t) = (-1)^k\,\overline{\left\{ c_k \begin{bmatrix} 0 & 1 \\ -1 & 0 \end{bmatrix} \right\}}\, \beta_f(\chi^{-1} + (k-2,1),g,\,t)$$

Thus the function $\chi_o \longrightarrow \delta_f(\chi,g,t)$ *is analytic and has a continuation to an analytic function on all of* $\mathbb{C}$.

Proof. From above we have for $\mathrm{Re}(\chi_o) > k-2$

$$\text{(5-80)} \qquad \delta_f(\chi,g,0) = \frac{2}{\Gamma(\chi_o-k+3)\Gamma(\chi_o-\frac{k}{2}+1)} \left\{ \int_{\mathbb{R}_+^* \times \mathbb{R}^{k-2}} \hat{f}(M_Q\, g\, A_a(r)\, N_a(Y)\, \tilde{v}_a) \chi^{-1}(r)|r|^{k-2} \frac{dr}{|r|}\, dY \right\} = \frac{2}{\Gamma(\chi_o-k+3)\Gamma(\chi_o-\frac{k}{2}+1)} R_{\hat{f}}^{\chi^{-1}+(k-2,1)}(M_Q g,0) .$$

But we recall that $\pi_{\mathfrak{m}}(w_o^{-1})\, f(X) = (-1)^k\, c_k\begin{bmatrix} 0 & 1 \\ -1 & 0 \end{bmatrix} \hat{f}(M_Q(X))$ for all $X \in \mathbb{R}^k$.

And then from Corollary 3 to Theorem 5.9, we deduce that

$$\text{(5-81)} \qquad \frac{2}{\Gamma(\chi_o-k+3)\Gamma(\chi_o-\frac{k}{2}+1)} R_{\hat{f}}^{\chi^{-1}+(k-2,1)}(M_Q\, g,\, 0) = (-1)^k \overline{\left\{ c_k \begin{bmatrix} 0 & 1 \\ -1 & 0 \end{bmatrix} \right\}}\, \sigma_f(\chi^{-1}+(k-2,1),g)$$

for all χ so that $\mathrm{Re}(\chi_o) > k-2$.

Q. E. D.

Again from the proof of Corollary 2 to Proposition 5.6, we see that for all $t \in \mathbb{R}$ and χ so that $\mathrm{Re}(\chi_o) < 0$, it is possible to find a function $\eta(r,Y)$ independent of t so that $|\eta|(r,Y) \in L^1(\mathbb{R}^* \times \mathbb{R}^{k-2}, \frac{dr}{|r|} \times dY)$ and $|\eta| \geq |f \circ \beta_a||r|^{\mathrm{Re}(\chi_o)-1}$ for all $Y \in \mathbb{R}^{k-2}$ and $r \in \mathbb{R}^*$. However noting that $t \longrightarrow f \circ \beta_a(r,Y,t)$ is a C^∞ function, it follows that $t \longrightarrow R_f^\chi(g,t)$ is a C^∞ function for $\mathrm{Re}(\chi_o) < 0$. Moreover using the disintegration of measure argument of Remark 5.5, we have that $t \longrightarrow R_f^\chi(g,t)$ $\in L^1(\mathbb{R})$ $(\mathrm{Re}(\chi_o) < 0)$.

From Remark 4.8, we know that $t \longrightarrow \theta_f^\chi(g,t,\tau) \in L^1(\mathbb{R})$ for $\mathrm{Re}(\chi_o) < \frac{k}{2} - 1$.

Corollary 6 to Theorem 5.9. For $f \in S(\mathbb{R}^k)$ *and* $\mathrm{Re}(\chi_o) < 0$

$$(5\text{-}82) \qquad R^{\chi}_{f*}(g,\ 2t) = \frac{\Gamma(1-\chi_o)}{2} \int_{-\infty}^{+\infty} \Theta^{\chi}_f(g,s,\tau)\ e^{2\pi i s t}\ ds\ .$$

Then the function $\chi_o \longrightarrow R^{\chi}_f(g,t)$ has a continuation to a meromorphic function $\chi_o \longrightarrow \tilde{R}^{\chi}_f(g,t)$ with possible simple poles at $\chi_o = 1, 2, 3, \ldots$. Moreover for $g \in O(Q)$ and $G \in S\tilde{\ell}_2$, if we define $W_f(\chi,g,G,t) = \frac{1}{\Gamma(1-\chi_o)} \tilde{R}^{\chi}_{\pi_{\mathfrak{m}}((Gw_o)^{-1})f}(g,-2t)$, then for $t \neq 0$ and all quasicharacters χ, $f \longrightarrow W_f(\chi, \ , \ , t)$ is a continuous $\tilde{G}_2(Q)$ intertwining map between the representation $\sigma_{\mathfrak{m}}$ of $\tilde{G}_2(Q)$ on $S(\mathbb{R}^k)$ and the representation L_χ of $\tilde{G}_2(Q)$ on $C^{\infty}_{\chi+(\frac{2-k}{2},1)}(O(Q)) \hat{\otimes} C^{\infty}(t)\left((\chi_o^2 + \chi_o(2-k) + (\frac{k^2}{4} - k)),\ \frac{i}{2}\tilde{\Phi}(k,Q,\varepsilon(\chi))\right)$, where $L_\chi(g,G) = \pi_{\chi+(\frac{2-k}{2},1)}(g) \hat{\otimes} \pi_t(G)$ (see §4) .

Proof. From above we have that the functions $t \longrightarrow \Theta^{\chi}_f(g,t,\tau)$ and $t \longrightarrow R^{\chi}_f(g,t) \in C^{\infty}(\mathbb{R}) \cap L^1(\mathbb{R})$ for $\mathrm{Re}(\chi_o) < 0$. Then we can apply the Fourier inversion formula and (5-47) of Remark 5.6 with $\varphi(t) = e^{i\pi t}$.

Then recalling Proposition 4.13 and Corollary 2 to Theorem 5.9, we know that the function $\chi_o \longrightarrow \int_{-\infty}^{+\infty} \Theta_f(g,s,\tau)\ e^{-2i\pi st} ds$ has an analytic continuation to all of $\mathbb{C}$. Thus the function $\chi_o \longrightarrow \frac{1}{\Gamma(1-\chi_o)} R^{\chi}_{f*}(g,t)$ $(\mathrm{Re}(\chi_o) < 0)$ has a continuation to an analytic function on all of $\mathbb{C}$. The last statements of the Corollary follow from Corollary 3 to Proposition 4.14.

Q. E. D.

We then consider the following diagram: $(t \neq 0)$

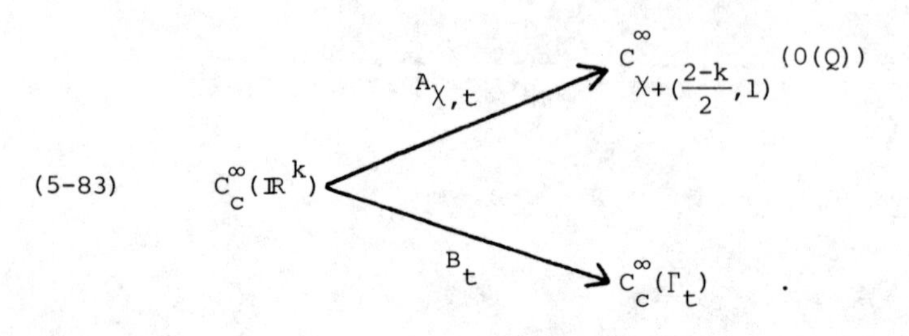

where $A_{\chi,t}(f)(g) = W_{\pi_{\mathfrak{m}}(w_o)}f(\chi, g, e_{\widetilde{s\ell}_2}, -\frac{t}{2})$ and $B_t(f)$ = the restriction of f to Γ_t. We note from Corollary 6 to Theorem 5.9 that $A_{\chi,t}$ is a continuous $O(Q)$ intertwining map. Moreover since Γ_t is closed in $\mathbb{R}^k$, we have that B_t is continuous and surjective (relative to the Schwartz topologies on $C_c^\infty(\mathbb{R}^k)$ and $C_c^\infty(\Gamma_t)$, respectively). Then for any $\varphi \in C_c^\infty(\Gamma_t)$, we define $c_{\chi,t}(\varphi) = A_{\chi,t}(F_\varphi)$, where $F_\varphi \in B_t^{-1}(\varphi)$. We observe that if $B_t(F_1) = B_t(F_2) = \varphi$, then $A_{\chi,t}(F_1) = A_{\chi,t}(F_2)$ (i.e. for $\mathrm{Re}(\chi_o) < 0$, $R_{F_1}^\chi(g,t) = R_{F_2}^\chi(g,t)$ and by analyticity the result holds for all $\chi_o \in \mathbb{C}$). We have the following:

<u>Theorem 5.10</u>. (1) *The map $c_{\chi,t}$ is a continuous $O(Q)$ intertwining map of $C_c^\infty(\Gamma_t)$ to $C^\infty_{\chi+(\frac{2-k}{k},1)}(O(Q))$ which completes the diagram in (5-83) (i.e. $c_{\chi,t} \circ B_t = A_{\chi,t}$ for all χ and $t \neq 0$). Moreover if $\chi \neq (k,\varepsilon)$, where k is any integer ≥ 1 and $\varepsilon = 1$ or -1, then $c_{\chi,t}$ is the unique, up to scalar multiple, intertwining map between $C_c^\infty(\Gamma_t)$ and $C^\infty_{\chi+(\frac{2-k}{2},1)}(O(Q))$.*

(2) *If $c_{\chi,t}(\varphi) = 0$ for all $\chi_o \in S$ where S is any subset of $\mathbb{C}$ having a nonempty interior then $\varphi \equiv 0$.*

(3) *If $U_t = \{M : C_c^\infty(\Gamma_t) \longrightarrow C_c^\infty(\Gamma_t) \mid M$ is continuous and $M\pi_{I_t}(g) = \pi_{I_t}(g)M$ for all $g \in O(Q)\}$ ("the algebra of all continuous operators on $C_c^\infty(\Gamma_t)$ which commute with the representation π_{I_t}"), then U_t is a commutative algebra of operators.*

<u>Proof</u>. From above we see that $c_{\chi,t}$ is well defined for all χ. We assume that $\varphi \in C_c^\infty(\Gamma_t)$ with $t > 0$. We note that the map $\mathbb{R}_+^* \times \Gamma_1 \xrightarrow{\rho} \{X \in \mathbb{R}^k \mid Q(X,X) > 0\}$ given by $\rho(t,\xi) = t\cdot\xi$ $(t \in \mathbb{R}_+^*, \xi \in \Gamma_1)$ is a diffeomorphism. Then it is possible to find $F_\varphi \in B_t^{-1}(\varphi)$ so that $\mathrm{supp}(F_\varphi) \subseteq \{X \in \mathbb{R}^K \mid Q(X,X) > 0\}$ and $F_\varphi(t\cdot\xi) = f_1(t)\varphi(\xi)$, where $f_1 \in C^\infty(\mathbb{R}^*)$, $\mathrm{supp}(f_1) \subseteq [\frac{1}{2}, 2]$ and $F_1(1) = 1$. Then

$F_\varphi(g\,A_a(r)\,N_a(Y)\,(\tilde{v}_a + \frac{1}{2}tv_a)) = f_1(t^{1/2})\varphi(g\,A_a(r)\,N_a(Y)\,(\tilde{v}_a + \frac{1}{2}v_a))$

from the definition of F_φ. Hence for $\mathrm{Re}(\chi_o) < 0$, $A_{\chi,t}(F_\varphi)(g) = \frac{1}{\Gamma(1-\chi_o)} R^{\chi}_{F_\varphi}(g,t) = \frac{f(t^{1/2})}{\Gamma(1-\chi_o)} R^{\chi}_{F_\varphi}(g,1)$, and by analyticity $A_{\chi,t}(F_\varphi) = f_1(t^{\frac{1}{2}})\, c_{\chi,1}(\varphi)$ for all $t \in \mathbb{R}$ and all χ. Thus if φ_α is a net converging to 0 in $C_c^\infty(\Gamma_t)$, we see easily that F_{φ_α} converges to 0 in $C_c^\infty(\mathbb{R}^k)$; thus $c_{\chi,1}(\varphi_\alpha)$ converges to 0 in $C^\infty_{\chi+(\frac{2-k}{2},1)}(O(Q))$. Hence $c_{\chi,1}$ is continuous. Repeating the same argument as above with $\Gamma_{t'}$, (t' fixed > 0) replacing Γ_1, we deduce that $c_{\chi,t}$ is continuous.

By using the same construction as in the proof of Corollary 3 to Theorem 5.9, we see that for all $\chi_o \neq k$ (k integer ≥ 1), $c_{\chi,1}$ is a nonzero map. Then the uniqueness of $c_{\chi,t}$ follows from conditions (A) and (B) of Proposition 3.6.

We then assume $c_{\chi,1}(\varphi) = 0$ for all $\chi \in S_1$, S_1 any subset of $\mathbb{C}$ having a nonempty interior.

Then $A_{\chi,1}(F_\varphi) = 0$ for all $\chi_o \in S_1$. But the map $\chi_o \longrightarrow A_{\chi,t}(F_\varphi)(g)$ is analytic in χ_o. Thus using the relation between $A_{\chi,t}$ and $c_{\chi,1}$ above, we deduce $A_{\chi,t}(F_\varphi) \equiv 0$ for all $\chi_o \in \mathbb{C}$ and all $t \in \mathbb{R}$. Then using (5-82) for $\mathrm{Re}(\chi_o) < 0$, $\Xi^{\chi}_{F_\varphi}(g,t,\tau) \equiv 0$ for any $g \in O(Q)$ and any $t \in \mathbb{R}$, $t \neq 0$. Again by analyticity, $\Xi^{\chi}_{F_\varphi}(\ ,\ ,\tau) \equiv 0$ for all $\chi_o \in \mathbb{C}$.

We then write $F_\varphi = F_\varphi^+ + F_\varphi^-$, $F_\varphi^{\pm}$ even (odd) part of F_φ. From (5-40) we observe that

$$\Xi^{\chi}_{F_\varphi}(g,t,\tau) = \begin{cases} \Xi^{\chi}_{F_\varphi^+}(g,t,\tau) & \text{if } \varepsilon(\chi) = 1 \\ \Xi^{\chi}_{F_\varphi^-}(g,t,\tau) & \text{if } \varepsilon(\chi) = -1 \end{cases}.$$

Again by analyticity of $\Xi^{\chi}_{F_\varphi}$, we then deduce from (5-53) that

$$\text{(5-84)} \qquad \int_0^\infty s^{\frac{\chi_o}{2}-1}\, \psi_{F_\varphi^+}(gv_a,\ t,\ s,\tau)\, ds \equiv 0\ ,$$

valid for all χ so that $0 < \mathrm{Re}(\chi_o) < 1$. From the definition of $\psi_{F_\varphi^+}$

in (5-51), we easily deduce that $\psi_{F_{\varphi}^{+}}(\ ,\ ,\ ,\tau) \in C^{\infty}(O(Q) \times \mathbb{R} \times \mathbb{R})$. Then by well known argument we have $\psi_{F_{\varphi}^{+}}(gv_a,t,s,\tau) = 0$ for all $g \in O(Q)$, $t, s, \in \mathbb{R}$. Thus

$$\int_{\mathbb{R}^k} F_{\varphi}^{+}(X)\, e^{-s\, Q(X,\xi)^2 + \pi i t\, Q(X,X)}\, dX \equiv 0 \tag{5-85}$$

for all $s, t \in \mathbb{R}$ and $\xi \in \Gamma_o - \{0\}$. But then differentiating with respect to s and t at $s = t = 0$ any number of times, we deduce that

$$\int_{\mathbb{R}^k} F_{\varphi}^{+}(X)\, Q(X,\xi)^{2j}\, Q(X,X)^{m}\, dX \equiv 0 \tag{5-86}$$

for any pair of integers $j, m \geq 0$.

However from the theory of separation of variables, we know that the complex linear span of the elements $Q(X,\xi)^{2j}\, Q(X,X)^{m}$ (as j, m range over all integers ≥ 0 and $\xi \in \Gamma_o - \{0\}$) coincides with the space of all even degree polynomials on $\mathbb{R}^k$.

Thus from (5-86), F_{φ}^{+} is orthogonal to all even degree polynomials on $\mathbb{R}^k$ relative to dX. But by simple application of the Theorem of Stone Weirstrass $F_{\varphi}^{+} \equiv 0$. Then a similar argument shows that $F_{\varphi}^{-} \equiv 0$ (using (5-52)). Thus we have (2) of Theorem 5.10.

We then let $Z \in U_t$ and define $c_{\chi,t}^{Z}(f) = c_{\chi,t}(Z\cdot f)$. Then it is clear that $c_{\chi,t}^{Z}$ is an $O(Q)$ intertwining map of $C_c^{\infty}(\Gamma_t)$ to $C^{\infty}_{\chi+\left|\frac{k-2}{2}\right|}(O(Q))$.

By the uniqueness property of (1) of this Theorem, we have for all χ so that $\chi_o \neq k$ (k integer), $c_{\chi,t}^{Z}(f) = \ell_2(\chi)\, c_{\chi,t}(f)$, where $\ell_2(\chi)$ is independent of $f \in C_c^{\infty}(\Gamma_t)$. Since $c_{\chi,t}$ is a nonzero map for all χ so that $\chi_o \neq$ integer, it follows that $\ell_2(\chi) \neq 0$ on an open dense subset of $\mathbb{C}$. Indeed we would have $c_{\chi,t}^{Z}(f) = c_{\chi,t}(Z\cdot f) = 0$ for all χ_o in a nonempty open set of $\mathbb{C}$ and hence by (2) of this Theorem, $Z\cdot f \equiv 0$.

Then we note that for A and $B \in U_t$, $c_{\chi,t}^{AB}(f) = c_{\chi,t}^{BA}(f)$ for all χ. But by (2) again, $AB - BA\,(f) = 0$ for all $f \in C_c^{\infty}(\Gamma_t)$. Thus $A\,B = B\,A$

on $C_c^\infty(\Gamma_t)$.

We note that the same argument works exactly for the case $t < 0$ above.

Q. E. D.

We let $\mathfrak{a}_Q = \{\varphi : O(Q) \times \mathbb{C} \to \mathbb{C} \mid$ (i) $\varphi(\ ,\lambda) \in C_\lambda^\infty(O(Q))$ for all λ, (ii) the function $(\lambda,g) \to D * \varphi(g,\lambda)$ is continuous and analytic in λ (for fixed g) for all $D \in \mathcal{U}(\mathfrak{S})\}$. Then for $D \in \mathcal{U}(\mathfrak{S})$, K_1 and K_2 compact subsets of $O(Q)$ and $\mathbb{C}$ respectively, we define on $\mathfrak{a}_Q$ a seminorm

$$(5\text{-}87) \qquad \nu_{D,K_1,K_2}(\varphi) = \sup_{\substack{g \in K_1 \\ \lambda \in K_2}} |D * \varphi(g,\lambda)| .$$

$\mathfrak{a}_Q$ equipped with this family of seminorms is a Frechet space; moreover we have the usual differentiable representation $\tilde{\Sigma}$ of $O(Q)$ on $\mathfrak{a}_Q$ given by $\tilde{\Sigma}(g)\ (\varphi(g',\lambda)) = \varphi(g^{-1}g',\lambda)$.

<u>Corollary 1 to Theorem 5.10</u>. <u>The map</u> $\varphi \rightsquigarrow c_{-\lambda-(\frac{1}{2}(2-k),1),t}(\varphi)(gv_a)$ <u>defines an injective and continuous</u> $O(Q)$ <u>intertwining map of</u> $C_c^\infty(\Gamma_t)$ <u>to</u> $\mathfrak{a}_Q$.

<u>Proof</u>. We know by simple adaption of the argument in Corollary 2 to Theorem 5.9 that for any $W \in \mathcal{U}(\mathfrak{S})$, the map $(g,G,\lambda) \to \Theta^\lambda_{\pi_{\mathfrak{m}}(G^{-1})\pi_{\mathfrak{m}}(W)f}(g,0,\tau)$ defines a continuous function on $O(Q) \times S\tilde{\ell}_2 \times \mathbb{C}$ and is analytic in λ (for fixed (g,G)). Then by letting $F_{\lambda+(\frac{1}{2}(2-k),1)}(G) = \Theta^\lambda_{\pi_{\mathfrak{m}}(G^{-1})\pi_{\mathfrak{m}}(W)f}(g,0,\tau)$ in the proof of Proposition 4.13, we see that, by reasoning similar to that in Proposition 4.13, the function $(g,\lambda) \rightsquigarrow A_{F_{\lambda+(\frac{1}{2}(2-k),1)}}(G,\lambda+(\frac{1}{2}(2-k),1),u)$ $(u \neq 0)$ is continuous and analytic in λ_o (for fixed g). But then noting that $c_{\lambda,t}(\varphi)(gv_a) = A_{\lambda,t}(F_\varphi)(gv_a) = W_{\pi_{\mathfrak{m}}(w_o)F_\varphi}(\lambda,g,e_{S\tilde{\ell}_2},-\frac{1}{2}t)$ and that $W_{\pi_{\mathfrak{m}}(w_o)F_\varphi}(\lambda,\ ,e_{S\tilde{\ell}_2},t)$ is the continuation of (5-82) (see Corollary 6 to Proposition 5.9), we deduce that $(g,\lambda) \rightsquigarrow c_{-\lambda-(\frac{1}{2}(2-k),1),t}(W * \varphi)(gv_a)$ determines an element of $\mathfrak{a}_Q$.

Similar reasoning as in the steps above also shows the continuity of the map. The injectivity of the map follows from (2) of Theorem 5.10.

Q. E. D.

We let $\varphi \in \mathcal{a}_Q$. Then we define

$$T_{\lambda,t}(\varphi)(h) = \frac{1}{\Gamma(\lambda_o+\frac{1}{2}(4-k))} \oint_{O(Q)/P_a} \varphi(hg,\lambda)\left[Q(gv_a,\xi_t)\right]^{\lambda+(\frac{1}{2}(2-k),1)} d\mu_k(\dot{g}) \tag{5-88}$$

for $h \in O(Q)$, λ a quasicharacter (i.e. $\lambda = (\lambda_o,\varepsilon)$) and with $\xi_t = \tilde{v}_a+\frac{1}{2}tv_a$.

Then we note that for $\mathrm{Re}(\lambda_o) \geq \frac{1}{2}(k-2)$

$$|T_{\lambda,t}(\varphi)(h)| \leq \left|\frac{1}{\Gamma(\lambda_o+\frac{1}{2}(4-k))}\right| \int_K |\varphi(k,\lambda)|\left[Q(kv_a,h\xi_t)\right]^{\mathrm{Re}(\lambda_o)+\frac{1}{2}(2-k)} dk \tag{5-89}$$

$$\leq \left|\frac{1}{\Gamma(\lambda_o+\frac{1}{2}(4-k))}\right| \left\{\sup_{k\in K}|\varphi(k,\lambda)|\right\} \left\{\sup_{k\in K}\left[Q(kv_a,h\xi_t)\right]^{\mathrm{Re}(\lambda_o)+\frac{1}{2}(2-k)}\right\}.$$

But then for any compact set L_1 of $O(Q)$ (L_2 of $\{\lambda|\mathrm{Re}(\lambda_o) \geq \frac{1}{2}(k-2)\}$ resp.), we see that

$$\sup_{k\in K}\left[Q(kv_a,h\xi_t)\right]^{\mathrm{Re}(\lambda_o)+\frac{1}{2}(2-k)} \leq c_{L_1,L_2}, \tag{5-90}$$

where c_{L_1,L_2} is a positive constant depending only on L_1 and L_2. Thus

$$|T_{\lambda,t}(\varphi)(h)| \leq \sup_{\lambda_o\in L}\left|\frac{1}{\Gamma(\lambda_o+\frac{1}{2}(4-k))}\right| c_{L_1,L_2}\, \nu_{I,K,L_2}(\varphi) \tag{5-91}$$

for λ so that $\mathrm{Re}(\lambda_o) \geq \frac{1}{2}(k-2)$.

Then we observe the following basic relation between $T_{\lambda,t}$ and $c_{\lambda,t}$: let $\beta \in C_c^\infty(\Gamma_t)$,

$$(\beta,T_{\lambda,t}(\varphi))_t = \int_{\Gamma_t} \beta(X)\, T_{\lambda,t}(\varphi)(X)\, d\mu_t(X) = \tag{5-92}$$

$$\frac{1}{\Gamma(\lambda_o+\frac{1}{2}(4-k))}\int_{O(Q)/O(Q)^{\xi_t}} \beta(g'\xi_t)\left\{\oint_{O(Q)/P_a} \varphi(g'gv_a,\lambda)\right.$$

$$\left.[Q(gv_a,\xi_t)]^{\lambda+(\frac{1}{2}(2-k),1)} d\mu_k(\dot g)\right\} d\mu_t(g'\cdot\xi_t) =$$

$$\frac{1}{\Gamma(\lambda_o+\frac{1}{2}(4-k))}\oint_{O(Q)/P_a} \varphi(gv_a,\lambda)\left\{\int_{O(Q)/O(Q)^{\xi_t}} \beta(g'\xi_t)\right.$$

$$\left.[Q(gv_a,g'\xi_t)]^{\lambda+(\frac{1}{2}(2-k),1)} d\mu_t(g'\xi_t)\right\} d\mu_k(\dot g) = (\varphi\,|\,c_{-\lambda-(\frac{1}{2}(2-k),1),t}(\beta))$$

(see (3-21)), valid for $\mathrm{Re}(\lambda_o) > \frac{1}{2}(k-2)$.

Proposition 5.11. *The map $\lambda \longrightarrow T_{\lambda,t}(\varphi)$ has a continuation to an analytic functional $\lambda \longrightarrow T^*_{\lambda,t}(\varphi)$ on $C_c^\infty(\Gamma_t)$ for all quasicharacters λ . And if for λ we have $\psi(\ ,\lambda) \equiv 0$, then $T^*_{\lambda,t}(\psi(\ ,\lambda))$ defines the zero distribution on $C_c^\infty(\Gamma_t)$. Hence $T^*_{\lambda,t}$ defines a continuous $O(Q)$ intertwining map of $C_\lambda^\infty(O(Q))$ to $C_c^\infty(\Gamma_t)'$.*

Proof. For any compact subset L_2 , with L_2^o = interior of $L_2 \neq \emptyset$, we know that $\lambda \longrightarrow c_{\lambda,t}(\beta)$ is an analytic function for $\lambda \in L_2$ since $\lambda \longrightarrow A_{\lambda,t}(F_\beta)$ has this property as shown in Corollary 1 to Theorem 5.10. Moreover since $\varphi \in \mathcal{a}_Q$ it follows that $|\varphi\cdot c_{\lambda,t}(\beta)|$ restricted to K is bounded independent of the value of $\lambda \in L_2$. Thus $\lambda \longrightarrow (\varphi\,|\,c_{-\lambda-(\frac{1}{2}(2-k),1),t}(\beta))$ is an analytic function for $\lambda \in \mathbb{C}$. Hence for any $\beta \in C^\infty(\Gamma_t)$, $\lambda \longrightarrow (\beta, T_{\lambda,t}(\varphi))$ is analytic for $\mathrm{Re}(\lambda_o) > \frac{1}{2}(k-2)$. Then since $\lambda_o \longrightarrow c_{\lambda,t}(\beta)$ is analytic for all $\lambda_o \in \mathbb{C}$, it follows by (5-89) that we can extend $T_{\lambda,t}(\varphi)$ to an analytic functional $T^*_{\lambda,t}(\varphi)$ for all $\lambda_o \in \mathbb{C}$. Again (5-92) implies that if $\psi(\ ,\lambda) \equiv 0$ for some quasicharacter λ , then $T^*_{\lambda,t}(\psi(\ ,\lambda))$ defines the zero distribution on $C_c^\infty(\Gamma_t)$. Moreover since $C_c^\infty(\Gamma_t)$ with the Schwartz topology is a reflexive space, it follows, using (5-92), that $T^*_{\lambda,t}$ is a continuous map of $C_\lambda^\infty(O(Q))$ to $C_c^\infty(\Gamma_t)'$.

Q. E. D.

Remark 5.9. By the analyticity property of $T^*_{\lambda,t}$ in Proposition 5.11, we have a power series expansion about $\lambda_o \in \mathbb{C}$, where $\lambda = (z,\varepsilon(\lambda))$ and $\tilde{\lambda}_o = (\lambda_o,\varepsilon(\tilde{\lambda}_o))$ with $\varepsilon(\lambda) = \varepsilon(\lambda_o)$:

(5-93) $$(\varphi, T^*_{\lambda,t}(\psi(\ ,\lambda))\ ) = \sum_{\ell=0}^{\ell=+\infty} \frac{1}{\ell!}\, r_{\ell,t}(\varphi|\psi(\ ,\tilde{\lambda}_o))(z-\lambda_o)^{\ell} ,$$

where the coefficient $r_{\ell,t}(\varphi|\psi(\ ,\tilde{\lambda}_o))$ is given by the Cauchy integral:

(5-94) $$r_{\ell,t}(\varphi|\psi(\ ,\tilde{\lambda}_o)) = \frac{1}{2\pi\sqrt{-1}} \int_{|z-\lambda_o|=a} (\varphi, T^*_{\lambda,t}(\psi(\ ,\lambda))\left(\frac{1}{z-\lambda_o}\right)^{\ell+1} dz$$

$$= \left(\frac{d}{dz}\right)^{\ell} (\varphi, T^*_{\lambda,t}(\psi(\ ,\lambda))\Big|_{z=\lambda_o}$$

(with a any positive number).

Corollary 1 to Proposition 5.11. *The map* $\psi(\ ,\lambda) \rightsquigarrow r_{\ell,t}(\ |\psi(\ ,\lambda_o))$ *is a continuous linear map of* $\mathfrak{a}_Q$ *to* $C_c^\infty(\Gamma_t)'$. *Moreover if* $\ell = 0$, *then* $(\varphi, T^*_{\lambda,t}(\psi(\ ,\lambda))) = r_{o,t}(\varphi|\psi(\ ,\lambda))$. *Then we define* $S^o_{\lambda,t}$ = Kernel $(T^*_{\lambda,t})$ *and for* $m \geq 1$, $S^m_{\lambda,t}$ = Kernel of the map $\psi(\ ,\lambda) \rightsquigarrow r_{m,t}(\ |\psi(\ ,\lambda))$ *when restricted to* $S^{m-1}_{\lambda,t}$. *Then* $S^m_{\lambda,t}$ *is an* $O(Q)$ *invariant subspace of* $C^\infty_\lambda(O(Q))$ *and the map* $(T^*_{\lambda,t})^{m+1}: \psi(\ ,\lambda) \rightsquigarrow r_{m+1,t}(\ |\psi(\ ,\lambda))$ *when restricted to* $S^m_{\lambda,t}$ *determines a continuous* $O(Q)$ *intertwining map of* $S^m_{\lambda,t}$ *to* $C_c^\infty(\Gamma_t)'$.

Proof. From the general theory of analytic functionals (see [6]), we know that $\frac{d}{d\lambda} T^*_{\lambda,t}(\psi(\ ,\lambda))$ determines a distribution on $C_c^\infty(\Gamma_t)$ and that the map $\psi(\ ,\lambda) \rightsquigarrow \frac{d}{d\lambda} T^*_{\lambda,t}(\psi(\ ,\lambda))$ is continuous.

Q. E. D.

We recall at this point for $\mathrm{Re}(\lambda) < 0$ and $f \in S(\mathbb{R}^k)$, $\psi \in \mathfrak{a}_Q$, and $\beta \in \mathfrak{a}_s^{+1}$ $(s = \tilde{\phi}(k,Q,\varepsilon(\lambda))$ in (5-70), the following relation:

(5-95) $$(\Theta^{\lambda}_f(\ ,\ ,\tau)|\psi(\ ,-\lambda-(\tfrac{1}{2}(2-k),1))\beta(\ ,\lambda_o - \tfrac{k}{2}+1)) =$$

$$\oint_{0(Q)/P_a \times \mathbb{R}} \Theta_f^{\lambda}(g,t,\tau)\psi(g,-\lambda-(\tfrac{1}{2}(2-k),1))\beta(t,\lambda_o - \tfrac{k}{2} + 1)\, d\mu_k(\dot{g})\, dt$$

$$= \frac{1}{\Gamma(1-\lambda_o)} \oint_{0(Q)/P_a \times \mathbb{R}} \psi(g,-\lambda-(\tfrac{1}{2}(2-k),1))\beta(t,\lambda_o - \tfrac{k}{2} + 1) \left\{ \int_{\mathbb{R}^k} f^*(X) \right.$$

$$\left. [Q(gv_a,X)]^{-\lambda} e^{-\pi i t Q(X,X)} dX \right\} d\mu_k(\dot{g})\, dt \, .$$

Now if we assume $\operatorname{supp}_t\left(\beta(\ ,\lambda_o - \frac{k}{2} + 1)\right)$ is compact for all λ , then we can perform the integration in (5-95) in any order and obtain

$$(5\text{-}95) = \frac{1}{\Gamma(1-\lambda_o)} \int_{\mathbb{R}} \beta(t,\lambda_o - \tfrac{k}{2} + 1) \left\{ \int_{\mathbb{R}^k} f^*(X) \left\{ \oint_{0(Q)/P_a} \psi(g,-\lambda-(\tfrac{1}{2}(2-k),1)) \right.\right.$$

$$(5\text{-}96) \qquad \left.\left. [Q(gv_a,X)]^{-\lambda} d\mu_k(\dot{g}) \right\} e^{-\pi i t Q(X,X)}\, dX \right\} dt$$

$$= \int_{\mathbb{R}} \beta(t,\lambda_o - \tfrac{k}{2} + 1) \left\{ \int_{\mathbb{R}^k} f^*(X) T^*_{-\lambda+(\frac{1}{2}(k-2),1),Q(X,X)}(\psi)(X) \right.$$

$$\left. e^{-\pi i t\, Q(X,X)} dX \right\} dt = \int_{\mathbb{R}^k} f^*(X) T^*_{-\lambda+(\frac{1}{2}(k-2),1),Q(X,X)}(\psi)(X)$$

$$\hat{\beta}(-\tfrac{1}{2}\, Q(X,X),\lambda_o - \tfrac{k}{2} + 1)\, dX \, ,$$

valid for $\operatorname{Re}(\lambda_o) < 0$.

But then we note that for $\operatorname{Re}(\lambda_o) > 0$ (using (5-60)), we deduce

$$(5\text{-}95) = \frac{\gamma(\lambda^{-1}+(1,1))}{\Gamma(1-\lambda_o)} \oint_{0(Q)/P_a \times \mathbb{R}} \psi(g,-\lambda-(\tfrac{1}{2}(2-k),1))\beta(t,\lambda_o - \tfrac{k}{2} + 1)$$

$$(5\text{-}97) \qquad \left\{ \int_{-\infty}^{+\infty} \lambda(r) \left[e^{-\pi i t Q}\cdot f^*\right]^{\wedge} (-M_Q(rgv_a)) \frac{dr}{|r|} \right\} d\mu_k(\dot{g})\, dt \, .$$

But assuming β has compact support property as above, we have, switching the orders of integration in (5-97)

$$(5\text{-}95) = \frac{\gamma(\lambda^{-1}+(1,1))}{\Gamma(1-\lambda_o)} \int_{-\infty}^{+\infty} \lambda(r) \left\{ \int_{\mathbb{R}} \beta(t,\lambda_o - \tfrac{k}{2}+1) \left\{ \int_K [e^{-\pi i t Q} f^*]^\wedge \right.\right.$$

(5-98) $$\left.\left.(-M_Q(rkv_a))\psi(k,-\lambda-(\tfrac{1}{2}(2-k),1))\, dk \right\} dt \right\} \frac{dr}{|r|} .$$

But we write in the definition of $[e^{-\lambda i t Q} f^*]^\wedge$ above and we have, again switching orders of integration,

(5-99) $$(5\text{-}95) = \frac{\gamma(\lambda^{-1}+(1,1))}{\Gamma(1-\lambda_o)} \int_{-\infty}^{+\infty} \lambda(r) \left\{ \int_{\mathbb{R}^k} G(r,X)\, dX \right\} \frac{dr}{|r|}$$

where

(5-100) $$G(r,X) = f^*(X)\hat{\beta}(-\tfrac{1}{2} Q(X,X),\lambda_o - \tfrac{k}{2} + 1) \int_K \psi(k,-\lambda-(\tfrac{1}{2}(2-k),1)) \, e^{-2\pi i r\, Q(kv_a,X)}\, dk \; .$$

Then we note that $h(r) = \int_{\mathbb{R}^k} G(r,X)\, dX$ is integrable relative to $r^{\mathrm{Re}(\lambda_o)-1}\, dr$. But we know that

(5-101) $$\int_{-\infty}^{+\infty} \lambda(r)h(r) \frac{dr}{|r|} = \lim_{\varepsilon\to 0} \int_{-\infty}^{+\infty} \lambda(r)\, e^{-\varepsilon|r|} \left\{ \int_{\mathbb{R}^k} G(r,X)\, dX \right\} \frac{dr}{|r|}$$

$$= \lim_{\varepsilon\to 0} \int_{\mathbb{R}^k} \left\{ \int_{-\infty}^{+\infty} G(r,X)\lambda(r)\, e^{-\varepsilon|r|} \frac{dr}{|r|} \right\} dX \, .$$

Thus we have from (5-101)

$$(5\text{-}95) = \frac{\gamma(\lambda^{-1}+(1,1))}{\Gamma(1-\lambda_o)} \lim_{\varepsilon\to 0} \int_{\mathbb{R}^k} f^*(X)\hat{\beta}(-\tfrac{1}{2} Q(X,X),\lambda_o - \tfrac{k}{2} +1)$$

(5-102) $$\left\{ \int_{-\infty}^{+\infty} e^{-\varepsilon|r|}\lambda(r) \left\{ \int_K \psi(k,-\lambda-(\tfrac{1}{2}(2-k),1))\, e^{-2\pi i r Q(kv_a,X)}\, dk \right\} \frac{dr}{|r|} \right\} dX \, .$$

<u>Proposition 5.12</u>. <u>The distribution</u> $T^*_{\lambda,t}(\psi(\ ,\lambda)) \in C_c^\infty(\Gamma_t)'$ <u>is equal to</u>

(5-103) $$\frac{\gamma(\lambda+(\frac{1}{2}(4-k),1))}{\Gamma(\lambda_o+\frac{1}{2}(4-k))} \lim_{\varepsilon \to 0} \int_{-\infty}^{+\infty} e^{-\varepsilon|r|}\lambda^{-1}(r)|r|^{\frac{1}{2}(k-2)} \left\{ \int_K \psi(k,\lambda) e^{-2\pi ir\, Q(kv_a,X)} dk \right\} \frac{dr}{|r|}$$

for $\mathrm{Re}(\lambda) < \frac{1}{2}(k-2)$ *and for* $X \in \Gamma_t$ $(t \neq 0)$ *where limit is taken relative to* $\varepsilon \to 0$ *weak topology in* $C_c^\infty(\Gamma_t)'$.

Proof. We choose f so that $f^* \circ \beta_a$ is of compact support in $A_a(r) \times N_a(Y) \times \mathbb{R}$ (see (5-25) and Lemma 5.25). Then for $\mathrm{Re}(\lambda_o) < 0$ we have from the right hand side of (5-96)

$$(5\text{-}95) = \int_{\mathbb{R}} \hat{\beta}(-\frac{1}{2}s,\lambda_o-\frac{k}{2}+1)(f^*|_{\Gamma_s}, T^*_{-\lambda+(\frac{1}{2}(k-2),1),s}(\psi))\, ds$$

(5-104) $$= \int_{\mathbb{R}} \hat{\beta}(-\frac{1}{2}s,\lambda_o-\frac{k}{2}+1)(\psi|c_{\lambda,s}(f^*))_\lambda\, ds .$$

But if f^* satisfies the above condition, it follows that, using (5-31), $R^\lambda_{f^*}$ is defined for all values of λ and is a bounded continuous function in t (for all λ). Thus since both terms of the integrand in (5-104) can be continued analytically in λ and using the rapid decrease of $\hat{\beta}$ (in s), we have

$$(5\text{-}95) = \int_{\mathbb{R}} \hat{\beta}(-\frac{1}{2}s,\lambda_o-\frac{k}{2}+1)(\psi|c_{\lambda,s}(f^*))_\lambda\, ds$$

(5-105) $$= \int_{\mathbb{R}} \hat{\beta}(-\frac{1}{2}s,\lambda_o-\frac{k}{2}+1)(f^*|_{\Gamma_s}, T^*_{-\lambda+(\frac{1}{2}(k-2),1),s}(\psi))\, ds$$

is valid for all λ (for β such that $\mathrm{supp}_t(\beta(\ ,\lambda_o-\frac{k}{2}+1))$ compact for all λ, and f^* as chosen above).

But from (5-102) we deduce that for $\mathrm{Re}(\lambda) > 0$,

(5-106) $$(5\text{-}95) = \frac{\gamma(\lambda^{-1}+(1,1))}{\Gamma(1-\lambda_o)} \lim_{\varepsilon \to 0} \int_{\mathbb{R}} \hat{\beta}(-\frac{1}{2}s,\lambda_o-\frac{k}{2}+1)(f^*|_{\Gamma_s}, L_{\lambda,\varepsilon}(\psi))_s\, ds$$

where

$$(5\text{-}107)\qquad L_{\lambda,\varepsilon}(\psi(\ ,-\lambda-(\tfrac{1}{2}(2-k),1)))(X) = \frac{\gamma(\lambda^{-1}+(1,1))}{\Gamma(1-\lambda_o)} \int_{-\infty}^{+\infty} e^{-\varepsilon|r|}\lambda(r)$$

$$\left\{ \int_K \psi(k,-\lambda-(\tfrac{1}{2}(2-k),1))e^{-2\pi ir\, Q(kv_a,X)}\,dk \right\} \frac{dr}{|r|} .$$

We observe at this point that the map:

$$(5\text{-}108)\qquad \psi(\ ,-\lambda-(\tfrac{1}{2}(2-k),1)) \rightsquigarrow \lim_{\varepsilon\to 0} L_{\lambda,\varepsilon}(\psi(\ ,-\lambda-(\tfrac{1}{2}(2-k),1)))$$

(limit taken in weak topology on $C_c^\infty(\Gamma_t)'$) is a continuous $O(Q)$ intertwining operator of $C^\infty_{\frac{1}{2}-\lambda-(\frac{1}{2}(2-k),1)}(O(Q))$ to $C_c^\infty(\Gamma_t)'$. Indeed if we choose $f^*(t\cdot\xi) = h(t^2)\delta(\xi)$ for $\xi\in\Gamma_{+1}$ with support$(f)\subseteq[\frac{1}{2},2]$ and $\delta\in C_c^\infty(\Gamma_{+1})$, then

$$(5\text{-}106) = \frac{\gamma(\lambda^{-1}+(1,1))}{\Gamma(1-\lambda_o)} \left\{ \int_{\mathbb{R}} \hat{\beta}(-\tfrac{1}{2}s,\lambda_o-\tfrac{k}{2}+1)h(s^{\frac{1}{2}})\,ds \right\}$$

$$(5\text{-}109)\qquad \lim_{\varepsilon\to 0}(\delta, L_{\lambda,\varepsilon}(\psi)) .$$

Then using the continuity and intertwining properties of the map $f \longrightarrow \theta_f^\lambda$, we deduce the continuity and intertwining properties of (5-108), except at the zeroes of $\frac{\gamma(\lambda^{-1}+(1,1))}{\Gamma(1-\lambda_o)}$. (i.e. choose h so that $\int_{\mathbb{R}} \hat{\beta}(-\frac{1}{2}s,\lambda_o-\frac{k}{2}+1)h(s^{\frac{1}{2}})\,ds \neq 0$).

Since β is an arbitrary function of compact support, it follows that for almost all s

$$(5\text{-}110)\qquad (f^*|_{\Gamma_s}, T^*_{-\lambda+(\frac{1}{2}(k-2),1),s}(\psi))_s = \lim_{\varepsilon\to 0}(f^*|_{\Gamma_s}, L_{\lambda,\varepsilon}(\psi))_s .$$

Thus it follows that support $\{T^*_{-\lambda+(\frac{1}{2}(k-2),1),s}(\psi) - \lim_{\varepsilon\to 0} L_{\lambda,\varepsilon}(\psi)\} \subseteq \{X\in\Gamma_s \mid Q(X,v_a) = 0\}$. But we know that both $T^*_{-\lambda+(\frac{1}{2}(k-2),1),s}(\psi)$ and $\lim_{\varepsilon\to 0} L_{\lambda,\varepsilon}(\psi)$ define $O(Q)$ intertwining maps of $C^\infty_{-\lambda+(\frac{1}{2}(k-2),1)}(O(Q))$ to

$C_c^\infty(\Gamma_t)'$. However if ψ is a K finite function of isotypic type in $C^\infty_{-\lambda+(\frac{1}{2}(k-2),1)}(O(Q))$, then $T^*_{-\lambda+(\frac{1}{2}(k-2),1),s}(\psi)$ and $\lim_{\varepsilon\to 0} L_{\lambda,\varepsilon}(\psi)$ both satisfy an elliptic equation as distributions (i.e. both distributions are eigenfunctions of the Casimir operators $\Omega_{O(Q)}$ and Ω_K of $O(Q)$ and K, respectively, and hence of $\Omega_{O(Q)} - \Omega_k$, which is elliptic). Thus for ψ a K finite function in $C^\infty_{-\lambda+(\frac{1}{2}(k-2),1)}(O(Q))$, we have that

$T^*_{-\lambda+(\frac{1}{2}(k-2),1)s}(\psi) = \lim_{\varepsilon\to 0} L_{\lambda,\varepsilon}(\psi)$ as distributions.

Then by the continuity of $T^*_{-\lambda+(\frac{1}{2}(k-2),1),s}$ and $\lim_{\varepsilon\to 0} L_{\lambda,\varepsilon}$ (i.e. given at those λ where $\dfrac{\gamma(\lambda^{-1}+(1,1))}{\Gamma(1-\lambda_o)}$ is not zero), we deduce that

$T^*_{-\lambda+(\frac{1}{2}(k-2),1),s}(\psi) = \lim_{\varepsilon\to 0} L_{\lambda,\varepsilon}(\psi(\ ,\ ))$ for all $\psi \in C^\infty_{-\lambda+(\frac{1}{2}(k-2),1)}(O(Q))$.

Q. E. D.

§6. DISCRETE SPECTRUM OF WEIL REPRESENTATION

Let $\Omega_{\pm} = \left\{ X \in \mathbb{R}^k \middle| \begin{array}{ll} Q(X,X) > 0 & \text{if } + \\ Q(X,X) < 0 & \text{if } - \end{array} \right\}$ and let $\mathcal{A}_{\pm} =$ $\left\{ \varphi \in L^2(\mathbb{R}^k) \middle| \varphi(X) = 0 \text{ if } \left\{ \begin{array}{l} X \in \Omega_+ \\ X \in \Omega_- \end{array} \right\} \right\}$. We consider the maps $L^2(\mathbb{R}^k) \underset{\psi_{\pm}}{\rightarrow} L^2(\Omega_{\pm}, dX)$ defined by $\psi_{\pm}(\varphi) = \varphi|_{\Omega_{\pm}}$ (restricted to $\Omega_{\pm}$). Then $\psi_{\pm}$ are continuous maps in the appropriate sense and $\operatorname{Ker}(\psi_{\pm}) = \mathcal{A}_{\pm}$. Moreover we have the decomposition $L^2(\mathbb{R}^k) = \mathcal{A}_+ \oplus \mathcal{A}_-$.

Let M_Q be the Lie subgroup, the direct product $a(r) \times O(Q)$ of $\tilde{G}_2(Q)$. Then we restrict $\pi_{\mathfrak{m}}$ to M_Q and easily see from §1 that the spaces $\mathcal{A}_+$ and $\mathcal{A}_-$ are stable by $\pi_{\mathfrak{m}}(M_Q)$. Moreover it is easy to see that the representation $\pi_{\mathfrak{m}}(M_Q)$ in $\mathcal{A}_-$ ($\mathcal{A}_+$ respectively) is unitarily equivalent to the unitarily induced representation of M_Q in $L^2(M_Q/O(Q)^{Z_+}, d\mu_1 \otimes \frac{dr}{r})$ $(L^2(M_Q/O(Q)^{Z_-}, d\mu_{-1} \otimes \frac{dr}{r})$, resp.) where $Z_{\pm} \in \Omega_{\pm}$ and $d\mu_{\pm 1}$ is an $O(Q)$ invariant measure on $O(Q)/O(Q)^{Z_{\pm}}$ $(\tilde{=}\Gamma_{\pm 1})$ and $\frac{dr}{r}$, Haar measure on $a(r)$. Indeed the unitary equivalence is given by sending $\varphi \rightsquigarrow L^{\pm}_{\varphi}(\gamma) = \pi_{\mathfrak{m}}(\gamma)^{-1}(\varphi)(Z_{\pm})$ with $\gamma \in M_Q$. We recall here the topological decomposition of the spaces $\Omega_{\pm}$ as $\Omega_+ \cong \mathbb{R}^*_+ \times \Gamma_{+1}$ and $\Omega_- \cong \mathbb{R}^*_+ \times \Gamma_{-1}$, with the pair (t,ξ) defining the point $t\cdot\xi \in \Omega_{\pm}$ $(t \in \mathbb{R}^*_+,\ \xi \in \Gamma_{+1}, \Gamma_{-1}$ resp.). Then we note that $dX|_{\Omega_{\pm}} \cong t^k\, d\mu_{\pm 1} \otimes \frac{dt}{t}$.

Then by using the regularity Theorem of C^∞ vectors in an induced representation, we deduce that the C^∞ vectors of M_Q with respect to $\pi_{\mathfrak{m}}$ in $\mathcal{A}_-$ ($\mathcal{A}_+$ resp.) are precisely all $\varphi \in \mathcal{A}_-$ ($\mathcal{A}_+$ resp.) so that $\varphi|_{\Omega_+}$ ($\varphi|_{\Omega_-}$ resp.) is a C^∞ function, and $\pi_{\mathfrak{m}}(D)\cdot\varphi \in \mathcal{A}_-$ ($\pi_{\mathfrak{m}}(D)\cdot\varphi \in \mathcal{A}_+$ resp.) for all $D \in \mathcal{U}(M_Q)$, the universal enveloping algebra of M_Q.

Then we also note at this point that relative to the coordinates on

$\Omega_{\pm}$ above, the differential operators $\partial(Q)$, Q, and E_1 become respectively:

$$(6\text{-}1) \qquad \partial(Q) = \begin{cases} \dfrac{\partial^2}{\partial t^2} + \dfrac{k-1}{t}\dfrac{\partial}{\partial t} - \dfrac{1}{t^2} W_\xi^+ & \text{for } \Omega_+ , \\[2ex] -\left(\dfrac{\partial^2}{\partial t^2} + \dfrac{k-1}{t}\dfrac{\partial}{\partial t} - \dfrac{1}{t^2} W_\xi^-\right) & \text{for } \Omega_- , \end{cases}$$

where $W_\xi^{\pm}$ is a multiple of the Laplace Beltrami operator on the hyperboloid Γ_{+1}, Γ_{-1} resp.

$$(6\text{-}2) \qquad E_1 = t\frac{\partial}{\partial t}$$

$$(6\text{-}3) \qquad Q = \begin{cases} t^2 & \text{for } \Omega_+ \\ -t^2 & \text{for } \Omega_- \end{cases} .$$

We let $\mathscr{L} = L^2(\mathbb{R}_+^* \times \Gamma_{\pm 1}, t^{k-1}\, dt \otimes d\mu_{\pm 1})$.

<u>Lemma 6.1</u>. <u>If</u> $\varphi \in \mathbb{F}_Q$ <u>then we may write for</u> $X = t\cdot\xi$ $(\xi \in \Gamma_{+1}, \Gamma_{-1}$ resp.) $\varphi(X) = E_\varphi^{\pm}(t,\xi)$, <u>where</u> $E_\varphi^{\pm} \in C^\infty(\mathbb{R}_+^* \times \Gamma_{\pm 1})$ <u>so that</u>

(1) $t^{2j} E_\varphi^{\pm}(t,\xi) \in \mathscr{L}$ <u>for all integers</u> $j \geq 0$.

(2) $(t\frac{d}{dt})^m E_\varphi^{\pm}(t,\xi) \in \mathscr{L}$ <u>for all integers</u> $m \geq 0$.

(3) $(\frac{\partial^2}{\partial t^2} + \frac{k-1}{t}\frac{\partial}{\partial t} - \frac{1}{t^2} W_\xi^{\pm})^s E_\varphi^{\pm}(t,\xi) \in \mathscr{L}$ <u>for all integers</u> $s \geq 0$.

(4) $D * E_\varphi^{\pm}(t,\xi) \in \mathscr{L}$ <u>for all</u> $D \in \mathscr{U}(\mathfrak{S})$, <u>the enveloping algebra of</u> $O(Q)$.

<u>Conversely any function</u> φ , <u>defined on</u> $\mathbb{R}^k$ <u>so that</u> $\varphi|_{\Omega_\pm}$ <u>is</u> C^∞ <u>and the associated</u> $E_\varphi^{\pm}$ <u>satisfies</u> (1) - (4) <u>above, belongs to</u> $\mathbb{F}_Q$.

<u>Proof</u>. We use the characterization of C^∞ vectors given at the beginning of §5. Moreover we note the comments preceding the Lemma.

Q. E. D.

From §5 we recall the Casimir operator $\omega_{s\ell_2}$ of $\widetilde{s\ell}_2(Q)$ and observe that

$$(6\text{-}4)\qquad \pi_{\mathfrak{M}}(\omega_{s\ell_2})_\infty = \begin{cases} W_\xi^+ + (\frac{k^2}{4} - k)\cdot I & \text{on } \Omega_+ \\ W_\xi^- + (\frac{k^2}{4} - k)\cdot I & \text{on } \Omega_- \end{cases}$$

We define $\mathbb{F}_Q(\lambda) = \{\varphi \in \mathbb{F}_Q \mid \pi_{\mathfrak{M}}(\omega_{s\ell_2})(\varphi) = \lambda\varphi\}$. It is then clear that $\mathbb{F}_Q(\lambda)$ is $\pi_{\mathfrak{M}}(\tilde{G}_2(Q))$ stable.

We note that the maximal compact subgroup of $\tilde{G}_2(Q)$ is $\underline{\tilde{K}} \times K$ where $\underline{\tilde{K}} = \{k(\theta,\varepsilon)\}$ given in §4 and K is the maximal compact subgroup of $O(Q)$ given in §2. Since the group K is isomorphic to $O(a) \times O(b)$, then $\lambda \in \hat{K}$ is characterized as the tensor product $\gamma_1 \otimes \gamma_2$ with $\gamma_1 \in \widehat{O(a)}$ and $\gamma_2 \in \widehat{O(b)}$. We now observe that every character $\rho \in \hat{\underline{\tilde{K}}}$ is of the form $\rho(k(\theta,\varepsilon)) = e^{i\frac{m}{2}\theta}(\mathrm{sgn}(\varepsilon))^m$ for any fixed integer m, as can be checked easily from the formulae in §1 and §4. We also recall the class one-representations of $\widehat{O(\ell)}$ are parametrized as nonegative integers. In particular the integer $s \geqq 0$ corresponds to the representation $[s]_\ell$ of $O(\ell)$ on the space of harmonic polynomials of degree s, $H^s(\mathbb{R}^\ell)$. Then we obtain the "other" classical realization of $[s]_\ell$ by restricting the functions in $H^s(\mathbb{R}^\ell)$ to the unit sphere $S^{\ell-1}$ in $\mathbb{R}^\ell$. If $d\omega_\ell$ is the $O(\ell)$ invariant measure on $S^{\ell-1}$, normalized according to (3-23) in §3, we let $Y_s(\ell,1),\ldots,$ $Y_s(\ell,d(s,\ell))$ (where $d(s,\ell)$ is the degree of the representation $[s]_\ell$) be a basis of $H^s(\mathbb{R}^\ell)$ so that it is orthonormal in $L^2(S^{\ell-1},d\omega_\ell)$. Moreover if $\chi_{\ell,[s]_\ell}$ is the character associated to $[s]_\ell$, then we choose $Y_s(\ell,1)$ so that

$$(6\text{-}5)\qquad \int_{O(\ell)} \chi_{\ell,[s]_\ell}(\xi_1\xi_2)\, d\sigma_\ell(\xi_1) = c\, Y_s(\ell,1)\,(\xi_2\cdot x_0)\,,$$

$$(\text{for } \xi_1 \in O(\ell))$$

where x_0 is a base vector in $\mathbb{R}^\ell$ with $\|x_0\| = 1$ and c, a specified constant. We observe that $Y_s(\ell,1)$ is the unique function of $H^s(\mathbb{R}^\ell)$ (up to scalar multiple) which is invariant by $O(\ell)^{X_0}$ = isotropy group of X_0 in $O(\ell)$. Moreover we know that $Y_s(\ell,1)(Z) = d\,\|Z\|^s C_s^{\frac{\ell-2}{2}}((\frac{Z}{\|Z\|},X_0))$ if $\ell > 2$, (where $C_s^{\frac{\ell-2}{2}}$ is the Gegenbauer polynomial function defined in [9,II], and d is a nonzero constant) and $Y_s(\ell,1)(Z) = d'\{(X_0+\sqrt{-1}\,X_1)^s + (X_0-\sqrt{-1}X_1)^s$ with d' a nonzero constant. (X_0, X_1 orthobasis of $\mathbb{R}^2$) .

We note that if $s = 1$, then $O(1)$ consists of 2 elements corresponding to the 2 nonequivalent representations of $\mathbb{Z}_2$, i.e. $\widehat{O(1)}$ consists of 2 elements, the trivial representation and the sign representation.

We let $E_{\mathfrak{m}}(\rho_m,\gamma_1 \otimes \gamma_2)$ be the isotypic component in $L^2(\mathbb{R}^k)$ relative to $\underline{\tilde{K}} \times K$ which transforms under $\pi_{\mathfrak{m}}$ according to the representation ρ_m of $\underline{\tilde{K}}$ and $\gamma_1 \otimes \gamma_2$ of K .

We recall the system of bipolar coordinates on the space $\mathbb{R}^k$. Indeed every point $X \in \mathbb{R}^k$ can be written in the form $r\omega_1 + s\omega_2$ with $r, s \geq 0$, and ω_1 and ω_2 belong to the unit spheres in $(e_1,\dots,e_a)$ and $(e_{a+1},\dots,e_n)$ respectively. Thus the representation $\pi_{\mathfrak{m}}$ restricted to K is equivalent to the representation of $O(a) \times O(b)$ on the tensor product $L^2(\mathbb{R}^a) \hat{\otimes} L^2(\mathbb{R}^b)$. Our problem is to determine $\lambda \in \mathbb{C}$ for which $\mathbb{F}_Q(\lambda) \neq 0$, the irreducibility of $\mathbb{F}_Q(\lambda)$ and the decomposition of $\mathbb{F}_Q(\lambda)$ when restricted to $\underline{\tilde{K}} \times K$.

Again returning to the regions $\Omega_{\pm}$, it is possible to coordinatize Ω_+ and Ω_- so that $\Omega_+ = \{t(\cosh(x)\,\omega_1 + \sinh(x)\,\omega_2) \mid t \in \mathbb{R}_+^* , x \in \mathbb{R}$, and ω_1,ω_2 as above$\}$ and $\Omega_- = \{t(\sinh(x)\,\omega_1 + \cosh(x)\,\omega_2) \mid t \in \mathbb{R}_+^*, x \in \mathbb{R} , \omega_1,\omega_2$ as above$\}$. Then relative to this system of coordinates and after computation:

$$W_\xi^+ = (\cosh x)^{1-a}(\sinh x)^{1-b}\frac{\partial}{\partial x}((\cosh x)^{a-1}(\sinh x)^{b-1}\frac{\partial}{\partial x}) + \frac{1}{\sinh^2 x}\nabla_{\omega_2} - \frac{1}{\cosh^2 x}\nabla_{\omega_1} \tag{6-6}$$

and

$$W_\xi^- = (\cosh x)^{1-b}(\sinh x)^{1-a}\frac{\partial}{\partial x}((\cosh x)^{b-1}(\sinh x)^{a-1}\frac{\partial}{\partial x})$$

$$+ \frac{1}{\sinh^2 x}\nabla_{\omega_1} - \frac{1}{\cosh^2 x}\nabla_{\omega_2} ,$$

where ∇_{ω_i} , $i = 1,2$ is the Laplace-Beltrami operator on the spheres S^{a-1} and S^{b-1} respectively.

Then we deduce from §5 that, taking $\begin{pmatrix} 0 & 1 \\ -1 & 0 \end{pmatrix}$ as the infinitesimal generator of the group $\underline{\tilde{K}}$, (here and in (6-8) and (6-9) below, we use $i = \sqrt{-1}$)

(6-7) $$\pi_{\mathfrak{m}}(\begin{pmatrix} 0 & 1 \\ -1 & 0 \end{pmatrix})_\infty = \pi i Q + \frac{1}{4\pi i}\partial(Q) .$$

Then relative to the coordinates in Ω_+ and Ω_- , we know that

(6-8) $$\pi i Q + \frac{1}{4\pi i}\partial(Q) = \begin{cases} \frac{1}{4\pi i}(\frac{\partial^2}{\partial t^2} + \frac{k-1}{t}\frac{\partial}{\partial t} - (4\pi^2 t^2 + \frac{1}{t^2}W_\xi^+) & \text{in } \Omega_+ \\ -\frac{1}{4\pi i}(\frac{\partial^2}{\partial t^2} + \frac{k-1}{t}\frac{\partial}{\partial t} - (4\pi^2 t^2 + \frac{1}{t^2}W_\xi^-) & \text{in } \Omega_- \end{cases} .$$

Moreover we recall that the operator $\pi_{\mathfrak{m}}(\omega_{(s\ell_2)})_\infty$ has a unique extension to a self adjoint operator (unbounded) on $L^2(\mathbb{R}^k)$; thus it follows that $\{\lambda \in \mathbb{C} | \mathbb{F}_Q(\lambda) \neq 0\} \subseteq \mathbb{R}$.

Lemma 6.2. Let $\lambda > -1$ *and* $\mathbb{F}_Q(\lambda) \neq 0$. *Then*

$$\lambda = \begin{Bmatrix} \nu^2 - \nu - 3/4 , & k \text{ odd} \\ \nu^2 - 1 , & k \text{ even} \end{Bmatrix} \text{ for } \nu \text{ integer} \begin{Bmatrix} \geqq 0 \\ \geqq 1 \end{Bmatrix} .$$

If $\mathbb{F}_Q^+(\lambda)$ ($\mathbb{F}_Q^-(\lambda)$ *respectively) is* $\{\varphi \in \mathbb{F}_Q(\lambda) | \varphi|_{\Omega-} = 0\}$ ($\{\varphi \in \mathbb{F}_Q(\lambda) | \varphi|_{\Omega+} = 0\}$ resp.), *then* $\mathbb{F}_Q^+(\lambda)$ ($\mathbb{F}_Q^-(\lambda)$ resp.) *is a closed* $\pi_{\mathfrak{m}}(\tilde{G}_2(Q))$ *invariant subspace of* $\mathbb{F}_Q(\lambda)$. *In addition the representations* $\pi_{\mathfrak{m}}$ *restricted to* $\mathbb{F}_Q^+(\lambda)$ *and* $\mathbb{F}_Q^-(\lambda)$ *are inequivalent.*

Proof. From the discussion above we take $s \in \frac{1}{2}\mathbb{Z}$ (we let $\tilde{s} = \sqrt{-1}\, s$) and integers s_1 and s_2 ; we consider the isotypic component (in $\mathbb{F}_Q(\lambda)$) $E_{\mathfrak{m}}(\lambda,s,s_1,s_2)$ of $\underline{\tilde{K}} \times K$ of type $(s, [s_1]_a \otimes [s_2]_b)$. Then if $\varphi \in E_{\mathfrak{m}}(\lambda,s,s_1,s_2)$ and $\varphi \neq 0$, we deduce that E_φ^+ (E_φ^- resp.) satisfies the 2nd order differential equation in t:

$$\text{(6-9)} \qquad \{\frac{d^2}{dt^2} + \frac{k-1}{t}\frac{d}{dt} + (-4\pi i\,\tilde{s} - 4\pi^2 t^2 - \frac{1}{t^2}(\lambda - \frac{k^2}{4} + k))\} E_\varphi^+(\ ,\xi) \equiv 0 \ .$$

$$(\{\frac{d^2}{dt^2} + \frac{k-1}{t}\frac{d}{dt} + (4\pi i\,\tilde{s} - 4\pi^2 t^2 - \frac{1}{t^2}(\lambda - \frac{k^2}{4} + k))\} E_\varphi^-(\ ,\xi) \equiv 0)$$

Then $\sigma_1(t) = t^{-k/2}E_\varphi^+(t^{-1},\xi)$ ($\sigma_2(t) = t^{-k/2}E_\varphi^-(t^{-1},\xi)$ resp.) satisfies a differential equation of type (4-22) (where we take $u = -1$ and relabeling λ in (4-22) as λ_1 to avoid confusion, we deduce that $\lambda_1 = \lambda$ above for both E_φ^+ and E_φ^- and $m = \begin{cases} -\tilde{s} \text{ for } E_\varphi^+ \\ \tilde{s} \text{ for } E_\varphi^- \end{cases}$. But $E_\varphi^{\pm}(\ ,\xi) \in L^2(\mathbb{R}_+^*, t^{k-1}dt)$ (for almost all $\xi \in \Gamma_{\pm 1}$) implies that $\sigma_j \in L^2(\mathbb{R}_+^*, \frac{dt}{t})$ $(j = 1,2)$.

Then we apply the methods of §4 to determine all σ which are solutions of (4-22) with $\sigma \in L^2(\mathbb{R}_+^*, \frac{dt}{t})$. First we note that $e^{-\pi t^{-2}} t^{-2i\tilde{s}} \phi_{\beta,-\tilde{s}}(t^{-2}) \in L^2([0,a], \frac{dt}{t})$ for $a > 0$ ($\beta = \frac{1}{2}\sqrt{1+\lambda}$ and this function is the only solution which has this property (up to scalar multiple). But then studying the behavior of the solutions of (4-22) at $t = \infty$ and again using the same arguments as in Proposition 4.6, we deduce that if σ satisfies (4-22) and $\sigma \in L^2(\mathbb{R}_+^*, \frac{dt}{t})$, then $M_{-1}^+(\lambda,\tilde{s}) \cap W_{-1}^+(\lambda,\tilde{s}) \neq (0)$. But from Remark 4.15, it follows that if $\lambda > -1$ and if $E_{\mathfrak{m}}(\lambda,s,s_1,s_2) \neq 0$, then we must have $\lambda = \begin{cases} \nu^2 - \nu - 3/4 & (\nu \text{ integer} \geq 0) \\ \nu^2 - 1 & (\nu \text{ integer} \geq 1) \end{cases}$. This means that if $\varphi \in E_{\mathfrak{m}}(\lambda,s,s_1,s_2)$ and $\varphi|_{\Omega_+} \neq 0$ ($\varphi|_{\Omega_-} \neq 0$ resp.) then

$$\tilde{s} = \begin{cases} \sqrt{-1}\,(\nu + 2j + \frac{1}{2}), \ j \text{ integer} \geq 0 \ \text{ with } \ \lambda = \nu^2 - \nu - 3/4 \\ \sqrt{-1}\,(\nu + 2j + 1), \ j \text{ integer} \geq 0 \ \text{ with } \ \lambda = \nu^2 - 1 \end{cases} .$$

$$(\tilde{s} = \begin{cases} -\sqrt{-1}\,(\nu + 2j' + \frac{1}{2}),\ j' \text{ integer } \geqq 0 \text{ with } \lambda = \nu^2 - \nu - 3/4 \\ -\sqrt{-1}\,(\nu + 2j' + 1),\ j' \text{ integer } \geqq 0 \text{ with } \lambda = \nu^2 - 1 \end{cases} \text{ resp.})$$

Thus if $\varphi|\Omega_+ \not\equiv 0$ ($\varphi|_{\Omega_-} \not\equiv 0$ resp.), it follows that $\varphi|_{\Omega_-} \equiv 0$ ($\varphi|_{\Omega_+} \equiv 0$ resp.).

But then we recall that $\pi_{\mathfrak{m}}((\begin{pmatrix}1 & 0\\ 0 & 1\end{pmatrix},\varepsilon)) = (\mathrm{sgn}\varepsilon)^k \cdot I$; hence $\mathbb{F}_Q(\lambda) \neq 0$

implies $\lambda = \begin{cases} \nu^2 - \nu - 3/4\ ,\ k \text{ odd } (\nu \text{ integer } \geqq 0) \\ \nu^2 - 1,\ k \text{ even } (\nu \text{ integer } \geqq 1) \end{cases}$.

Then we observe that $\mathbb{F}^+_Q(\lambda)$ and $\mathbb{F}^-_Q(\lambda)$ are closed subspaces of $\mathbb{F}_Q(\lambda)$ and that $\mathbb{F}^+_Q(\lambda) \cap \mathbb{F}^-_Q(\lambda) = \{0\}$ (recall $\mathbb{F}_Q \subseteq L^2(\mathbb{R}^k)$) . But from the last paragraph we deduce that $\mathbb{F}_Q(\lambda)_{\underline{\tilde{K}} \times K}$ (set of $\underline{\tilde{K}} \times K$ finite vectors) is the direct sum of $\mathbb{F}^+_Q(\lambda)_{\underline{\tilde{K}} \times K}$ and $\mathbb{F}^-_Q(\lambda)_{\underline{\tilde{K}} \times K}$. But since $\mathbb{F}_Q(\lambda)_{\underline{\tilde{K}} \times K}$ is dense in $\mathbb{F}_Q(\lambda)$, it follows that $\mathbb{F}^+_Q(\lambda)_{\underline{\tilde{K}} \times K}$ ($\mathbb{F}^-_Q(\lambda)_{\underline{\tilde{K}} \times K}$ resp.) is either <u>dense</u> or <u>zero</u> in $\mathbb{F}^+_Q(\lambda)$ ($\mathbb{F}^-_Q(\lambda)$ resp.). Thus we deduce that in any case $\mathbb{F}^+_Q(\lambda)$ and $\mathbb{F}^-_Q(\lambda)$ are stable under $\pi_{\mathfrak{m}}(\underline{\tilde{K}} \times K)$.

This then implies that $\mathbb{F}^+_Q(\lambda)$ and $\mathbb{F}^-_Q(\lambda)$ are stable by $\pi_{\mathfrak{m}}(\tilde{G}_2(Q))$. Indeed from above the spaces are stable under M_Q and $n(y)$ (see §4). But in the last paragraph we proved $\mathbb{F}^+_Q(\lambda)$ and $\mathbb{F}^-_Q(\lambda)$ are stable under $\underline{\tilde{K}}$. Since $\underline{\tilde{K}}$, M_Q , and $n(y)$ generate $\tilde{G}_2(Q)$, we are done.

Q. E. D.

<u>Remark 6.1</u>. We have seen in the course of the proof of Lemma 6.2 that if $E^{\pm}_{\varphi} \in E_{\mathfrak{m}}(\lambda,s,s_1,s_2)$, then $E^{\pm}_{\varphi}(t,\xi) = \rho^{\pm}_{\varphi}(t)\beta^{\pm}_{\varphi}(\xi)$, where $\rho^{\pm}_{\varphi} \in C^{\infty}(\mathbb{R}^*_+)$ and $\beta^{\pm}_{\varphi} \in C^{\infty}(\Gamma_{\underline{+}1})$. (i.e. a separation of variables).

At this point we recall the representation of $\tilde{S\ell}_2$ on the spaces $\overline{H}_\nu$ and $\overline{D}_\nu$, described in Proposition 4.15. We then note that by conjugation the matrix $\begin{pmatrix}1 & 0\\ 0 & -1\end{pmatrix} = A$ defines a Lie group outer automorphism of $PS\ell_2(\mathbb{R})$ ($\cong S\ell_2(\mathbb{R})/\mathbb{Z}_2$) . Moreover it is easy to see that this operation extends to an automorphism of $\tilde{S\ell}_2$ (since $\tilde{S\ell}_2$ is a covering of $PS\ell_2(\mathbb{R})$) . Then we define the unitary irreducible representations of $\tilde{S\ell}_2$ on $\overline{H}^A_\nu$ and

$\overline{D}^A_\nu$ by $\pi_u(g^A)|_{\overline{H}_\nu}$ and $\pi_u(g^A)|_{\overline{D}_\nu}$ respectively (g^A is the automorphism on $\widetilde{s\ell}_2$ induced by A) .

Lemma 6.3. Let $\lambda > -1$ *and* $\mathbb{F}_Q(\lambda) \neq 0$. *Then if* $\mathbb{F}^+_Q(\lambda) \neq 0$ ($\mathbb{F}^-_Q(\lambda) \neq 0$, *resp.) and we choose* s *and* s_1 , s_2 *so that* $0 \neq E_{\mathfrak{m}}(\lambda,s,s_1,s_2) \subseteq \mathbb{F}^+_Q(\lambda)$ ($\subseteq \mathbb{F}^-_Q(\lambda)$, resp.), *then* $\dim E_{\mathfrak{m}}(\lambda,s,s_1,s_2) = \dim([s_1]_a \otimes [s_2]_b)$. *Moreover if* $0 \neq E_{\mathfrak{m}}(\lambda,s,s_1,s_2) \subseteq \mathbb{F}^+_Q(\lambda)$ ($\subseteq \mathbb{F}^-_Q(\lambda)$, resp.), *then* (recall $\tilde{s} = \sqrt{-1}\, s$)

$$\text{(6-10)} \qquad \tilde{s} = \begin{cases} \sqrt{-1}\,(\nu + 2j + \frac{1}{2}) , & j \text{ integer} \geq 0 \text{ with } \lambda = \nu^2 - \nu - \frac{3}{4} , \text{ (k odd)} \\ \sqrt{-1}\,(\nu + 2j + 1) , & j \text{ integer} \geq 0 \text{ with } \lambda = \nu^2 - 1 \text{ (k even)} \end{cases}$$

(resp.

$$\tilde{s} = \begin{cases} -\sqrt{-1}\,(\nu + 2j' + \frac{1}{2}), & j' \text{ integer} \geq 0 \text{ with } \lambda = \nu^2 - \nu - \frac{3}{4} \text{ (k odd)} \\ -\sqrt{-1}\,(\nu + 2j' + 1), & j' \text{ integer} \geq 0 \text{ with } \lambda = \nu^2 - 1 \text{ (k even)} \end{cases}) .$$

Thus if we define $\mathbb{F}^{\pm}_Q(\lambda,s_1,s_2) = \{\varphi \in \mathbb{F}^{\pm}_Q(\lambda) \mid E^{\pm}_{\varphi}(\ ,\xi)$ transforms in ξ according to the rep. $[s_1]_a \otimes [s_2]_b$ of $K\}$, *then* $\mathbb{F}^+_Q(\lambda,s_1,s_2)$ *is a closed subspace of* $\mathbb{F}_Q(\lambda)$ *and is stable under* $\pi_{\mathfrak{m}}(\widetilde{s\ell}_2(Q))$. *Moreover the representation of* $\widetilde{s\ell}_2(Q)$ *on* $\mathbb{F}^+_Q(\lambda,s_1,s_2)$ ($\mathbb{F}^-_Q(\lambda,s_1,s_2)$, resp.) *is equivalent to the direct sum* $\dim([s_1]_a \otimes [s_2]_b)$ *times of the representation of* $\widetilde{s\ell}_2$ *on*

$$\begin{cases} (\overline{D}^A_\nu)_\infty \\ (\overline{H}^A_\nu)_\infty \end{cases} \text{(on} \begin{cases} (\overline{D}_\nu)_\infty \\ (\overline{H}_\nu)_\infty \end{cases} \text{resp.)}.$$

Proof. If $\varphi \in E_{\mathfrak{m}}(\lambda,s,s_1,s_2)$, then using Remark 6.1 we may write $\beta^+_\varphi(\xi) = \Upsilon^+(x)\, F^+_1(\omega_1)\, F^+_2(\omega_2)$ ($\beta^-_\varphi(\xi) = \Upsilon^-(x)\, F^-_1(\omega_1)\, F^-_2(\omega_2)$, resp.), where $F^{\pm}_1$ and $F^{\pm}_2$ satisfy $\nabla_{\omega_1}(F^{\pm}_1) = -\frac{1}{2}\, s_1(s_1 + a - 2)\, F^{\pm}_1$ and $\nabla_{\omega_2}(F^{\pm}_2) = -\frac{1}{2}\, s_2(s_2 + b - 2)\, F^{\pm}_2$, and $\Upsilon^{\pm}(x)$ satisfy

$$\text{(6-11)} \qquad (\cosh x)^{1-a}(\sinh x)^{1-b} \frac{d}{dx}((\cosh x)^{a-1}(\sinh x)^{b-1} \frac{d}{dx}(\Upsilon^+))$$

$$+ \{ -\frac{s_2(s_2+b-2)}{\sinh^2 x} + \frac{s_1(s_1+a-2)}{\cosh^2 x} + (-k + \frac{k^2}{4} - \lambda)\} \Upsilon^+ \equiv 0 ;$$

$$(\cosh x)^{1-b}(\sinh x)^{1-a} \frac{d}{dx}((\cosh x)^{b-1}(\sinh x)^{a-1} \frac{d}{dx}(\Upsilon^-))$$

$$+ \{- \frac{s_1(s_1+a-2)}{\sinh^2 x} + \frac{s_2(s_2+b-2)}{\cosh^2 x} + (-k+\frac{k^2}{4} - \lambda)\} \gamma^- \equiv 0 .$$

Then $\gamma^+(x) = \rho^+((\tanh x)^2)$ and $\gamma^-(x) = \rho^-((\tanh x)^2)$, where ρ^+ and ρ^- satisfy the 2nd order equations:

$$4y(1-y)\frac{d^2}{dy^2} + \{(2a-8)y+2b\} \frac{d}{dy} + \{-s_2(s_2+b-2) \frac{1}{y} + s_1(s_1+a-2) + (-k+\frac{k^2}{4} - \lambda) \frac{1}{1-y}\} (\rho^+) \equiv 0 ; \tag{6-12}$$

$$4y(1-y) \frac{d^2}{dy^2} + \{(2b-8)y + 2a\} \frac{d}{dy} + \{-s_1(s_1+a-2) \frac{1}{y} + s_2(s_2+b-2) + (-k + \frac{k^2}{4} - \lambda) \frac{1}{1-y}\} (\rho^-) \equiv 0 ,$$

where $0 < y < 1$. Then $\beta^{\pm}_{\varphi} \in L^2(\Gamma_{\pm 1}, d\mu_{\pm 1}(\xi))$ if and only if $\gamma^+ \in L^2(\mathbb{R}_+, (\cosh x)^{a-1}(\sinh x)^{b-1}dx)$ and $\gamma^- \in L^2(\mathbb{R}_+ , (\cosh x)^{b-1} (\sinh x)^{a-1} dx)$. But this is equivalent to the fact that $\rho^+ \in L^2((0,1), y^{\frac{1}{2}b-1}(1-y)^{-\frac{1}{2}k}dy)$ and $\rho^- \in L^2((0,1), y^{\frac{1}{2}a-1}(1-y)^{-\frac{1}{2}k}dy)$.

Then the indicial equation of (6-12) at $y = 1$ is given by $X^2 + (1 - \frac{1}{2}k)X + \frac{1}{4}\lambda^* = 0$, with $\lambda^* = -\lambda + \frac{1}{4}k^2 - k$. But the roots of this equation are $\frac{1}{2}((\frac{1}{2}k-1) \pm \sqrt{1+\lambda})$. However from Lemma 6.2

$$\sqrt{1+\lambda} = \begin{cases} \frac{1}{2}, \text{ if } \nu = 0 \text{ or } 1 \text{ and } k \text{ odd} \\ \nu - \frac{1}{2}, \text{ if } \nu > 1 \text{ and } k \text{ odd} \\ \nu, \ k \text{ even} \end{cases} \tag{6-13}$$

But then there exists a fundamental system $\{\alpha, \beta\}$ of solutions to (6-12) so that $\alpha(y) = (1-y)^{\frac{1}{4}k - \frac{1}{2} + \frac{1}{2}\sqrt{1+\lambda}}\varphi_1(y)$ and $\beta(y) = (1-y)^{\frac{1}{4}k - \frac{1}{2} - \frac{1}{2}\sqrt{1+\lambda}}\varphi_2(y) + \gamma \log(1-y)\alpha(y)$, where φ_1 is analytic at $y = 1$ with $\varphi_1(1) = 1$, φ_2 is analytic at $y = 1$ with $\varphi_2(1) = 1$ if $\sqrt{1+\lambda} \neq 0$, and $\varphi_2(1) = 0$ if $\sqrt{1+\lambda} = 0$, and γ, a specified constant. Then with $\sqrt{1+\lambda} > 0$, we see that $\alpha \in L^2((c,1),(1-y)^{-k/2}dy)$ with $0 < c < 1$. Moreover if $\sqrt{1+\lambda}$ is a positive integer, then $\log(1-y)\alpha(y) \in L^2((c,1),(1-y)^{-k/2}dy)$, and $(1-y)^{\frac{1}{4}k-\frac{1}{2}-\frac{1}{2}\sqrt{1+\lambda}}\varphi_2 \notin L^2((c,1),(1-y)^{-k/2}dy)$. Thus $\beta \notin L^2((c,1),(1-y)^{-k/2}dy)$.

Thus in any case we have deduced that <u>if</u> $E_{\mathfrak{m}}(\lambda,s,s_1,s_2) \neq 0$, then $\dim E_{\mathfrak{m}}(\lambda,s,s_1,s_2) = \dim([s_1]_a \otimes [s_2]_b)$.

Now we assume that $E_{\mathfrak{m}}(\lambda,s,s_1,s_2) \neq 0$ and $E_{\mathfrak{m}}(\lambda,s,s_1,s_2) \subseteq \mathbb{F}_Q^+(\lambda)$. By the argument given in Lemma 6.2 we easily deduce (6-10).

We now show that if $E_{\mathfrak{m}}(\lambda,s,s_1,s_2) \subseteq \mathbb{F}_Q^+(\lambda)$, then $E_{\mathfrak{m}}(\lambda,s \pm 2,s_1,s_2) \subseteq \mathbb{F}_Q^+(\lambda)$. Indeed if $\varphi \in E_{\mathfrak{m}}(\lambda,s,s_1,s_2)$, then

$$\text{(6-14)} \qquad \rho^+_{\pi_{\mathfrak{m}}(N_\pm)(\varphi)}(t) = t\frac{d}{dt} + \tfrac{1}{2}k \mp \sqrt{-1}\,(\pi\sqrt{-1}\,t^2 - (s - \pi\sqrt{-1}\,t^2))\,\rho^+_\varphi \ ,$$

(+ sign (- sign, resp.) corresponds to N_- (N_+ , resp.)). Then using the relation $\sigma_1(t) = t^{-k/2}\rho^+_\varphi(t^{-1})$ from the proof of Lemma 6.2, we deduce that

$$\text{(6-15)} \qquad \rho^+_{\pi_{\mathfrak{m}}(N_\pm)\varphi}(u^{-1}) = u^{k/2}\pi^A_{-1}(N_\pm)\,(\sigma_1 e^{s\theta})e^{-(\tilde{s}\mp 2\sqrt{-1})\theta} \ ,$$

(here we use the definition of $\pi_{-1}(N_\pm)$ given in (4-16) and (4-17)). Then for any element φ of $E_{\mathfrak{m}}(\lambda,s,s_1,s_2)$, we define a map to $\left\{\begin{matrix}\overline{D}^A_\nu \\ \overline{H}^A_\nu\end{matrix}\right.$ by $\varphi \rightsquigarrow t^{-k/2}\rho^+_\varphi(t^{-1})e^{\tilde{s}\theta}$. By what is observed in (6-15), this map is an <u>infinitesimal intertwining isomorphism</u> between the representation of $\pi_{\mathfrak{m}}(\mathcal{U}(s\ell_2))$ on $\bigoplus_{j=0}^{j=\infty} E_{\mathfrak{m}}(\lambda, \begin{cases}(\nu+2j+\tfrac{1}{2}), & k \text{ odd}\\ (\nu+2j), & k \text{ even}\end{cases}, s_1,s_2)$ and the representation of $\mathcal{U}(s\ell_2)$ on the set of $\tilde{\underline{K}}$ finite vectors in $\dim([s_1]_a \otimes [s_2]_b)\cdot \left\{\begin{matrix}\overline{D}^A_\nu \\ \overline{H}^A_\nu\end{matrix}\right.$ respectively. Thus we deduce that there is a $\mathcal{U}(s\ell_2)$ intertwining equivalence between the set of $\tilde{\underline{K}}$ finite vectors in $\mathbb{F}_Q^+(\lambda,s_1,s_2)$ and the set of $\tilde{\underline{K}}$ finite vectors in $\dim([s_1]_a \otimes [s_2]_b) \left\{\begin{matrix}\overline{D}^A_\nu & (k \text{ odd}) \\ \overline{H}^A_\nu & (k \text{ even})\end{matrix}\right.$. But the closure $\overline{\mathbb{F}_Q^+(\lambda,s_1,s_2)}$ of $\mathbb{F}_Q^+(\lambda,s_1,s_2)$ in the Hilbert space $L^2(\mathbb{R}^k)$ is a <u>closed</u> subspace of $L^2(\mathbb{R}^k)$, and is $\pi_{\mathfrak{m}}(S\tilde{\ell}_2(Q))$ <u>invariant</u> since $\mathbb{F}_Q^+(\lambda,s_1,s_2)$ is $\pi_{\mathfrak{m}}(S\tilde{\ell}_2(Q))$ invariant (follows from Lemma 6.2). But the set of $\tilde{\underline{K}}$ finite vectors of $\overline{\mathbb{F}_Q^+(\lambda,s_1,s_2)}$ coincides with the set of $\tilde{\underline{K}}$ finite vectors in $\mathbb{F}_Q^+(\lambda,s_1,s_2)$. Thus the unitary representations of $\tilde{s\ell}_2$ on $\overline{\mathbb{F}_Q^+(\lambda, s_1, s_2)}$ and on $\dim([s_1]_a \otimes [s_2]_b) \begin{cases} \bar{D}_\nu^A \\ \bar{H}_\nu^A \end{cases}$ are equivalent. Thus the differentiable representations of $\tilde{s\ell}_2$ on $\mathbb{F}_Q^+(\lambda, s_1, s_2)$ and on $\dim([s_1]_a \otimes [s_2]_b) \cdot \begin{cases} (\bar{D}_\nu^A)_\infty \\ (\bar{H}_\nu^A)_\infty \end{cases}$ are equivalent.

We repeat the same arguments for the case $\mathbb{F}_Q^-(\lambda, s_1, s_2)$.

Q. E. D.

Corollary 1 to Lemma 6.3. If $\lambda \leqq -1$, then $\mathbb{F}_Q(\lambda) \equiv 0$.

Proof. We look at the subspace $E_{\mathfrak{m}}(\lambda, s, s_1, s_2)$. By the argument in Lemma 6.3, if $\varphi \in E_{\mathfrak{m}}(\lambda, s, s_1, s_2)$, then $\varphi|_{\Gamma_{+1}} = \gamma^+(x)\ F_1^+(\omega_1)\ F_2^+(\omega_2)$ and γ^+ satisfies the differential equation (6-11). But then by the arguments in Lemma 6.3, we have for $\lambda < -1$ that there is a fundamental system $\{\alpha, \beta\}$ of (6-12) so that $\alpha(y) = (1-y)^{\frac{1}{2}(\frac{k}{2} - 1) + \frac{1}{2}\sqrt{1+\lambda}}\ \varphi_1(y)$ and $\beta(y) = (1-y)^{\frac{1}{2}(\frac{k}{2} - 1) - \frac{1}{2}\sqrt{1+\lambda}}\ \varphi_2(y)$ with φ_1 and φ_2 analytic at $y = 1$ and $\varphi_1(1) = \varphi_2(1) = 1$. Then we may write $A\alpha + B\beta = A(1-y)^{\frac{1}{4}k - \frac{1}{2} + \frac{1}{2}\sqrt{1+\lambda}} + B(1-y)^{\frac{1}{4}k - \frac{1}{2} - \frac{1}{2}\sqrt{1+\lambda}} + G(y)$, with $G(y) \in L^2((c,1),(1-y)^{-k/2}dy)$. Then $A\alpha + B\beta \in L^2((c,1),(1-y)^{-k/2}dy)$ if and only if $A(1-y)^{\frac{1}{4}k - \frac{1}{2} + \frac{1}{2}\sqrt{1+\lambda}} + B\ (1-y)^{\frac{1}{4}k - \frac{1}{2} - \frac{1}{2}\sqrt{1+\lambda}} \in L^2((c,1),(1-y)^{-k/2}dy)$. However by an argument similar to that used in Proposition 4.6, this is possible only if $A = 0$ and $B = 0$. The case $\lambda = -1$ follows exactly by the same reasoning used in a similar case in Proposition 4.6. Thus $\varphi|_{\Omega_+} \equiv 0$. Similar reasoning shows $\varphi|_{\Omega_-} \equiv 0$.

Q. E. D.

Corollary 2 to Lemma 6.3. *Let* $\mathbb{F}_Q(\lambda) \neq 0$. *Let* S_{ξ_0} *be any compact neighborhood of a point* $\xi_0 \in \Gamma_0$. *Then if a nonzero* $\varphi \in \mathbb{F}_Q(\lambda)$ *and* φ *is* $\tilde{K} \times K$ *finite, then the quantity*

$$\text{(6-16)} \qquad \sup_{X \in S_{\xi_0} - (S_{\xi_0} \cap \Gamma_0)} \left| \varphi(X) \left| Q(X,X) \right|^{\alpha} \right|$$

<u>is finite for all real numbers</u> α <u>so that</u>

$$\alpha \geqq \begin{cases} \frac{1}{2}(\frac{k-1}{2} - \nu) \text{ if } \lambda = \nu^2 - \nu - 3/4 \quad (k \text{ odd}) \\ \frac{1}{2}(\frac{k-2}{2} - \nu) \text{ if } \lambda = \nu^2 - 1 \quad (k \text{ even}) \end{cases}$$

<u>Proof</u>. We may assume from Lemma 6.2 that $\varphi \in \mathbb{F}^+_Q(\lambda)$ or $\varphi \in \mathbb{F}^-_Q(\lambda)$. Then it follows from Lemma 6.3 that if $\varphi \in E_{\mathcal{M}}(\lambda,s,s_1,s_2)$, then

$$\rho^{\pm}_{\varphi}(t) = e^{-\pi t^2}\psi^{\pm}(t^{-2}) \begin{cases} t^{\nu+2j+\frac{1}{2}-\frac{1}{2}k} & (k \text{ odd and } \nu^2 - \nu - 3/4 = \lambda) \\ t^{\nu+2j+1-\frac{1}{2}k} & (k \text{ even and } \nu^2 - 1 = \lambda) \end{cases},$$

where $\psi^{\pm}$ is a polynomial of one variable of degree j. Thus since the $\underline{\tilde{K}} \times K$ finite functions are finite linear combinations of elements belonging to the various $E_{\mathcal{M}}(\lambda,s,s_1,s_2)$, we then deduce the results of the Corollary.

Q. E. D.

<u>Remark 6.2</u>. If φ satisfies the hypotheses of Corollary 2 to Lemma 6.3 and

if $\begin{cases} \frac{k-1}{2} - \nu < 0, \ k \text{ odd} \\ \frac{k-2}{2} - \nu < 0, \ k \text{ even} \end{cases}$, then φ <u>extends to a continuous function on all of</u> $\mathbb{R}^k$ <u>and vanishes identically on</u> Γ_0.

We now define the "extreme vectors" for the metaplectic representation. Namely we consider the family of equations:

$$\text{(6-17)} \qquad (A_s) \begin{cases} \pi_{\mathcal{M}}(K)_\infty \varphi = is\varphi \\ \pi_{\mathcal{M}}(N_+)_\infty \varphi = 0 \end{cases} \qquad (B_s) \begin{cases} \pi_{\mathcal{M}}(K)_\infty \varphi = is\varphi \\ \pi_{\mathcal{M}}(N_-)_\infty \varphi = 0 \end{cases}$$

(with $i = \sqrt{-1}$).

Then we define ($s \in \frac{1}{2}\mathbb{Z}$)

$$\text{(6-18)} \qquad \mathcal{A}^+_s = \{\varphi \in \mathbb{F}_Q \mid \varphi \text{ satisfies system } (A_s)\}$$

$$\mathcal{A}^-_s = \{\varphi \in \mathbb{F}_Q \mid \varphi \text{ satisfies system } (B_s)\}$$

We observe that $\mathcal{A}_s^+$ and $\mathcal{A}_s^-$ are <u>closed</u> subspaces of $\mathbb{F}_Q$ and invariant by $\pi_{\mathfrak{M}}(O(Q))$. Moreover for $\psi \in \mathcal{A}_s^+$ ($\in \mathcal{A}_s^-$ resp.),

$$\pi_{\mathfrak{M}}(\omega_{O(Q)})(\psi) = -\pi_{\mathfrak{M}}(\omega_{(s\ell_2)})(\psi) + (\tfrac{1}{4}k^2 - k)\psi$$

$$= \begin{cases} -(s^2-2s+k-\tfrac{1}{4}k^2)\psi & \text{for } \psi \in \mathcal{A}_s^+ \\ -(s^2+2s+k-\tfrac{1}{4}k^2)\psi & \text{for } \psi \in \mathcal{A}_s^- \end{cases}$$

(see Remark 5.2). Then it follows that the space $\pi_{\mathfrak{M}}(s\tilde{\ell}_2(Q))(\mathcal{A}_s^{\pm}) \subseteq \mathbb{F}_Q(\lambda^{\pm}(s))$, where

$$\lambda^{\pm}(s) = \begin{cases} s^2 - 2s & \text{for } + \\ s^2 + 2s & \text{for } - \end{cases}$$

. Then for all $s \in \tfrac{1}{2}\mathbb{Z}$, we have $\lambda^{\pm}(s) > -1$, and then we can apply Lemma 6.2 and Lemma 6.3. In particular we deduce that for $\lambda^{\pm}(s) > -1$ the set of K finite vectors in $\mathcal{A}_s^+$ ($\mathcal{A}_s^-$ resp.) is the direct sum of $E_{\mathfrak{M}}(\lambda^+(s),s,s_1,s_2)$ ($E_{\mathfrak{M}}(\lambda^-(s),s,s_1,s_2)$ resp.) as s_1, s_2 range over integers describing the representations of $O(a)$ and $O(b)$ given above.

<u>Theorem 6.4</u>. <u>The representation</u> $\pi_{\mathfrak{M}}$ <u>of</u> $\tilde{G}_2(Q)$ <u>restricted to</u> $\mathbb{F}_Q^+(s^2-2s)$ ($\mathbb{F}_Q^-(s^2+2s)$ resp.) <u>is equivalent to the representation of</u> $\tilde{G}_2(Q)$ <u>on the tensor product</u>

$$\begin{Bmatrix} (\bar{D}^A_{s-\frac{1}{2}})_\infty, & k \text{ odd} \\ (\bar{H}^A_{s-1})_\infty, & k \text{ even} \end{Bmatrix} \hat{\otimes} \mathcal{A}_s^+ \quad \left(\begin{Bmatrix} (\bar{D}_{-(s+\frac{1}{2})})_\infty, & k \text{ odd} \\ (\bar{H}_{-(s+1)})_\infty, & k \text{ even} \end{Bmatrix} \hat{\otimes} \mathcal{A}_s^-, \text{ resp.} \right)$$

<u>provided</u> $\mathbb{F}_Q^+(s^2-2s) \neq 0$ ($\mathbb{F}_Q^-(s^2+2s) \neq 0$ resp.) <u>with</u> $s > 0$ ($s < 0$, resp.).

<u>Proof</u>. From above we know that the set of $K = O(a) \times O(b)$ finite vectors in $\mathcal{A}_s^+$ ($\mathcal{A}_s^-$ resp.) is the algebraic direct sum $\oplus E_{\mathfrak{M}}(s^2 \mp 2s,s,s_1,s_2)$ over all s_1 and s_2 where $E_{\mathfrak{M}}(\ ,\ ,s_1,s_2) \neq 0$. But from Lemma 6.3 we recall that $\mathbb{F}_Q^+(s^2-2s,s_1,s_2)$ is $s\tilde{\ell}_2(Q)$ equivalent to $\dim([s_1]_a \otimes [s_2]_b)$ times

$$\begin{Bmatrix} (\bar{D}^A_{s-\frac{1}{2}})_\infty & (k \text{ odd}) \\ (\bar{H}^A_{s-1})_\infty & (k \text{ even}) \end{Bmatrix}.$$

Thus the closure of $\pi_{\mathfrak{M}}(s\tilde{\ell}_2)(\mathcal{A}_s^+)$ in $\mathbb{F}_Q$ (in $\mathbb{F}_Q$

topology) equals $\mathbb{F}^{+}_{Q}(s^2-2s)$ and the statement of the theorem follows. We note that the case $\mathbb{F}^{-}_{Q}(s^2+2s)$ is argued similarly.

Q. E. D.

We recall at this point the Fock—Bargmann model of the metaplectic representation. First let $\mathbb{C}^k$ be complex k space so that the elements of $\mathbb{C}^k$ are of the form $X + \sqrt{-1}\, Y$, where $X, Y \in \mathbb{R}^k$. Then we recall the conjugation operation $\overline{}$ on $\mathbb{C}^k$ given by $X + \sqrt{-1}\, Y \to X - \sqrt{-1}\, Y$. Moreover we consider the complex bilinear form on $\mathbb{C}^k$ given by $[Z_1,Z_2] = [X_1,Y_1] - [X_2,Y_2] + \sqrt{-1}\,\{[X_1,Y_2] + [X_2,Y_1]\}$, where $Z_j = X_j + \sqrt{-1}\, Y_j$, $j = 1, 2$ and $[\ ,\ ]$ is defined on $\mathbb{R}^k$ in §1. Then we let $d\lambda_k(Z)$ be the measure on $\mathbb{C}^k$ given by $d\lambda_k(Z) = (\frac{1}{\pi})^k e^{-[Z,\bar{Z}]} dX\, dY$ $(Z=X+\sqrt{-1}\, Y)$. Then we let F_k be the space of analytic functions f on $\mathbb{C}^k$ provided with the inner product

$$(6\text{-}19) \qquad (f_1|f_2) = \int_{\mathbb{C}^k} \overline{f_1(Z)}\; f_2(Z)\; d\lambda_k(Z)\ .$$

Then F_k is a Hilbert space with this inner product. We also define the kernel $A\colon \mathbb{C}^k \times \mathbb{R}^k \to \mathbb{C}$ given by

$$(6\text{-}20) \qquad A(Z,X) = (\tfrac{1}{\pi})^{\frac{1}{4}k}\, e^{-\frac{1}{2}([Z,Z]+[X,X])+\sqrt{2}[Z,X]}\ .$$

Then following [19] we define a unitary isomorphism A^* of $L^2(\mathbb{R}^k)$ onto F_k by

$$(6\text{-}21) \qquad A^*(\varphi)(Z) = \int_{\mathbb{R}^k} A(Z,X)\varphi(X)\, dX$$

We know that for $\varphi \in L^2(\mathbb{R}^k)$ the integral is absolutely convergent and $A^*(\varphi)$ is an analytic function in $\mathbb{C}^k$ and is an element of F_k. The inverse $(A^*)^{-1}$ to A^* is given by

$$(6\text{-}22) \qquad (A^*)^{-1}(f)(X) = \underset{r\to\infty}{\text{l.i.m.}} \int_{[Z,\bar{Z}]<r} \overline{A(Z,X)}\; f(Z)\; d\lambda_k(Z)$$

(l.i.m. = limit in L^2 sense).

We also recall from [19] that if we take the linear topological space

$$(6\text{-}23) \qquad F' = \{f\colon \mathbb{C}^k \to \mathbb{C} \,|\, f \text{ holomorphic and } \|f\|_\rho =$$

$$\int_{\mathbb{C}^k} |f(Z)|^2 e^{-\rho[Z,\bar{Z}]} d\lambda_k(Z) < \infty \quad \text{for some real number } \rho\}$$

with the family of seminorms $\| \ \|_\rho$ for <u>all real</u> ρ, then $(A^*)^{-1}$ <u>extends</u> to an isomorphic linear mapping of $F' \to S(\mathbb{R}^k)'$ relative to the pairing $<(A^*)^{-1}(f), \varphi> = (f, A^*(\varphi))$, where

$$(6\text{-}24) \qquad (f, A^*(\varphi)) = \int_{\mathbb{C}^k} \overline{f(Z)}\, A^*(\varphi)(Z)\, d\lambda_k(Z) .$$

Then relative to the coordinate system used in §2 and §5, we compute for $\varphi \in S(\mathbb{R}^k)$ that

$$(6\text{-}25) \qquad A^*(x_j \cdot \varphi) = \left(\frac{1}{\sqrt{2}}\right)\left(z_j + \frac{\partial}{\partial z_j}\right) A^*(\varphi)$$

$$A^*\left(\frac{\partial}{\partial x_j} \varphi\right) = \left(\frac{-1}{\sqrt{2}}\right)\left(z_j - \frac{\partial}{\partial z_j}\right) A^*(\varphi)$$

Thus it follows that

$$(6\text{-}26) \qquad A^*(\partial(Q) \cdot \varphi) = \tfrac{1}{2}\{Q + \partial(Q) - 2\sum_{j=1}^{k} \varepsilon_j z_j \frac{\partial}{\partial z_j} + (b-a)\cdot I\}\, A^*(\varphi)$$

$$A^*(Q \cdot \varphi) = \tfrac{1}{2}\{Q + \partial(Q) + 2\sum_{j=1}^{k} \varepsilon_j z_j \frac{\partial}{\partial z_j} + (a-b)\cdot I\}\, A^*(\varphi)$$

$$A^*(E_1 \cdot \varphi) = \tfrac{1}{2}\{\sum_{j=1}^{k} \frac{\partial^2}{\partial z_j^2} - \sum_{j=1}^{k} z_j^2 - k \cdot I\}\, A^*(\varphi) ,$$

where Q is extended to a function on $\mathbb{C}^k$ and $\partial(Q)$ similarly is obtained from Q by replacing z_j by $\frac{\partial}{\partial z_j}$. Moreover $\varepsilon_j = \begin{cases} 1 & \text{if } 1 \leq j \leq a \\ -1 & \text{if } a+1 \leq j \leq k \end{cases}$. We note at this point that if P is any polynomial-coefficient, differential operator on $\mathbb{C}^k$, then P defines an unbounded operator on F_k with Domain $(P) = \{\psi \in F_k | P(\psi) \in F_k\}$.

<u>Lemma 6.5</u>. <u>The space</u> $A^*(\pi_{\mathfrak{m}}(a(\sqrt{2\pi}))\mathcal{A}_s^{\pm})$ <u>is characterized as</u> $\{f \in F_k | f$ satisfies system $\begin{Bmatrix} (R_s^+) \\ (R_s^-) \end{Bmatrix}\}$ <u>where</u>

$$\text{(6-27)}\qquad R_s^+ : \begin{cases} \sum_{j=1}^{k} \varepsilon_j z_j \frac{\partial}{\partial z_j}(f) = \{s - \tfrac{1}{2}(a-b)\} f , \\ \{\sum_{j=1}^{a} \frac{\partial^2}{\partial z_j^2} - \sum_{j=a+1}^{k} z_j^2\} (f) = 0 \end{cases}$$

$$R_s^- : \begin{cases} \sum_{j=1}^{k} \varepsilon_j z_j \frac{\partial}{\partial z_j}(f) = \{s - \tfrac{1}{2}(a-b)\} f , \\ \{\sum_{j=a+1}^{k} \frac{\partial^2}{\partial z_j^2} - \sum_{j=1}^{k} z_j^2\} (f) = 0 . \end{cases}$$

Proof. We first note that if $L = \pi_{\mathfrak{m}}(a(\sqrt{2\pi}))$, then $L^{-1}(\pi i Q)L = \tfrac{1}{2} iQ$ and $L^{-1}(-\frac{1}{4\pi i}\partial(Q))L = \tfrac{1}{2} i\, \partial(Q)$.

We know that if $\varphi \in \mathbb{F}_Q$, then $\varphi \in$ domain $(\pi_{\mathfrak{m}}(\xi))$ for any ξ in the enveloping algebra of $\tilde{G}_2(Q)$. But $\pi_{\mathfrak{m}}(\xi)$ is a polynomial-coefficient differential operator on $\mathbb{R}^k$. Similarly $(A^*)\pi_{\mathfrak{m}}(\xi)(A^*)^{-1}$ is a polynomial-coefficient differential operator on $\mathbb{C}^k$. Thus $A^*(\varphi) \in$ Domain $(A^*\pi_{\mathfrak{m}}(\xi)(A^*)^{-1})$. Then by computation we deduce that $\mathcal{A}_s^+$ ($\mathcal{A}_s^-$ resp.) is carried to R_s^+ (R_s^-, resp.).

Q. E. D.

We look first for all solutions to the systems $R_s^{\pm}$ in the ring $\mathcal{F}_k$ of all <u>formal power series</u> in $\mathbb{C}^k$. Here the problem is purely algebraic. Indeed we know that $K = O(a) \times O(b)$ acts reductively on $\mathcal{F}_k$ and the representations of K which occur in $\mathcal{F}_k$ are exactly the $[s_1]_a \otimes [s_2]_b$ discussed above. We let $\mathcal{F}_k^{\pm}(s,s_1,s_2) = \{\varphi \in F_k \mid \varphi$ transforms according to $[s_1]_a \otimes [s_2]_b$ and φ satisfies $R_s^{\pm}\}$.

We let $\Delta_1 = \sum_{j=1}^{a} z_j^2$ and $\Delta_2 = \sum_{j=a+1}^{k} z_j^2$. Then we have

<u>Lemma 6.6. The space $\mathcal{F}_k^+(s,s_1,s_2) = \{0\}$ ($\mathcal{F}_k^-(s,s_1,s_2) = \{0\}$ resp.) unless</u> $s_1 - s_2 - (s - \tfrac{1}{2}(a-b)) = 2\mu,\ \mu \in \mathbb{Z}_+$ $(s_2 - s_1 + s - \tfrac{1}{2}(a-b)) = 2\mu$, $\mu \in \mathbb{Z}_+$

resp.). Otherwise if $s_1 - s_2 - (s - \frac{1}{2}(a-b)) = 2\mu_+$ with $\mu_+ \in \mathbb{Z}_+$ ($s_2 - s_1 + (s - \frac{1}{2}(a-b)) = 2\mu_-$ with $\mu_- \in \mathbb{Z}_+$, resp.) then

(i) $\mathscr{F}_k^+(s,s_1,s_2)$ consists of all functions of the form

$$H_1^{s_1}(z_1,\dots,z_a)\cdot H_2^{s_2}(z_{a+1},\dots,z_k)\cdot\Delta_2^{\mu_+}\{\sum_{j=0}^{\infty} a_j^+(\Delta_1\Delta_2)^j\} \tag{6-28}$$

where $H_1^{s_1}$ and $H_2^{s_2}$ are harmonic polynomials of degree s_1 and s_2 in the variables $(z_1,\dots,z_a)$ and $(z_{a+1},\dots,z_k)$ respectively. Moreover

$$a_j^+ = \frac{\Gamma(s_1+\frac{1}{2}a)}{4^j\ j!\ \Gamma(s_1+\frac{1}{2}a+j)} .$$

(ii) $\mathscr{F}_k^-(s,s_1,s_2)$ consists of all functions of the form

$$\tilde{H}_2^{s_2}(z_{a+1},\dots,z_k)\ \Delta_1^{\mu_-}\ \tilde{H}_1^{s_1}(z_1,\dots,z_a)\ \{\sum_{j=0}^{\infty} a_j^-(\Delta_1\Delta_2)^j\} \tag{6-29}$$

where $\tilde{H}_1^{s_1}$ and $\tilde{H}_2^{s_2}$ are harmonic polynomials of degrees s_1 and s_2 in the variables $(z_1,\dots,z_a)$ and $(z_{a+1},\dots,z_k)$ respectively. Moreover

$$a_j^- = \frac{\Gamma(s_2+\frac{1}{2}b)}{4^j\ j!\ \Gamma(s_2+\frac{1}{2}b+j)}$$

Proof. First we know that every element $\xi \in \mathscr{F}_k$ can be written in the form $\sum_{\substack{m\geq 0\\ n\geq 0}} \xi_{(m,n)}$ where $\xi_{(m,n)}$ is an element of $\mathscr{P}^m(z_1,\dots,z_a) \otimes \mathscr{P}^n(z_{a+1},\dots,z_k)$ (i.e. $\mathscr{P}^m(z_1,\dots,z_a)$ and $\mathscr{P}^n(z_{a+1},\dots,z_k)$ are the spaces of homogeneous polynomials of degree m and degree n, respectively, in $(z_1,\dots,z_a)$ and $(z_{a+1},\dots,z_k)$.). Then we note that $\sum_{j=1}^{k} \varepsilon_j z_j \frac{\partial}{\partial z_j}(\xi_{(m,n)}) = (m-n)\xi_{(m,n)}$. Thus we must have $m - n = s - \frac{1}{2}(a-b)$. But if ξ transforms according to $[s_1]_a \otimes [s_2]_b$, then $m = s_1 + 2\lambda_m$ and $n = s_2 + 2\sigma_n$, where λ_m, σ_n are nonnegative integers (indeed the representation $[s_1]_a$ occurs in the space of polynomials $\bigoplus_{s=0}^{\infty} \mathscr{P}^s(z_1,\dots,z_a)$ once in each $\mathscr{P}^{s_1+2j}$ for $j \geq 0$ and zero otherwise).

We thus assume $\xi = \sum_j \xi_{(s_1+2\lambda_j, s_2+2\sigma_j)}$ where $s_1 - s_2 + 2(\lambda_j - \sigma_j) = s - \frac{1}{2}(a-b)$; we can further assume that $\lambda_j = j + \lambda_1$ and $\sigma_j = j + \sigma_1$. Thus if $\partial(\Delta_1)\xi = \Delta_2 \cdot \xi$ then $\partial(\Delta_1)(\xi_{(s_1+2\lambda_1, s_2+2\sigma_1)}) = 0$ and $\partial(\Delta_1)(\xi_{(s_1+2\lambda_{j+1}, s_2+2\sigma_{j+1})}) = \Delta_2\, \xi_{(s_1+2\lambda_j, s_2+2\sigma_j)}$ for $j \geq 0$. This implies that $\lambda_1 = 0$ (i.e. the $(z_1,\ldots,z_a)$ part of $\xi_{(s_1+2\lambda_1, s_2+2\sigma_1)}$ must be harmonic). Moreover if $\xi_{(s_1+2j, s_2+2\sigma_1+2j)} = \Delta_1^j \Delta_2^{\sigma_1+j} V_j$ (where V_j is an element of $H_1^{s_1}(z_1,\ldots,z_a) \otimes H_2^{s_2}(z_{a+1},\ldots,z_k))$, then $\partial(\Delta_1)(\Delta_1^j\Delta_2^{\sigma_1+j} V_j) = \Delta_1^{j-1}\Delta_2^{\sigma_1+j} V_{j-1}$ implies that $\partial(\Delta_1)(\Delta_1^j)\Delta_2^{\sigma_1+j} V_j + 2\{\sum_{i=1}^{a} \frac{\partial}{\partial z_i}(\Delta_1^j)\frac{\partial}{\partial z_i}(\Delta_2^{\sigma_1+j} V_j)\} = \Delta_1^{j-1}\Delta_2^{\sigma_1+j} V_{j-1}$. But by computation $\partial(\Delta_1)(\Delta_1^j) = 4j(j-1+\frac{1}{2}a)\Delta_1^{j-1}$ and hence we deduce that $\{4j(j-1+\frac{1}{2}a) + 4js_1\}\Delta_1^{j-1}\Delta_2^{\sigma_1+j} V_j = \Delta_1^{j-1}\Delta_2^{\sigma_1+j} V_{j-1}$. Thus we have $V_j = \frac{1}{4j(j-1+\frac{1}{2}a+s_1)} V_{j-1}$. This implies (i) of the Lemma. A similar argument works for (ii).

Q. E. D.

Remark 6.3. If $b = 1$, then since $s_2 = 0$ or $s_2 = 1$, we deduce from Lemma 6.6 that if $s < 0$ then $\mathscr{F}_k^-(s,s_1,s_2) = 0$ for all triples (s,s_1,s_2) (if $a > 2$) and for all triples $(s,s_1,s_2) \neq (-\frac{1}{2},0,1)$ (if $a = 2$).

Corollary 1 to Lemma 6.6. If $b = 1$, then for $s < 0$ the space $\mathcal{A}_s^- = 0$. Hence it follows (for $b = 1$) that $\mathbb{F}_Q^-(s^2+2s) = 0$ for $s < 0$.

Proof. From Remark 6.3 and Lemma 6.6 it follows that $\mathcal{A}_s^- = 0$ except when $a = 2$ and $s = -\frac{1}{2}$. Then $\mathcal{A}_{-\frac{1}{2}}^-$ (if nonzero) determines a one-dimensional representation of $O(Q)$ (i.e. $\mathcal{A}_{-\frac{1}{2}}^- = E_{\mathfrak{M}}(-3/4, -\frac{1}{2},0,1)$ in such a case). Thus $\varphi \in \mathcal{A}_{-\frac{1}{2}}^-$ determines a constant when restricted to Γ_t for $t < 0$. That is $\varphi(t\cdot\xi) = c(t)$ and hence

$$\infty > \int_{Q(X,X)<0} |\varphi(X)|^2\, dX = \{\int_0^\infty |c(t)|^2 t^{k-1} dt\}\{\int_{\Gamma_{-1}} d\mu_{-1}(\xi)\} \tag{6-30}$$

which is clearly impossible.

Q. E. D.

For any $s \in \frac{1}{2}\mathbb{Z}$ we let $V_s^{\pm} = \oplus \mathcal{F}_k^{\pm}(s,s_1,s_2)$ (the algebraic direct sum). Then $V_s^{\pm}$ is stable by the infinitesimal action of $\mathcal{U}(\mathfrak{G})$, ($\mathfrak{G}$, the Lie algebra of $O(Q)$) given by transferring the infinitesimal representation $\pi_{\mathfrak{m}}$ of $O(Q)$ to $\mathcal{F}_k$ via A^* given above.

We observe at this point if ψ is a harmonic polynomial of degree m on $\mathbb{C}^s$ ($s \geq 2$), then for any nonzero element $A \in \mathbb{C}^s$, $\frac{\partial\psi}{\partial(A)}$ (defined by $\frac{\partial\psi}{\partial(A)}(W) = \frac{d}{dt}\psi(W+tA)\big|_{t=0}$) is a harmonic polynomial of degree $m-1$. Moreover if z_i is a coordinate direction on $\mathbb{C}^s$, then we have the decomposition

$$(6\text{-}31) \qquad z_i \cdot \psi = W_{z_i}(\psi) + c \cdot \Delta_s \cdot \frac{\partial\psi}{\partial z_i}$$

where $\Delta_s = z_1^2 + \dots + z_s^2$ and $W_{z_i}(\psi)$ is a nonzero harmonic polynomial of degree $m+1$ and $c = \frac{1}{s+2m-2}$ (with $m > 0$). If $s = 1$, then we use the convention that $W_z(1) = z$ and $W_z(z) = 0$ for the harmonics 1 and z on $\mathbb{C}$.

We note, moreover, that (see (5-3)) for $\varphi \in S(\mathbb{R}^k)$

$$(6\text{-}32) \qquad A^*(N_{1,a+1}(\varphi)) = c'(z_1 z_{a+1} - \frac{\partial}{\partial z_1}\frac{\partial}{\partial z_{a+1}}) A^*(\varphi) ,$$

(c' some nonzero constant), where $N_{1,a+1}$ restricted to $S(\mathbb{R}^k)$ is just the differential of $\pi_{\mathfrak{m}}(A_a(e^x))$ at $x = 0$.

<u>Lemma 6.7</u>. <u>Let</u> $H_1^{s_1} \cdot H_2^{s_2} \Delta_2^{\mu_+} L_{s_1}(\Delta_1,\Delta_2)$ <u>be a nonzero element of</u> $\mathcal{F}_k^+(s,s_1,s_2)$ <u>with</u> $L_{s_1}(\Delta_1,\Delta_2) = \sum_{j=0}^{\infty} a_j^+(\Delta_1\Delta_2)^j$ (see Lemma 6.6). <u>Then</u>

$$(6\text{-}33) \qquad (z_1 z_{a+1} - \frac{\partial}{\partial z_1}\frac{\partial}{\partial z_{a+1}})(H_1^{s_1} H_2^{s_2} \Delta_2^{\mu_+} L_{s_1}(\Delta_1,\Delta_2)) =$$

$$\frac{(s_1+\frac{1}{2}(a-2)-\mu_+)}{(s_1+\frac{1}{2}a)} \{W_{z_1}(H_1^{s_1}) W_{z_{a+1}}(H_2^{s_2}) \Delta_2^{\mu_+} L_{s_1+1}(\Delta_1,\Delta_2)\}$$

$$+ (-2\mu_+)\{\frac{\partial H_1^{s_1}}{\partial z_1} W_{z_{a+1}}(H_2^{s_2})\Delta_2^{\mu_+-1} L_{s_1-1}(\Delta_1,\Delta_2)\} +$$

$$\frac{(s_1-s_2+\frac{1}{2}(a-b)-\mu_+)}{(s_1+\frac{1}{2}a)(b+2s_2-2)}\{W_{z_1}(H_1^{s_1})\frac{\partial H_2^{s_2}}{\partial z_{a+1}}\Delta_2^{\mu_++1} L_{s_1+1}(\Delta_1,\Delta_2)\}$$

$$+ \frac{(2-2\mu_+-2s_2-b)}{(b+2s_2-2)}\{\frac{\partial H_1^{s_1}}{\partial z_1}\frac{\partial H_2^{s_2}}{\partial z_{a+1}}\Delta_2^{\mu_+} L_{s_1-1}(\Delta_1,\Delta_2)\},$$

where in case $s_1 = 0$ *or* $s_2 = 0$ *the term containing* $\frac{\partial H_1^{s_1}}{\partial z_1}$ *or* $\frac{\partial H_2^{s_2}}{\partial z_{a+1}}$ *does not appear*.

Proof. The computation is done directly; we do term by term differentiation of the series $L_s(\Delta_1,\Delta_2)$ and make use of (6-31) given above.

Q. E. D.

Remark 6.4. If $\Delta_1^{\mu_-}\tilde{H}_1^{s_1}\tilde{H}_2^{s_2}\tilde{L}_{s_2}(\Delta_1,\Delta_2)$ is a nonzero element of $\mathscr{F}_k^-(s,s_1,s_2)$ with $\tilde{L}_{s_2}(\Delta_1,\Delta_2) = \sum_{j=0}^{\infty} a_j^-(\Delta_1,\Delta_2)^j$ (given by 6-29), then by a computation similar to Lemma 6.7

$$(6\text{-}34) \qquad (z_1 z_{a+1} - \frac{\partial}{\partial z_1}\frac{\partial}{\partial z_{a+1}})(\Delta_1^{\mu_-}\tilde{H}_1^{s_1}\tilde{H}_2^{s_2}\tilde{L}_{s_2}(\Delta_1,\Delta_2)) =$$

$$\frac{(s_2+\frac{1}{2}(b-2)-\mu_-)}{(s_2+\frac{1}{2}b)}\{\Delta_1^{\mu_-} W_{z_1}(\tilde{H}_1^{s_1}) W_{z_{a+1}}(\tilde{H}_2^{s_2})\tilde{L}_{s_2+1}(\Delta_1,\Delta_2)\}$$

$$+ (-2\mu_-)\{\Delta_1^{\mu_--1} W_{z_1}(\tilde{H}_1^{s_1})\frac{\partial \tilde{H}_2^{s_2}}{\partial z_{a+1}}\tilde{L}_{s_2-1}(\Delta_1,\Delta_2)\} +$$

$$\frac{(s_2-s_1+\frac{1}{2}(b-a)-\mu_-)}{(s_2+\frac{1}{2}b)(a+2s_1-2)}\{\Delta_1^{\mu_-+1}\frac{\partial \tilde{H}_1^{s_1}}{\partial z_1} W_{z_{a+1}}(\tilde{H}_2^{s_2})\tilde{L}_{s_2+1}(\Delta_1,\Delta_2)\}$$

$$+ \frac{(2-2\mu_--2s_1-a)}{(a+2s_1-2)}\{\Delta_1^{\mu_-}\frac{\partial \tilde{H}_1^{s_1}}{\partial z_1}\frac{\partial \tilde{H}_2^{s_2}}{\partial z_{a+1}}\tilde{L}_{s_2-1}(\Delta_1,\Delta_2)\},$$

where again in case $s_1 = 0$ or $s_2 = 0$ the term containing $\frac{\partial \tilde{H}_1^{s_1}}{\partial z_1}$ or

$\frac{\partial \tilde{H}_2^{s_2}}{\partial z_{a+1}}$ vanishes.

Theorem 6.8. If $s \geq 1$ ($s < 0$ resp.), then the infinitesimal representation of $\mathscr{U}(\mathfrak{S})$ induced by $A^* \cdot \pi_{\mathfrak{m}}$ above on V_s^+ (V_s^- resp.) is algebraically irreducible. Thus the representation $\pi_{\mathfrak{m}}$ of $O(Q)$ on $\mathcal{A}_s^+$ (on $\mathcal{A}_s^-$ for $s < 0$, resp.) is irreducible (i.e. the only closed subspaces of $\mathcal{A}_s^+$ ($\mathcal{A}_s^-$ resp.) are $\mathcal{A}_s^+$ ($\mathcal{A}_s^-$) and $\{0\}$).

Proof. We note that if $H_1^{s_1} H_2^{s_2} \Delta_2^{\mu_+} L_{s_1}(\Delta_1, \Delta_2) \in \mathscr{F}_k^+(s, s_1, s_2)$, then $W_{z_1}(H_1^{s_1}) W_{z_{a+1}}(H_2^{s_2})\Delta_2^{\mu_+} L_{s_1+1}(\Delta_1, \Delta_2) \in \mathscr{F}_k^+(s, s_1+1, s_2+1)$. Similarly we see that each of the terms $\{\ldots\}$ in (6-33) (if they are nonzero) also belong to V_s^+.

Then the algebraic irreducibility of V_s^+ under $\mathscr{U}(\mathfrak{S})$ will follow provided each of the coefficients of $\{\ldots\}$ in (6-33) does not vanish (i.e. $O(Q)$ is generated by $K = O(a) \times O(b)$ and $A_a(r)$). This means that we must have $s_1 + \frac{1}{2}(a-2) \neq \mu_+$, $s_1 - s_2 + \frac{1}{2}(a-b) \neq \mu_+$, and $1-(s_2+\frac{1}{2}b) \neq \mu_+$. But we recall that $s_1 - s_2 = 2\mu_+ + (s - \frac{1}{2}(a-b))$, $\mu_+ \geq 0$ and the hypothesis on s in the Theorem. Then we obtain easily the inequalities required.

The proof for the irreducibility of V_s^- follows in the same way with the hypothesis on s given in Theorem.

Q. E. D.

Corollary 1 to Theorem 6.8. The representation $\pi_{\mathfrak{m}}$ of $\tilde{G}_2(Q)$ on $\mathbb{F}_Q^+(s^2-2s)$ with $s \geq 1$ is irreducible. And the representation $\pi_{\mathfrak{m}}$ of $\tilde{G}_2(Q)$ on $\mathbb{F}_Q^-(s^2+2s)$ for $s < 0$ is irreducible.

Proof. We recall Theorem 6.4 and use Theorem 6.8.

Q. E. D.

We let $U_{\pm 1}$ be the Hilbert space $L^2(\Gamma_{\pm 1}, d\mu_{\pm 1})$ and $\pi_{\pm 1}$, the corresponding representation of $O(Q)$ on $U_{\pm 1}$. We know that $(U_{\pm 1})_\infty = \{\varphi \in C^\infty(\Gamma_{\pm 1}) \mid D * \varphi \in U_{\pm 1}$ for all $D \in \mathscr{U}(\mathfrak{S})\}$.

We let $U_{\underline{+1}}(\lambda) = \{\varphi \in (U_{\underline{+1}})_\infty | W_\xi^{\pm} * \varphi = (\lambda + k - \frac{1}{4}k^2)\varphi\}$. Then $U_{\underline{+1}}(\lambda)$ is a closed subspace of $(U_{\underline{+1}})_\infty$ and is $\pi_{\underline{+1}}(O(Q))$ stable.

<u>Theorem 6.9</u>. <u>The restriction map of</u> $\varphi \in (\mathcal{A}_s^+)_K$ $(\varphi \in (\mathcal{A}_s^-)_K$, resp.) <u>to</u> $\varphi|_{\Gamma_{+1}}$ $(\varphi|_{\Gamma_{-1}}$ resp.) <u>is a bijective</u> $\mathcal{U}(\mathfrak{S})$ <u>intertwining map between</u> $(\mathcal{A}_s^+)_K$ $((\mathcal{A}_s^-)_K$ resp.) <u>and</u> $U_{+1}(s^2-2s)_K$ $(U_{-1}(s^2+2s)_K$ resp.). <u>Thus the representation of</u> $O(Q)$ <u>on</u> $\mathcal{A}_s^+$ $(\mathcal{A}_s^-$ resp.) <u>is equivalent to the representation of</u> $O(Q)$ <u>on</u> $U_{+1}(s^2-2s)$ $(U_{-1}(s^2+2s)$, resp.). <u>And if</u> $s \geqq 1$ <u>the representation</u> π_{+1} <u>of</u> $O(Q)$ <u>on</u> $U_{+1}(s^2-2s)$ <u>is irreducible</u>. <u>Similarly if</u> $s < 0$ <u>the representation</u> π_{-1} <u>of</u> $O(Q)$ <u>on</u> $U_{-1}(s^2+2s)$ <u>is irreducible</u>.

<u>Proof</u>. We know from Lemma 6.3 explicitly the form of a $K = O(a) \times O(b)$ isotypic component of $\mathbb{F}_Q^+(s^2-2s)$. But noting the separation of variables of elements of $E_{\mathfrak{M}}(s^2-2s,s,s_1,s_2)$ given in Remark 6.1, it follows that the restriction map of $(\mathcal{A}_s^+)_K$ to $U_{+1}(s^2-2s)_K$ is $\mathcal{U}(\mathfrak{S})$ intertwining and injective. Then if $\psi \in U_{+1}(s^2-2s)$ is of isotypic type $[s_1]_a \otimes [s_2]_b$ for K, we can assume that ψ has the form $\psi(\xi) = \gamma^+(x)\, F_1^+(\omega_1)\, F_2^+(\omega_2)$, where γ^+, F_1^+ and F_2^+ satisfy the same conditions as given in the proof of Lemma 6.3. Then we choose $\rho^+(t)$ to satisfy (6-9) and so that $\alpha(t) = t^{-\frac{1}{2}k} \rho^+(t^{-1}) \in L^2(\mathbb{R}_+^*, \frac{dt}{t})$. This is possible by the choice of s and the argument in Lemma 6.2. Thus $\rho^+(t)\psi(\xi)$ defines a function $\varphi \in E_{\mathfrak{M}}(s^2-2s,s,s_1,s_2)$.

Thus $(\mathcal{A}_s^+)_K$ and $U_{+1}(s^2-2s)_K$ are $\mathcal{U}(\mathfrak{S})$ equivalent. This implies that the closures of $\mathcal{A}_s^+$ in $L^2(\mathbb{R}^k)$ and of $U_{+1}(s^2-2s)$ in $L^2(\Gamma_{+1}, d\mu_{+1})$ are unitarily equivalent representations of the connected component of $O(Q)$. But it is also clear from above that the respective closures of $\mathcal{A}_s^+$ and $U_{+1}(s^2-2s)$ are K equivalent. But since $O(Q)$ is generated by K and the connected component of $O(Q)$, we have, in fact, $O(Q)$ unitary equivalence. We then recall that $\mathcal{A}_s^+$ and $U_{+1}(s^2-2s)$ are the C^∞ vectors for the respective $O(Q)$ unitary representations. Thus $\mathcal{A}_s^+$ and $U_{+1}(s^2-2s)$ give equivalent $O(Q)$ representations.

A similar argument works in the case of $\mathcal{A}_s^-$.

Q. E. D.

The problem at hand is to determine for which s the relation $\mathcal{A}_s^+ \neq 0$ ($\mathcal{A}_s^- \neq 0$, resp.) holds. From the proof of Lemma 6.3 we recall that if $\varphi \in E_{\mathfrak{m}}(s^2-2s,s,s_1,s_2)$, then φ has the form $\varphi(t\cdot\xi) = \rho_\varphi^+(t)\ \beta_\varphi^+(\xi)$ with $\beta_\varphi^+(\xi) = \gamma^+(x)\ F_1^+(\omega_1)\ F_2^+(\omega_2)$ and $\gamma^+(x) = \rho^+(\tanh^2 x)$, where ρ^+ satisfies (6-12). But (6-12) is related to the hypergeometric equation. Indeed if we let $\rho^+(u) = u^L\ (1-u)^M\ \delta_+(u)$ with $L = \frac{1}{2}s_2$ and $M = \frac{1}{2}\{(\frac{1}{2}k - 1) + \sqrt{1+\lambda}\}$ (with $\sqrt{1+\lambda}$ having the special form given by (6-13)), then δ_+ satisfies the hypergeometric equation

$$\text{(6-35)} \qquad \left(z(1-z)\frac{d^2}{dz^2} + \{C-(A+B+1)z\}\frac{d}{dz} - A\,B\right)\delta_+ \equiv 0$$

for $0 < z < 1$ where $A = \frac{1}{2}\left[\frac{1}{2}k + (s_1+s_2) + \sqrt{1+\lambda} - 1\right]$, $B = \frac{1}{2}\left[b + s_2-s_1 + \sqrt{1+\lambda} - \frac{1}{2}k + 1\right]$, and $C = s_2 + \frac{1}{2}b$. Then from Lemma 6.6, we note that s_1 and s_2 satisfy the relationship $s_1-s_2 = 2\mu_+ + \frac{1}{2}(b-a) + \begin{cases} \nu + \frac{1}{2}, & k \text{ odd} \\ \nu + 1, & k \text{ even}\end{cases}$ for some nonnegative integer μ_+ (see Theorem 6.4 for the relation between ν and s). Then using the special form of $\sqrt{1+\lambda}$ given by (6-13), we deduce that with $s > 1$, $B = \frac{1}{2}\left[b + \frac{1}{2}(a-b) - 2\mu_+ - \frac{1}{2}k\right] = -\mu_+$. Hence it follows that there is a nonzero solution V of (6-35) which is a polynomial. We note that if $s = \frac{1}{2}$, then $B = -\mu_+ + \frac{1}{2}$.

<u>Theorem 6.10</u>. <u>(Non-Vanishing Criterion)</u>. <u>We assume</u> $s \equiv \frac{1}{2}k \bmod 1$. <u>Moreover the space</u> $E_{\mathfrak{m}}(s^2-2s,s,s_1,s_2) \neq 0$ <u>if and only if</u> $s > 1$ <u>and</u> $s_1-s_2 = (s - \frac{1}{2}(a-b)) + 2\mu_+$ <u>for a nonnegative integer</u> μ_+. <u>Thus</u> $\mathcal{A}_s^+ \neq 0$ <u>and</u> $\mathbb{F}_Q^+(s^2-2s) \neq 0$ <u>for</u> $s > 1$. <u>Similarly the space</u> $E_{\mathfrak{m}}(s^2+2s,s,s_1,s_2) \neq 0$ <u>if and only if</u> $s < -1$ <u>and</u> $s_2-s_1 = -(s - \frac{1}{2}(a-b)) + 2\mu_-$ <u>for a nonnegative integer</u> μ_-. <u>Thus</u> $\mathcal{A}_s^- \neq 0$ <u>and</u> $\mathbb{F}_Q^-(s^2+2s) \neq 0$ <u>for</u> $s < -1$ (provided $b > 1$ <u>with Corollary</u> 1 <u>to Lemma</u> 6.6 <u>in mind</u>).

Proof. If $s > 1$ and $s_1 - s_2 = (s - \frac{1}{2}(a-b)) + 2\mu_+$ as above. Then applying the reasoning given before the Theorem, we can construct an element $\rho^+(u) = u^L (1-u)^M V(u) \in L^2((0,1), u^{\frac{1}{2}b-1} (1-u)^{-\frac{1}{2}k} du)$. Thus $E_{\mathfrak{m}}(s^2-2s, s, s_1, s_2) \neq 0$. Conversely the characterization of $E_{\mathfrak{m}}(s^2 - 2s, s, s_1, s_2) \neq 0$ is given in Lemma 6.3 and Lemma 6.6. If $s = \frac{1}{2}$ then we must show that $E_{\mathfrak{m}}(-3/4, \frac{1}{2}, s_1, s_2) = 0$. We go back to (6-12) and determine that the indicial equation of (6-12) at $y = 0$ is given by $X^2 + (\frac{1}{2}b-1)X + \frac{1}{4}(-s_2(s_2+b-2)) = 0$. The roots are $\frac{1}{2}s_2$ and $1 - \frac{1}{2}b - \frac{1}{2}s_2$. Hence it follows that there exists a solution of (6-12) of the form $y^{\frac{1}{2}s_2} L(y)$ with L analytic at 0 and $L(0) = 1$. Thus if there is a nontrivial solution σ of (6-12) so that $\sigma \in L^2((0,1), y^{\frac{1}{2}b-1}(1-y)^{-\frac{1}{2}k})$, then σ must be of the form $y^{\frac{1}{2}s_2}(1-y)^{\frac{1}{4}k-\frac{1}{4}}G(y)$, with G an analytic function in y on all of $\mathbb{C}$. But with $\lambda = -3/4$ above, we see that G is a solution of (6-35). Hence we know that G must be a nonzero multiple of ${}_2F_1(A,B,C,y)$, the hypergeometric series (see [9,I]), for y in a small neighborhood of 0, and A, B, and C as above (i.e. since the indicial roots of (6-35) at $z = 0$ are 0 and $1 - s_2 - \frac{1}{2}b$, then by separate examination of the cases, the only solution of (6-35) analytic at 0 is a scalar multiple of ${}_2F_1(A,B,C,z)$.). But ${}_2F_1(A,B,C,y)$ has an infinite radius of convergence only if A or B is a nonpositive integer. However this is impossible from the comments above. Thus it follows that $E_{\mathfrak{m}}(-3/4, \frac{1}{2}, s_1, s_2) = 0$.

Similar reasoning as above is used to deduce the results for the spaces $E_{\mathfrak{m}}(s^2 + 2s, s, s_1, s_2)$.

Q. E. D.

Corollary 1 to Theorem 6.10. Let φ *be a* $\tilde{K} \times K$ *finite vector in* $\mathbb{F}^+_Q(s^2 - 2s)$ ($\mathbb{F}^-_Q(s^2 + 2s)$, resp.).

(i) *Then for* $s \geq \frac{1}{2}k$ ($s \leq -\frac{1}{2}k$, resp.), $\sup_{X \in \mathbb{R}^k} (1 + \|X\|^2)^\beta |\varphi(X)| < \infty$ *for all* $\beta \in \mathbb{R}$ *so that* $\beta \leq \frac{1}{2}(s + \frac{1}{2}k - 2)$ ($\beta \leq \frac{1}{2}(-s+\frac{1}{2}k - 2)$, resp.).

(ii) φ <u>extends to a continuous function on</u> $\mathbb{R}^k - \{0\}$ <u>which vanishes on</u> $\Gamma_0 - \{0\}$ <u>when</u> $s > 1$ ($s < -1$ resp.).

<u>Proof</u>. We let $\varphi \in E_{\mathfrak{m}}(s^2 - 2s, s, s_1, s_2)$ where s_1 and s_2 satisfy the conditions given in Theorem 6.10. Then $\varphi(t \cdot \xi) = \rho_{\varphi}^+(t)\gamma^+(x)H^{s_1}(\tilde{\omega}_1)H^{s_2}(\tilde{\omega}_2)$. Also $\gamma^+(x) = \rho^+((\tanh x)^2)$ and from the paragraph preceding Theorem 6.10 we have that $\rho^+(u) = u^{\frac{1}{2}s_2}(1-u)^{\frac{1}{4}k - \frac{1}{2} + \frac{1}{2}\sqrt{1+\lambda}}\beta^*(u)$, with β^* a polynomial in u, and $\sqrt{1+\lambda}$ given by (6-13).

Then if $r = \|X_+\|$ and $s = \|X_-\|$, we have $u = \frac{s^2}{r^2}$ and $t = r^2 - s^2$. Then $1 + \|X\|^2 = 1 + r^2 + s^2 = 1 + t(\frac{1+u}{1-u}) = [(1-u) + t(1+u)](1-u)^{-1}$ with $0 < u < 1$ and $t > 0$. Thus using the argument in Corollary 2 to Lemma 6.3,

$$(6\text{-}36) \qquad (1 + \|X\|^2)^{\beta}|\varphi(X)| \leq C\, e^{-\pi t^2}\{(1-u) + t(1+u)\}^{\beta}|\psi^+(t^{-2})| \\ (1-u)^{\frac{1}{2}\sqrt{1+\lambda} - \frac{1}{2} + \frac{1}{4}k - \beta} u^{\frac{1}{2}s_2}|\beta^*(u)| \begin{cases} t^{\nu + 2j + \frac{1}{2} - \frac{1}{2}k} & (k \text{ odd}) \\ t^{\nu + 2j + 1 - \frac{1}{2}k} & (k \text{ even}) \end{cases},$$

where ψ^+ is a polynomial in one variable of degree j and C, a positive constant. Then the quantity on the right hand side of (6-36) is bounded provided $s \geq \frac{1}{2}k$ and $\frac{1}{2}\sqrt{1+\lambda} + \frac{1}{4}k - \frac{1}{2} \geq \beta$. Then using the expression (6-13) we deduce (i) of the Corollary.

To study the continuity of φ on $\Gamma_0 - \{0\}$, we must determine $\lim_{Y \to \xi^*} \varphi(Y)$ where $\xi^* \in \Gamma_0 - \{0\}$ and $Y \in \Omega_+$ (recall $\varphi \in \mathbb{F}_{\Omega}^+(s^2 - 2s)$ vanishes on Ω_-). Then from above we know that $t^2 = r^2 - s^2 = r^2(1-u)$ and moreover, as $Y \to \xi^* \in \Gamma_0 - \{0\}$, we have $t \to 0$ and $u \to 1$. Also since $\xi^* \neq 0$ we observe that $\lim_{Y \to \xi^*} r \neq 0$. Thus to determine the behavior of $\varphi(Y)$ as $Y \to \xi^*$ we need examine only $\psi^+(t^{-2})\beta^*(u)(1-u)^{\frac{1}{2}\sqrt{1+\lambda} - \frac{1}{2} + \frac{1}{4}k}\begin{cases} t^{\nu+2j+\frac{1}{2}-\frac{1}{2}k} & (k \text{ odd}) \\ t^{\nu+2j+1-\frac{1}{2}k} & (k \text{ even}) \end{cases}$ as $t \to 0$ and $u \to 1$. But then since $\psi^+(t^{-2})$ is a polynomial of degree j and β^* a polynomial in u, we need only study

$$(1-u)^{\frac{1}{2}\sqrt{1+\lambda} - \frac{1}{2} + \frac{1}{4}k}\begin{cases} t^{\nu - \frac{1}{2}(k-1)}, & k \text{ odd} \\ t^{\nu - \frac{1}{2}(k-2)}, & k \text{ even} \end{cases}.$$ Using the relation between t,

u, and r above, we see that this expression $\to 0$ as $u \to 1$ when $s > 1$ (recall relation between ν and s given in Theorem 6.4).

Also the arguments for $\varphi \in \mathbb{F}_Q^-(s^2 + 2s)$ are given in a similar way.

Q. E. D.

We let ω_1 and ω_2 be elements of S^{a-1} and S^{b-1} in the planes $(e_1,\ldots,e_a)$ and $(e_{a+1},\ldots,e_k)$ respectively. Then we define $F_\gamma(Z,\omega_1,\omega_2) = e^{\gamma Q(Z,\omega_1)Q(Z,\omega_2)}$ for $\gamma \in \mathbb{R}$. If $|\gamma| < 1$, we deduce that $F_\gamma(Z,\omega_1,\omega_2) \in F_k$. We then recall the group $A_a(r)$ defined in §2.

<u>Lemma 6.11</u>. <u>Let</u> $k \in K$ <u>so that</u> $k(e_1) = \omega_1$ <u>and</u> $k(e_{a+1}) = \omega_2$. <u>Then for</u> $|\gamma| < 1$,

$$(6\text{-}37)\qquad (A^*)^{-1}(F_\gamma(\ ,\omega_1,\omega_2)) = c(1-\gamma^2)^{-\frac{1}{2}} e^{-\frac{1}{2}\left[A_a(\rho_\gamma)k^{-1}X,A_a(\rho_\gamma)k^{-1}X\right]},$$

<u>where</u> $\rho_\gamma = \left(\frac{1+\gamma}{1-\gamma}\right)^{\frac{1}{2}}$ <u>and</u> c <u>a nonzero constant</u>.

<u>Proof</u>. It suffices to compute $(A^*)^{-1}(F_\gamma(\ ,e_1,e_{a+1}))$. But then the integral

$$(6\text{-}38)\qquad \int_{\mathbb{C}^k} \overline{A(Z,X)}\, F_\gamma(Z,e_1,e_{a+1})\, d\lambda_k(Z) = I_\gamma(X)$$

converges absolutely for $|\gamma| < 1$. Indeed $I_\gamma(X)$ can be written in the form

$$(6\text{-}39)\qquad \left\{\int_{\mathbb{C}^2} e^{\gamma z_1 z_{a+1}} e^{\sqrt{2}\,(x_1\bar{z}_1+x_{a+1}\bar{z}_{a+1})} e^{-\frac{1}{2}(\bar{z}_1^2+\bar{z}_{a+1}^2)} e^{-(|z_1|^2+|z_{a+1}|^2)}\, dz_1\, dz_{a+1}\right\}$$
$$\left\{\int_{\mathbb{C}^{k-2}} e^{\sqrt{2}\,(\sum_{i\neq 1,a+1}^{k} x_i\bar{z}_i)} e^{-(\sum_{i\neq 1,a+1}^{k}(|z_i|^2+\frac{1}{2}\bar{z}_i^2))} \prod_{i\neq 1,a+1} dz_i\right\}.$$

However from [19] we know that the second part of $I_\gamma(X)$ is given by $c_1 e^{-\frac{1}{2}(\sum_{i\neq 1,a+1}^{k} y_i^2)}$ with $c_1 \neq 0$. Thus we must compute the first integral in (6-39) which is a two dimensional problem. Then if $\lambda_1, \lambda_2 \in \mathbb{C}$ with

$\mathrm{Re}(\lambda_1) > 0$, we recall the relation

$$(6\text{-}40) \qquad \int_{\mathbb{R}} e^{-\lambda_1 x^2 + 2\lambda_2 x}\, dx = \left(\frac{\pi}{\lambda_1}\right)^{\frac{1}{2}} e^{\lambda_2^2/\lambda_1}$$

where $\mathrm{Re}\left(\left(\frac{\pi}{\lambda_1}\right)^{\frac{1}{2}}\right) > 0$. Then applying this relation 4 times and making appropriate change of variables, we deduce that the first integral in (6-39) equals

$$(6\text{-}41) \qquad c_2 \cdot (1-\gamma^2)^{-\frac{1}{2}} e^{-\frac{1}{2}\left\{\frac{1+\gamma^2}{1-\gamma^2}(x_1^2 + x_{a+1}^2) + \frac{4\gamma}{1-\gamma^2} x_1 x_{a+1}\right\}} .$$

But then we note that $A_a\left(\left(\frac{1+\gamma}{1-\gamma}\right)^{\frac{1}{2}}\right) v_a = \left(\frac{1+\gamma}{1-\gamma}\right)^{\frac{1}{2}} v_a$ and $A_a\left(\left(\frac{1+\gamma}{1-\gamma}\right)^{\frac{1}{2}}\right)\tilde{v}_a = \left(\frac{1-\gamma}{1+\gamma}\right)^{\frac{1}{2}} \tilde{v}_a$. Then following the same reasoning as in [19], (p. 139), we have $I_\gamma \in L^2(\mathbb{R}^k)$ and $(A^*)^{-1}(F_\gamma(\ ,e_1,e_{a+1})) = c\, I_\gamma(X)$ for $X \in \mathbb{R}^k$.

Q. E. D.

We recall at this point a version of the Gegenbauer Addition Theorem: if L is any harmonic polynomial of degree t (in m variables) and $z \in \mathbb{C}$, then we have

$$(6\text{-}42) \qquad \int_{S^{m-1}} e^{z[\tilde{\omega}_1,\omega_1]} L(\omega_1)\, d\sigma_m(\omega_1) = c\, L(\tilde{\omega}_1) \sum_{j=0}^{j=\infty} a_j\, z^{2j+t} ,$$

where c is a nonzero constant and $a_j = \dfrac{\Gamma(t+\frac{1}{2}m)}{4^j\, j!\,\Gamma(j+t+\frac{1}{2}m)}$.

We also recall that every point $x \in \mathbb{R}^k$ can be written in the form $X = X_+ + X_-$, where $X_+ \in \langle e_1,\ldots,e_a\rangle$, $X_- \in \langle e_{a+1},\ldots,e_n\rangle$. We then let G_2^+ be the rotation group of the plane $\langle e_1,e_2\rangle$. If $g_2^+(\theta)$ is the generator of G_2^+ , then $g_2^+(\theta)$: $\begin{cases} e_1 \to \cos\theta\, e_1 + \sin\theta\, e_2 . \\ e_2 \to -\sin\theta\, e_1 + \cos\theta\, e_2 . \end{cases}$

We let $K_1 = O(a)$, $K_2 = O(b)$ so that $K = K_1 \times K_2$. Moreover we define $\tilde{M} = M_a \cap K$, $\tilde{M}_i = K_i \cap M_a$.

Then we recall the disintegration of Haar measure $d_1(k)$ on K_1 so that if $\varphi \in L^1(K_1,d_1(k))$, then there exist suitably normalized Haar measures so that

(6-43) $$\int_{K_1} \varphi(k)\, d_1(k) = \int_{\tilde{M}_1\times\tilde{M}_1\times G_2^{\pm}(\theta)} \varphi(m\, g_2^{+}(\theta)m')\, d_1(m)\, d_1(m')\, d^{+}(\theta) \; .$$

Explicitly we have

(6-44) $$d^{+}(\theta) = \begin{cases} (\sin\theta)^{a-2}\, d\theta \text{ with range } [0,\pi] \text{ if } a > 2 \\ d\theta \qquad\qquad \text{with range } [-\pi,\pi] \text{ if } a = 2 \end{cases} \; .$$

<u>Remark 6.5</u>. The discussion from here to Theorem 6.13 will involve another way (more geometric) to prove the nonvanishing criterion of Theorem 6.10 in the case $b = 1$. We include this method because it will be useful in the construction of automorphic forms attached to the Weil representation.

We want to determine for which s, s_1, and s_2 the space $\mathscr{F}_k^{+}(s,s_1,s_2) \subseteq F_k$. From Lemma 6.6 and (6-42) we have expressed $G = Q_1(Z_+)\, z_k^{s_2+2\mu^{+}} \sum_{j=0}^{\infty} a_j^{+} \Delta_1^{j}\, z_k^{2j} \in \mathscr{F}_k^{+}(s,s_1,s_2)$ in the form

(6-45) $$G(Z) = z_k^{\frac{1}{2}(k-2)-s} L(Z) \; ,$$

where $L(Z) = \lim_{\gamma\to 1} \int_{S^{a-1}} e^{\gamma Q(Z,\omega_+)Q(Z,\omega_-)}\, Q_1(\omega_+)\, d\,\omega_+$,

with limit taken pointwise.

However, we know that $\| e^{Q(Z,\omega_+)Q(Z,\omega_-)} \|_{\rho} < \infty$ if and only if $\rho > 0$. Thus it is easy to see $\lim_{\gamma\to 1}$ can be taken in the topology of F' , and we deduce that $L \in F'$.

Hence if $G \in F_k$, it follows from (6-24) and (6-25) that

(6-46) (i) $$(A^*)^{-1}(G) = \left(x_k - \frac{\partial}{\partial x_k}\right)^{\frac{1}{2}(k-2)-s} (A^*)^{-1}(L) \, , \text{ if } \tfrac{1}{2}(k-2)-s \geq 0 \; ,$$

(ii) $$\left(x_k - \frac{\partial}{\partial x_k}\right)^{s-\frac{1}{2}(k-2)} (A^*)^{-1}(G) = (A^*)^{-1}(L) \, , \text{ if } s-\tfrac{1}{2}(k-2) > 0$$

where equality and differentiation are taken in the distribution sense.

Then from (6-46) we see that $(A^*)^{-1}(L)$ must be determined explicitly. This is done in Proposition 6.12.

Proposition 6.12. Let Q_1 be a harmonic polynomial homogeneous of degree t_1 in the variables $\langle x_1,\ldots,x_a\rangle$, with $a = k-1$. Then

$$\text{(6-47)}\qquad \lim_{\gamma\to 1} \int_{K_1} I_\gamma(k\cdot X)\, Q_1(ke_1)\, d_1(k) =$$

$$\begin{cases} 0 & \text{if } Q(X,X) < 0 \\ d\cdot Q(X,X)^{\frac{1}{2}(a-3)} e^{-\frac{1}{2}Q(X,X)} \|X_+\|^{2-(a+t_1)} Q_1(X_+)\xi(X) & \text{if } Q(X,X) > 0 \end{cases}$$

where
$$\xi(X) = \begin{cases} C_{t_1}^{\frac{1}{2}(a-2)}\left(\dfrac{X_-}{\|X_+\|}\right) & \text{if } a > 2 \\ \operatorname{Re}\left\{\dfrac{X_-}{\|X_+\|} + \sqrt{-1}\,\dfrac{Q(X,X)^{\frac{1}{2}}}{\|X_+\|}\right\}^{t_1} & \text{if } a = 2 \end{cases}$$
and d, a non-zero constant.

Proof. We let $W_+(\theta,x) = g_2^+(\theta)\left[\cosh(x)\, e_1 + \sinh(x)\, e_{a+1}\right]$ and $W_-(\theta,x) = g_2^+(\theta)\left[\sinh(x)\, e_1 + \cosh(x)\, e_{a+1}\right]$. Then

$$\left[A_a(\rho_\gamma)W_+(\theta,x), A_a(\rho_\gamma)W_+(\theta,x)\right] = -\frac{1}{1-\gamma^2}(\cos\theta\cosh(x)\gamma + \sinh x)^2 - 1 \quad \text{and}$$

$$\left[A_a(\rho_\gamma)W_-(\theta,x), A_a(\rho_\gamma)W_-(\theta,x)\right] = -\frac{1}{1-\gamma^2}(\cos\theta\sinh(x)\gamma + \cosh(x))^2 + 1 .$$

Then using (6.43) with $\varphi_1 = I_\gamma(k\cdot X)Q_1(ke_1)$, we deduce for X hyperbolic, i.e. $X = t\,\tilde{k}\,W_+(0,x)$, $t > 0$, $x \in R$, and $\tilde{k} \in K_1$

$$\text{(6-48)}\qquad \int_{K_1} I_\gamma(t\,k\,\tilde{k}\,W_+(0,x))\, Q_1(ke_1)\, d_1(k) =$$

$$\int_{K_1} I_\gamma(t\,k''\,W_+(0,x))\, Q_1(\tilde{\mathbf{k}}(k'')^{-1}(e_1))\, d_1(k'') =$$

$$\int_{\tilde{M}_1\times\tilde{M}_1} \left\{ \int I_\gamma(t\,m\,W_+(\theta,x))\, Q_1(\tilde{k}(m')^{-1} g_2^+(-\theta)e_1)\, d^+(\theta) \right\} d_1(m)\, d_1(m') .$$

Thus it suffices first to study the behavior of the inner integral of (6-48) as $\gamma \to 1$. By making the change of variables $z = \gamma\cosh(x)\cos\theta + \sinh(x)$, we see that the inner integral of (6-48) is a nonzero multiple of

(6-49) $$(1-\gamma^2)^{-\frac{1}{2}}|\gamma \cosh x|^{2-a} e^{-\frac{1}{2}t^2} \int_{A(\gamma)}^{B(\gamma)} Q_1^*(\ldots) e^{-\frac{t^2}{1-\gamma^2} z^2}$$

$$\left[(z-A(\gamma))(B(\gamma)-z)\right]^{\frac{1}{2}(a-3)} dz .$$

where $A(\gamma) = \sinh(x)-\gamma \cosh(x)$, $B(\gamma) = \sinh(x) + \gamma \cosh(x)$ and

(6-50) $$Q_1^*(\ldots) = \begin{cases} Q_1(\tilde{k}(m')^{-1}(\cos\theta\, e_1-\sin\theta\, e_2), \text{ if } a > 2 \\ Q_1(\tilde{k}(m')^{-1}(\cos\theta\, e_1-\sin\theta\, e_2)) + \\ Q_1(\tilde{k}(m')^{-1}(\cos\theta\, e_1+\sin\theta\, e_2)) , \text{ if } a = 2 \end{cases}$$

where θ is expressed as a function of z, γ, and x by the change of variables above.

Then we choose γ so that $A(\gamma) < 0$ and $B(\gamma) > 0$. We divide the interval $[A(\gamma), B(\gamma)]$ into 3 parts: $[A(\gamma),-\varepsilon]$, $[-\varepsilon,\varepsilon]$, and $[\varepsilon,B(\gamma)]$ with $\varepsilon > 0$. Moreover we note that Q_1^* restricted to S^{a-1} is bounded. Thus we have

(6-51) $$e^{-\frac{1}{2}t^2} \frac{|\gamma\cosh x|^{2-a}}{\sqrt{1-\gamma^2}} \left| \int_{A(\gamma)}^{-\varepsilon} Q_1^*(\ldots) e^{-\frac{t^2}{1-\gamma^2} z^2} \right.$$

$$\left. \left[(z-A(\gamma))(B(\gamma)-z)\right]^{\frac{1}{2}(a-3)} dz \right|$$

$$\leqq D_1 e^{-\frac{1}{2}t^2} \frac{|\gamma\cosh x|^{2-a}}{\sqrt{1-\gamma^2}} \left[\max_{z\in[A(\gamma),-\varepsilon]} (B(\gamma)-z)^{\frac{1}{2}(a-3)}\right]\left[\varepsilon+A(\gamma)\right]^{\frac{1}{2}(a-1)}$$

$$e^{-\frac{t^2}{1-\gamma^2}\varepsilon^2}$$

where D_1 is a positive constant. But the last expression $\to 0$ as $\gamma \to 1$. A similar statement works for the range of integration $[\varepsilon,B(\gamma)]$. Thus we must study

(6-52) $$e^{-\frac{1}{2}t^2} \frac{|\gamma\cosh x|^{2-a}}{\sqrt{1-\gamma^2}} \int_{-\varepsilon}^{+\varepsilon} Q_1^*(\ldots) e^{-\frac{t^2}{1-\gamma^2} z^2} \left[(z-A(\gamma))(B(\gamma)-z)\right]^{\frac{a-3}{2}} dz .$$

If we let $z = \sqrt{1-\gamma^2}w$, then (6-52) becomes

$$(6\text{-}53)\qquad e^{-\frac{1}{2}t^2}|\gamma \cosh x|^{2-a} \int_{-\varepsilon(1-\gamma^2)^{-\frac{1}{2}}}^{\varepsilon(1-\gamma^2)^{-\frac{1}{2}}} Q_1^*(\ldots)e^{-t^2w^2}((1-\gamma^2)^{\frac{1}{2}}w-A(\gamma))^{\frac{1}{2}(a-3)}$$

$$(B(\gamma) - (1-\gamma^2)^{\frac{1}{2}}w)^{\frac{1}{2}(a-3)}dw .$$

But if $-\varepsilon(1-\gamma^2)^{-\frac{1}{2}} \le w \le \varepsilon(1-\gamma^2)^{-\frac{1}{2}}$, then $-\varepsilon - A(\gamma) \le (1-\gamma^2)^{\frac{1}{2}}w - A(\gamma) \le$

$\varepsilon-A(\gamma)$ and $B(\gamma)-\varepsilon \le B(\gamma) - (1-\gamma^2)^{\frac{1}{2}}w \le B(\gamma)+\varepsilon$. Then the integrand of (6-53) is bounded by $L\left[(\varepsilon+e^{-x})(\varepsilon+e^{x})\right]^{\frac{1}{2}(a-3)}e^{-t^2w^2}$ if $a \ge 3$, where L is some positive constant. Then we know that there exist λ_1,λ_2 so that $\lambda_1 > 1$ with $-e^{-x} \le A(\gamma) \le -\lambda_1\varepsilon$ and $\lambda_2 > 1$ with $\lambda_2\varepsilon \le B(\gamma) \le e^{x}$. Thus if $a = 2$ the integrand of (6-53) is bounded by $S\left[(\lambda_1-1)(\lambda_2-1)\right]^{-\frac{1}{2}}\frac{1}{\varepsilon}e^{-t^2w^2}$ with S a positive constant.

Thus in any case the integrand of (6-53) is bounded by an integrable function independent of γ and so we can apply Lebesque dominated convergence and we deduce as $\gamma \to 1$

$$(6\text{-}54)\qquad \int I_\gamma(t\ m\ W_+(\theta,x))Q_1(\tilde{k}(m')^{-1}g_2^+(-\theta)e_1)d^+(\theta) \to$$

$$d\ e^{-\frac{1}{2}t^2}(\cosh x)^{2-a}\left\{\int_{-\infty}^{+\infty} e^{-t^2w^2}\,dw\right\} l(k,m,x) ,$$

where

$$l(k,m,x) = \begin{cases} Q_1(\tilde{k}(m')^{-1}(\tanh(x)e_1 - \frac{1}{\cosh x}e_2)) , & \text{if } a \ge 3 \\ Q_1(\tilde{k}(m')^{-1}(\tanh(x)e_1 + \frac{1}{\cosh x}e_2)) + & \\ \qquad Q_1(\tilde{k}(m')^{-1}(\tanh(x)e_1 - \frac{1}{\cosh x}e_2)) , & \text{if } a = 2 \end{cases}$$

and d is a nonzero constant.

Then we repeat the same argument for the case X elliptic and deduce that

$$(6\text{-}55)\qquad \int_{K_1} I_\gamma(k\cdot X)Q_1(ke_1)d_1(k) = \int_{\tilde{M}_1\times\tilde{M}_1}\left\{\int I_\gamma(tm\ W_-(\theta,x))\right.$$

$$Q_1(\tilde{k}(m')^{-1}g_2^+(-\theta)e_1)d^+(\theta)\Big\} \; d_1(m)d_1(m')$$

(with $X = t\,\tilde{k}\,W_-(0,x))$.

Next we study the behavior of the inner integral of (6-55) as $\gamma \to 1$. Making the change of variable $z = \gamma \cos\theta \sinh(x) + \cosh(x)$ we see that this integral is a nonzero multiple of

$$(6\text{-}56) \qquad (1-\gamma^2)^{-\frac{1}{2}} e^{\frac{1}{2}t^2} |\gamma \sinh(x)|^{2-a} \int_{\tilde{A}(\gamma)}^{\tilde{B}(\gamma)} Q_1^*(\ldots) e^{-\frac{t^2}{1-\gamma^2} z^2} \Big[(z-\tilde{A}(\gamma))(\tilde{B}(\gamma)-z)\Big]^{\frac{a-3}{2}} dz ,$$

where $\tilde{A}(\gamma) = \cosh(x) - \gamma \sinh(x)$, $\tilde{B}(\gamma) = \cosh(x) + \gamma \sinh(x)$ and the argument inside Q_1^* is expressed by representing θ as a function of z , x , and γ (see (6-50) above). But then the integral in (6-56) is bounded by

$$(6\text{-}57) \qquad D\, e^{-\frac{t^2}{1-\gamma^2}(\tilde{A}(\gamma))^2} \left\{ \int_{\tilde{A}(\gamma)}^{\tilde{B}(\gamma)} (z-\tilde{A}(\gamma))^{\frac{a-3}{2}} (\tilde{B}(\gamma)-z)^{\frac{a-3}{2}} dz \right\} ,$$

where D is a positive constant. If $a \geqq 3$ then the integral in (6-57) is bounded by a constant independent of γ (i.e. $\max_{\gamma\in[0,1]} \tilde{A}(\gamma) = \cosh(x)$ and $\max_{\gamma\in[0,1]} \tilde{B}(\gamma) = e^x$) . If $a \geqq 3$ then the expression (6-57) tends to 0 as $\gamma \to 1$.

If $a = 2$ we see that the integration in (6-57) can be divided into two intervals $[\tilde{A}(\gamma)\, , \cosh x]$ and $[\cosh x, \tilde{B}(\gamma)]$ so that

$$(6\text{-}58) \qquad \int_{\tilde{A}(\gamma)}^{\tilde{B}(\gamma)} (z-\tilde{A}(\gamma))^{-\frac{1}{2}} (\tilde{B}(\gamma)-z)^{-\frac{1}{2}} dz \leqq$$

$$\frac{1}{2}\left[\frac{\tilde{B}(\gamma)-\cosh x}{\cosh x - \tilde{A}(\gamma)}\right]^{\frac{1}{2}} + \frac{1}{2}\left[\frac{\cosh x - \tilde{A}(\gamma)}{\tilde{B}(\gamma)-\cosh x}\right]^{\frac{1}{2}} = \gamma .$$

Hence (6-57) $\to 0$ as $\gamma \to 1$.

Finally we must study an integral of the form

(6-59) $$\psi_{Q_1}(k,k') = \int_{M_1} Q_1(k\, m\, k'\, e_1)\, dm \ , \ k \ , \ k' \in K_1$$

But ψ_{Q_1} can be interpreted as the <u>mean value operator</u> of the spherical harmonic $\xi \to Q_1(\xi \cdot e_1)$ for $\xi \in K_1$. Then it follows from [13] that if $Q_1(e_1) \neq 0, \psi_{Q_1}(k,k') = c_{Q_1} Q_1(ke_1)\psi_{Q_1}(e,k')$ where c_{Q_1} is a nonzero constant given by $1/Q(e_1)$.

Moreover from representation theory we know that for any $\eta \in S^{a-1}$

(6-60) $$\int_{\tilde{M}_1} Q_1(m \cdot \eta)\, d_1(m) = \langle Q_1, Y_{t_1}(a,1)\rangle\, Y_{t_1}(a,1)\,(\eta)\ ,$$

where $\langle\ ,\ \rangle$ is the inner product on S^{a-1} given by the K_1- invariant measure on S^{a-1} and $Y_{t_1}(a,1)$ defined on (6-5). Thus it follows that

(6-61) $$\psi_{Q_1}(e,k') = \langle Q_1, Y_{t_1}(a,1)\rangle\, Y_{t_1}(a,1)(k'e_1)\ .$$

But we know that $Q_1(e_1) = 0$ if and only if $\langle Q_1, Y_{t_1}(a,1)\rangle = 0$ (i.e. $Q_1(e_1) = c\,\langle Q_1, Y_{t_1}(a,1)\rangle$ where c is a nonzero constant, independent of Q_1).

Thus it follows by a limiting argument that for any Q_1

(6-62) $$\psi_{Q_1}(k,k') = s\, Q_1(ke_1) Y_{t_1}(a,1)\ (k'e_1)$$

(s a nonzero constant independent of Q_1).

Thus we have that

(6-63) $$\lim_{\gamma \to 1} \int_{K_1} I_\gamma(k \cdot X) Q_1(ke_1) d_1(k) \to$$

$$d\frac{1}{t}\, e^{-\frac{1}{2}t^2} (\cosh x)^{2-a} Q_1(\tilde{k}e_1) \begin{cases} Y_{t_1}(a,1)(\tanh(x)e_1 - \dfrac{1}{\cosh x} e_2)\ , & \text{if } a > 2 \\ Y_{t_1}(a,1)(\tanh(x)e_1 - \dfrac{1}{\cosh x} e_2) + Y_{t_1}(a,1)(\tanh(x)e_1 + \dfrac{1}{\cosh x} e_2) & \text{if } a = 2 \end{cases}$$

where $d \neq 0$, a nonzero constant.

Q. E. D.

Corollary 1 to Proposition 6.12. *The statement in Proposition 6-12 remains valid if pointwise convergence is replaced by uniform convergence on compact subsets of* $\Omega_+ \cup \Omega_-$.

Proof. Let L be a compact subset of Ω_+ of the form

$\{ t\,\tilde{k}\,(\cosh(x)e_1 + \sinh(x)e_{a+1}) \mid r_1 < t < r_2 ,\ |x| < B \text{ and } k \in \tilde{K} \}$.

Then it is clear that the integrands of (6-51) and (6-53) as functions of (Υ, X, z) and (Υ, X, w) are bounded by integrable functions of z and w where $(\Upsilon, X) \in [\frac{1}{2}, 1] \times L$. Thus (6-54) holds uniformly for all $X \in L$. Since the outer integral of (6-48) is concentrated on a compact set $\tilde{M}_1 \times \tilde{M}_1$, it follows that (6-47) holds uniformly on compact subsets of Ω_+ . A similar argument works on compact subsets of Ω_- .

Q. E. D.

Remark 6.6. From Corollary 1 to Proposition 6.12, we deduce that $(A^*)^{-1}(L)$ as a distribution restricted to $\Omega_+ \cup \Omega_-$ is given by (6-47).

We must show that $(A^*)^{-1}(G) \in L^2(\mathbb{R}^k)$, using (6-46) and Proposition 6.12. Since $(A^*)^{-1}(G)$ must vanish on Ω_- (see Lemma 6.2) and $(A^*)^{-1}(G)$ is a C^∞ vector for the representation $\pi_{\mathfrak{m}}$ (see Lemma 6.1), then we can consider equations (i) and (ii) of (6-46) as differential equations involving C^∞ functions valid for all points in Ω_+ . We note that $(A^*)^{-1}(L)$ vanishes on Ω_- and is a C^∞ function on Ω_+ by Proposition 6.12.

Then we first solve (i) of (6-46). Indeed using Proposition 6.12, we must show ($r = \|X_+\|$ and $\mathcal{A} = X_-$)

$$(6\text{-}64) \qquad e^{-\frac{1}{2}(r^2 - \mathcal{A}^2)}\, r^{3-k} \left(\frac{\partial}{\partial \mathcal{A}}\right)^m \left\{ (r^2 - \mathcal{A}^2)^{\frac{1}{2}(k-4)} F\left(\frac{\mathcal{A}}{r}\right) \right\}$$

$$\in L^2(\mathbb{R}_+^* \times \mathbb{R} ,\ r^{k-2}\, dr\, d\mathcal{A}) ,$$

where

$$(6\text{-}65) \qquad F\left(\frac{s}{r}\right) = \begin{cases} C_{s_1}^{\frac{1}{2}(k-3)}\left(\frac{s}{r}\right), & \text{if } a > 2 \\ \operatorname{Re}\left\{\frac{s}{r} + \sqrt{-1}\,\frac{(r^2 - s^2)^{\frac{1}{2}}}{r}\right\}^{s_1}, & \text{if } a = 2 \end{cases}$$

and $m = \frac{1}{2}(k-2)-s$. But from Theorem 6.4 we know that $s > 0$. However if we take the coordinates $w = \frac{s}{r}$, $t = (r^2 - s^2)^{\frac{1}{2}}$, then (6-64) becomes

$$(6\text{-}66) \qquad e^{-\frac{1}{2}t^2}\, t^{3-k}(1-w^2)^{\frac{1}{2}(k-3)} \left\{ \frac{(1-w^2)^{\frac{1}{2}}}{t} \frac{\partial}{\partial w} + \frac{w}{(1-w^2)^{\frac{1}{2}}} \frac{\partial}{\partial t} \right\}^m$$

$$\left\{t^{k-4}F(w)\right\} \in L^2(\mathbb{R}_+^* \times [-1,\, 1],\, t^{k-1}\,dt \otimes (1-w^2)^{-\frac{1}{2}k}\,dw) .$$

But since $F(w)$ is a <u>2 variable polynomial</u> in w and $(1-w^2)^{\frac{1}{2}}$, then it follows by an induction argument on m that

$$(6\text{-}67) \qquad \left(\frac{(1-w^2)^{\frac{1}{2}}}{t} \frac{\partial}{\partial w} + \frac{w}{(1-w^2)^{\frac{1}{2}}} \frac{\partial}{\partial t}\right)^m (t^{k-4}\,F(w)) =$$

$$\sum c_{\alpha_1\alpha_2\alpha_3}\, t^{\alpha_1} w^{\alpha_2} (1-w^2)^{\frac{1}{2}\alpha_3} ,$$

where $\alpha_1, \alpha_2, \alpha_3$ range over all integers; only finitely many of $c_{\alpha_1\alpha_2\alpha_3} \neq 0$ and

$$(6\text{-}68) \qquad \begin{aligned} k-4-m &= \min \alpha_1 \text{ so that } c_{\alpha_1\alpha_2\alpha_3} \neq 0 \\ 0 &\leq \min \alpha_2 \text{ so that } c_{\alpha_1\alpha_2\alpha_3} \neq 0 \\ -m &= \min \alpha_3 \text{ so that } c_{\alpha_1\alpha_2\alpha_3} \neq 0 . \end{aligned}$$

But $t^{\alpha_1} w^{\alpha_2} (1-w^2)^{\frac{1}{2}\alpha_3} \in L^2(\mathbb{R}_+^* \times [-1,1], t^{k-1}\,dt \otimes (1-w^2)^{-\frac{1}{2}k}\,dw)$ if and only if $\alpha_1 > -\frac{1}{2}k$, $\alpha_2 > -\frac{1}{2}$ and $\alpha_3 > \frac{1}{2}k-1$. Thus (6-64) is valid <u>if and only if</u> $s > 1$. Thus if $\frac{1}{2}(k-2) \geq s$, then $\mathcal{F}_k^+(s,s_1,s_2) \subseteq F_k$ if and only if $s > 1$.

Then to handle case (ii) of (6-46), we may assume $(A^*)^{-1}(G) \in L^2(\mathbb{R}^k)$ has the form

$$(6\text{-}69) \qquad Q_1(\omega_+)\, e^{-\frac{1}{2}(r^2 - s^2)}\, r^{3-k}\varphi(r, s)$$

where $\varphi(r,s)$ satisfies the equation

$$(6\text{-}70)\qquad \left(\frac{\partial}{\partial s}\right)^m (\varphi(r,s)) = (r^2 - s^2)^{\frac{1}{2}(k-4)} F\left(\frac{s}{r}\right) ,$$

where $m = s-\frac{1}{2}(k-2)$. Then since $\varphi \in C^\infty(\mathbb{R}_+^* \times \mathbb{R})$, we have

$$(6\text{-}71)\qquad \varphi(r,s) = \int_0^s (r^2-v^2)^{\frac{1}{2}(k-4)} F\left(\frac{v}{r}\right) (s-v)^{m-1}\, dv + \sum_{j=0}^{m-1} c_j(r)\, s^j$$

where c_j are arbitrary functions of r .

Next we evaluate the first term on the left hand side of (6-71). By change of variable $v = rw$, we have

$$(6\text{-}72)\qquad \int_0^s (r^2-v^2)^{\frac{1}{2}(k-4)} F\left(\frac{v}{r}\right) (s-v)^{m-1}\, dv =$$

$$r^{m+k-4} \int_0^{s/r} (1-w^2)^{\frac{1}{2}(k-4)} F(w) \left(\frac{s}{r} - w\right)^{m-1} dw .$$

If $k \geq 4$ then by Rodriques formula for Gegenbauer polynomials we have

$$(6\text{-}73)\qquad F(w) = c\, (1-w^2)^{\frac{1}{2}(4-k)} \left(\frac{d}{dw}\right)^{s_1} \left\{ (1-w^2)^{s_1+\frac{1}{2}(k-4)} \right\}$$

with c a nonzero constant. Then recalling from Lemma 6.6 that $s_1-m = s_2+2\mu^+ \geq 0$, it follows by application of Taylor's Theorem that

$$(6\text{-}74)\qquad \frac{1}{(m-1)!} \int_0^{s/r} \left(\frac{d}{du}\right)^m \left(\frac{d}{du}\right)^{s_1-m} \left\{ (1-u^2)^{s_1+\frac{1}{2}(k-4)} \right\} \left(\frac{s}{r} - u\right)^{m-1} du$$

$$= \tilde{H}\left(\frac{s}{r}\right) - \sum_{\nu=0}^{m-1} \frac{1}{\nu!} (\tilde{H})^{(\nu)}(0) \left(\frac{s}{r}\right)^\nu , \text{ where}$$

$$\tilde{H}(w) = \left(\frac{d}{du}\right)^{s_1-m} \left\{ (1-w^2)^{s_1+\frac{1}{2}(k-4)} \right\} .$$

But the smallest power of $(1-w^2)$ in $\tilde{H}(w)$ is $(1-w^2)^{m+\frac{1}{2}(k-4)}$. Then we observe that

$$(6\text{-}75)\qquad e^{-\frac{1}{2}(r^2-s^2)} r^{m-1} \tilde{H}\left(\frac{s}{r}\right) \in L^2(\mathbb{R}_+^* \times \mathbb{R} ,\ r^{k-2}\, dr \otimes ds) .$$

Thus if we let $c_j(r) = \delta_j r^{-j}$ in (6-71), where δ_j is an appropriately chosen constant so that $\varphi(r,\mathcal{A}) = \tilde{H}(\frac{\mathcal{A}}{r})$, then

$Q_1(\omega_+)e^{-\frac{1}{2}(r^2-s^2)}r^{3-k}\varphi(r,\mathcal{A}) \in L^2(\Omega_+)$ for $k \geqq 4$.

Then to handle case (i) of (6-46) for $k = 3$, we note that

$$\text{(6-76)} \qquad \int_0^{\mathcal{A}/r} (1-u^2)^{-\frac{1}{2}}\mathrm{Re}\left\{u+\sqrt{-1}\,(1-u^2)^{\frac{1}{2}}\right\}^{s_1}(\tfrac{\mathcal{A}}{r} - u)^{m-1}\,du =$$

$$\int_{\pi/2}^{\beta} \cos(s_1\theta)(\cos\beta - \cos\theta)^{m-1}\,d\theta \quad ,$$

where $\cos\beta = \frac{\mathcal{A}}{r}$ $(0 \leqq \beta \leqq \pi)$. Then we recall the Mehler Dirichlet formula:

$$\text{(6-77)} \qquad C^m_{s_1-m}(\cos\beta)\,(\sin\beta)^{2m-1} = d\int_0^{\beta} \cos(s_1\theta)(\cos\beta - \cos\theta)^{m-1}d\theta$$

(d, a nonzero constant depending on s_1 and m), i.e. see [9,I, p. 159].

Thus it follows that the left hand side of (6-76) is

$$\text{(6-78)} \qquad \frac{1}{d}\left[C^m_{s_1-m}(\tfrac{\mathcal{A}}{r})(1-(\tfrac{\mathcal{A}}{r})^2)^{m-\frac{1}{2}} - C^m_{s_1-m}(0)\right] .$$

Thus in (6-71) if we let $c_o(r) = \frac{1}{d}C^m_{s_1-m}(0)\,r^{m-1}$ and $c_j(r) = 0$ if $j > 0$ so that

$$\text{(6-79)} \qquad \varphi(r,\mathcal{A}) = r^{m-1}(1-(\tfrac{\mathcal{A}}{r})^2)^{m-\frac{1}{2}}C^m_{s_1-m}(\tfrac{\mathcal{A}}{r}) ,$$

then $Q_1(\omega_+)e^{-\frac{1}{2}(r^2-\mathcal{A}^2)}\varphi(r,\mathcal{A}) \in L^2(\Omega_+)$ for $k = 3$.

<u>Theorem 6.13</u>. <u>Let</u> $G \in \mathscr{F}_k(s,s_1,s_2)$ <u>be given by</u> (6-45).

(i) <u>Then</u> $(A^*)^{-1}(G) = (x_k - \frac{\partial}{\partial x_k})^{\frac{1}{2}(k-2)-s}(A^*)^{-1}(L) \in L^2(\mathbb{R}^k)$

<u>where</u> $\frac{k-2}{2} \geqq s > 1$ <u>and</u> $s_1 - s_2 = s - \frac{1}{2}(k-2) + 2\mu_+$ (μ_+ integer $\geqq 0$). <u>Moreover</u> <u>if</u> $X = r\omega_1 + \mathcal{A}\cdot e_n \in \Omega_+$, <u>with</u> $\omega_1 \in S^{a-1}$, <u>then</u>

$$\text{(6-80)} \qquad (A^*)^{-1}(G)\,(r,\mathcal{A},\omega_1) = c\,e^{-\frac{1}{2}(r^2-\mathcal{A}^2)}r^{3-k}(\tfrac{\partial}{\partial\mathcal{A}})^{\tilde{m}}$$

$$\left\{(r^2-\mathcal{A}^2)^{\frac{1}{2}(k-4)}F(\tfrac{\mathcal{A}}{r})\right\}Q_1(\omega_1)$$

with F *given by* (6-73) *and* $\tilde{m} = \frac{1}{2}(k-2)-s$ *and* c , *a nonzero constant.*

(ii) *If* $s > \frac{1}{2}(k-2)$, *then* $(A^*)^{-1}(G) \in L^2(\mathbb{R}^k)$ *when* s_1 *and* s_2 *satisfy* $s_1 \geq s-\frac{1}{2}(k-2)$ *and* $s_1-s_2 = s-\frac{1}{2}(k-2)+2\mu_+$ (μ_+ integer ≥ 0) . *Moreover if* $X = r\omega_1 + \mathscr{s} e_n \in \Omega_+$ (*as in* (i) *above*) *then*

$$(A^*)^{-1}(G)(r,\mathscr{s},\omega_1) = d\, e^{-\frac{1}{2}(r^2-\mathscr{s}^2)} r^{3-k} \varphi(r,\mathscr{s})\, Q_1(\omega_1) \tag{6-81}$$

with

$$\varphi(r,\mathscr{s}) = \begin{cases} \tilde{H}(\frac{\mathscr{s}}{r}) & \text{if } k > 3 \\ r^{m-1}(1-(\frac{\mathscr{s}}{r})^2)^{m-\frac{1}{2}} C^m_{s_1-m}(\frac{\mathscr{s}}{r}) , & \text{if } k = 3 \end{cases} \tag{6-82}$$

where $\tilde{H}$ *given by* (6-74), $m = s-\frac{1}{2}(k-2)$, *and* d *a nonzero constant.*

Proof. We need only show that $Q_1(\omega_+)e^{-\frac{1}{2}(r^2-\mathscr{s}^2)} r^{3-k}\varphi(r,\mathscr{s})$ (with φ chosen as in (6-75) for $k \geq 4$ and as in (6-79) for $k = 3$) is, in fact, an element of $\mathscr{A}_s^+$. From the discussion preceding (6-69), we know à priori that there is a function φ_1 so that $Q_1(\omega_1)e^{-\frac{1}{2}(r^2-\mathscr{s}^2)} r^{3-k}\varphi_1(r,\mathscr{s}) \in \mathscr{A}_s^+$. This means, in particular, that $\varphi - \varphi_1$ must belong to the kernel of the operator $(\frac{\partial}{\partial \mathscr{s}})^m$. Thus a function of the form $(\sum_{j=0}^{m-1} \alpha_j \mathscr{s}^j) e^{-\frac{1}{2}(r^2-\mathscr{s}^2)} r^{3-k} \in L^2(\mathbb{R}_+^* \times \mathbb{R} , r^{k-2}dr \otimes d\mathscr{s})$, where α_j are given functions of r . But this is only possible if all $\alpha_j \equiv 0$.

Q. E. D.

In the following discussion we return to the general case of the quadratic form Q with arbitrary signature. We recall the representation π_λ of $O(Q)$ in $C^\infty_\lambda(O(Q))$ (See §3). Since the representation π_λ is determined by its restriction to $K/K \cap M_a$, it follows that the K isotypic components appearing in $C^\infty_\lambda(O(Q))$ are of the type $[s_1]_a \otimes [s_2]_b$ with s_1 (s_2 resp.) a class one representation of $O(a)$ ($O(b)$ resp.). But since $\varphi \in C^\infty_\lambda(O(Q))$ has a parity condition (see §3), we deduce that $[s_1]_a \otimes [s_2]_b$ occurs in $C^\infty_\lambda(O(Q))$ if and only if $s_1+s_2 \equiv \tilde{\varepsilon}(\lambda) \bmod 2$, where

$$\tilde{\varepsilon}(\lambda) = \left\{ \begin{array}{ll} 0 & \text{if } \varepsilon(\lambda) = 1 \\ 1 & \text{if } \varepsilon(\lambda) = -1 \end{array} \right\} .$$

Our problem is to take the element $\tilde{f}(t(\omega_1+\omega_2)) = |t|^{\frac{1}{2}(2-k)-\lambda_o} H^{s_1}(\omega_1) H^{s_2}(\omega_2)$ for H^{s_1}, H^{s_2} harmonic polynomials of degrees s_1 and s_2 on $\mathbb{R}^a$ and $\mathbb{R}^b$, respectively, and compute $T^*_{\lambda,t}(\tilde{f})$ using Proposition 5.12.

First we note if $X = A\tilde{\omega}_1+B\tilde{\omega}_2$, with $A \geqq 0$ and $B \geqq 0$, then

$$Q(\omega_1+\omega_2, A\tilde{\omega}_1+B\tilde{\omega}_2) = A(\omega_1,\tilde{\omega}_1) - B(\omega_2,\tilde{\omega}_2) , \tag{6-83}$$

with (,) the usual inner product in Euclidean space $\mathbb{R}^k$. Then from (6-42) we have

$$\int_{S^{a-1}\times S^{b-1}} H^{s_1}(\omega_1) H^{s_2}(\omega_2)\, e^{-2\pi i r A(\omega_1,\tilde{\omega}_1)}\, e^{2\pi i r B(\omega_2,\tilde{\omega}_2)}\, d\omega_1 d\omega_2 \tag{6-84}$$

$$= \left\{ \int_{S^{a-1}} H^{s_1}(\omega_1) e^{-2\pi i r A(\omega_1,\tilde{\omega}_1)}\, d\omega_1 \right\} \cdot$$

$$\left\{ \int_{S^{b-1}} H^{s_2}(\omega_2) e^{2\pi i r B(\omega_2,\tilde{\omega}_2)}\, d\omega_2 \right\}$$

$$= c\, A^{s_1} B^{s_2} r^{s_1+s_2} (J_{\frac{1}{2}a+s_1-1}(2\pi r A) / (2\pi r A)^{\frac{1}{2}a+s_1-1})$$

$$(J_{\frac{1}{2}b+s_2-1}(2\pi r B) / (2\pi r B)^{\frac{1}{2}b+s_2-1})\, H^{s_1}(\tilde{\omega}_1)\, H^{s_2}(\tilde{\omega}_2) ,$$

where J_ν is the Bessel function of index ν (i.e. the series in (6-42) is equal to $(J_{\frac{1}{2}m+t-1}(z) / z^{\frac{1}{2}m+t-1})$).

We recall at this point the well-known Weber-Schlafheitlin integral (see [9,II], p. 51):

$$W(a,b,\nu_1,\nu_2,\lambda) = \int_0^\infty J_{\nu_1}(ax)\, J_{\nu_2}(bx)\, x^\lambda\, dx. \tag{6-85}$$

where the integral is absolutely convergent for the following domain:

$D = \{(a,b,\nu_1,\nu_2,\lambda) \mid a > 0, b > 0, 1 > \mathrm{Re}(\lambda)$, and $\mathrm{Re}(\nu_1+\nu_2+\lambda) > -1\}$.

Proposition 6.14. *If* $X = A\tilde{\omega}_1+B\tilde{\omega}_2$ *with* $X \notin \Gamma_o$, *then*

$$T^*_{\lambda,t}(\tilde{f})(X) = c\,\frac{\gamma(\lambda+(\tfrac12(4-k),1))}{\Gamma(\lambda_o+\tfrac12(4-k))}\,A^{1-\frac12 a}B^{1-\frac12 b}\tag{6-86}$$

$$W(2\pi A, 2\pi B, \tfrac12 a+s_1-1, \tfrac12 b+s_2-1, -\lambda_o)\; H^{s_1}(\tilde{\omega}_1)\; H^{s_2}(\tilde{\omega}_2)$$

where $T^*_{\lambda,t}(\tilde{f})$ *is given by* (5-103) *and* c , *a nonzero constant. The equality holds for all* λ *so that* $\tfrac12(k-2) > \mathrm{Re}(\lambda_o) > -1$.

Proof. In (5-103) we write

$$\int_{-\infty}^{+\infty} \lambda^{-1}(r)\,|r|^{\frac12(k-2)}\beta_\psi(r,X)\frac{dr}{|r|} = \int_0^\infty |r|^{-\lambda_o+\frac12 k-2}\beta_\psi(r,X)\,dr\tag{6-87}$$

$$+\varepsilon(\lambda)\int_0^\infty |r|^{-\lambda_o+\frac12 k-2}\beta_\psi(-r,X)\,dr\ ,$$

where $\beta_\psi(r,X) = \int_K \psi(k,\lambda)e^{-2\pi i r Q(kv_a,X)}\,dk$. But $\beta_\psi(r,X) = \varepsilon(\lambda)\beta_\psi(-r,X)$ by the parity condition put on ψ . Thus

$$(6\text{-}87) = 2\int_0^\infty |r|^{-\lambda_o+\frac12 k-2}\beta_\psi(r,X)\,dr$$

$$= c\,A^{1-\frac12 a}B^{1-\frac12 b}\,H^{s_1}(\omega_1)H^{s_2}(\omega_2)\int_0^\infty J_{\frac12 a+s_1-1}(2\pi rA)\tag{6-88}$$

$$J_{\frac12 b+s_2-1}(2\pi rB)\;r^{-\lambda_o}\,dr\ .$$

But from above we know that the latter integral is absolutely convergent for λ so that $\mathrm{Re}(\lambda_o) > -1$ and $\tfrac12 k+s_1+s_2-\mathrm{Re}(\lambda_o) > 1$. But then using the expression in (6-84), we deduce our result.

Q. E. D.

We recall at this point Sonine and Schlafheitlin's computation of

$W(a,b,\nu_1,\nu_2,\lambda)$ above. Indeed we let $2\alpha = \nu_1+\nu_2+\lambda+1$, $2\beta = -\nu_1+\nu_2+\lambda+1$, and $\gamma = \nu_2+1$. Then

$$(6\text{-}89)\quad W(a,b,\nu_1,\nu_2,\lambda) = \begin{cases} 2^{\alpha+\beta-\gamma}\dfrac{\Gamma(\alpha)}{\Gamma(\gamma)\Gamma(1-\beta)}\, b^{\gamma-1}a^{-(\alpha+\beta)}\,{}_2F_1(\alpha,\beta,\gamma,\dfrac{b^2}{2}), & \text{if } b < a \\ 2^{\alpha+\beta-\gamma}\dfrac{\Gamma(\alpha)}{\Gamma(\gamma-\alpha)\Gamma(\alpha-\beta+1)}\, b^{\gamma-1-2\alpha}a^{\alpha-\beta} \\ \qquad {}_2F_1(\alpha,\alpha-\gamma+1,\alpha-\beta+1,\dfrac{a}{b^2}), & \text{if } b > a \end{cases}$$

which is valid for $(a,b,\nu_1,\nu_2,\lambda) \in D$.

<u>Proposition 6.15</u>. <u>If</u> $X = A\tilde{\omega}_1 + B\tilde{\omega}_2$ <u>with</u> $X \in \Gamma_t(t\neq 0)$, <u>then</u>

$$(6\text{-}90)\quad T^*_{\lambda,t}(\tilde{f})(X) =$$

$$\begin{cases} c_+(\lambda)A^{1-\frac{1}{2}k-s_2+\lambda_o}B^{s_2}\,{}_2F_1(\alpha_o,\beta_o,\tfrac{1}{2}b+s_2,\dfrac{B^2}{A^2})H^{s_1}(\tilde{\omega}_1)H^{s_2}(\tilde{\omega}_2), & \text{if } B < A \\ c_-(\lambda)A^{s_1}B^{1-\frac{1}{2}k-s_1+\lambda_o}\,{}_2F_1(\alpha_1,\beta_1,\tfrac{1}{2}a+s_1,\dfrac{A^2}{B^2})H^{s_1}(\tilde{\omega}_1)H^{s_2}(\tilde{\omega}_2), & \text{if } B > A \end{cases}$$

<u>where</u>

$$(6\text{-}91)\quad c_+(\lambda) = \frac{\gamma(\lambda+(\frac{1}{2}(4-k),1))\Gamma(\frac{1}{2}(\frac{1}{2}k+s_1+s_2-1-\lambda_o))}{\Gamma(\lambda_o+\frac{1}{2}(4-k))\Gamma(s_2+\frac{1}{2}b)\Gamma(\frac{1}{2}(\lambda_o+s_1-s_2+1+\frac{1}{2}(a-b)))}\pi^{\lambda_o-1}$$

<u>and</u>

$$(6\text{-}92)\quad c_-(\lambda) = \frac{\gamma(\lambda+(\frac{1}{2}(4-k),1))\Gamma(\frac{1}{2}(\frac{1}{2}k+s_1+s_2-1-\lambda_o))}{\Gamma(\lambda_o+\frac{1}{2}(4-k))\Gamma(s_1+\frac{1}{2}a)\Gamma(\frac{1}{2}(\lambda_o+s_2-s_1+1+\frac{1}{2}(b-a)))}\pi^{\lambda_o-1}$$

<u>and</u> $\alpha_o = \frac{1}{2}(\frac{1}{2}k+s_1+s_2+1-\lambda_o) = \alpha_1$ <u>and</u> $\beta_o = \frac{1}{2}(\frac{1}{2}(b-a)-s_1+s_2+1-\lambda_o)$ <u>and</u> $\beta_1 = \frac{1}{2}(\frac{1}{2}(a-b)+s_1-s_2+1-\lambda_o)$, <u>which is valid for all</u> λ. <u>Moreover</u> $c_+(\lambda)$ <u>and</u> $c_-(\lambda)$ <u>are analytic functions in</u> λ_o.

<u>Proof</u>. We recall from Proposition 5.11 that $\lambda \to T^*_{\lambda,t}(\tilde{f})(X)$ as a distribution on $C_c^\infty(\Gamma_t)$ extends to an analytic function on all of $\mathbb{C}$. But using

(6-89) we see that the right hand side of (6-90) extends to a meromorphic function in λ (for fixed a, b, ν_1, ν_2) on $\mathbb{C}$. Moreover for all such λ, the right hand side of (6-90) is also a C^∞ function of $X \in \Omega_+ \cup \Omega_-$. Thus the right hand side of (6-90) defines, in the obvious way, for all such λ an element of $C_c^\infty(\Gamma_t)'$ for $t \neq 0$. Thus we deduce that (6-90) holds for all λ except possibly the poles given by $\frac{\gamma(\lambda+(\frac{1}{2}(4-1),1))}{\Gamma(\lambda_o+\frac{1}{2}(4-k))}$ and the poles which arise in the continuation of the right hand side of (6-89).

Then we note that $\lambda_o \to \frac{\gamma(\lambda+(\frac{1}{2}(4-k),1))}{\Gamma(\lambda_o+\frac{1}{2}(4-k))}$ is analytic with simple zeroes at $\frac{1}{2}k+j$, where $j \in \{2\mathbb{Z}\}$ if $\varepsilon(\lambda) = -1$ and $j \in \{2\mathbb{Z}+1\}$ if $\varepsilon(\lambda) = 1$ (see 5-59). Moreover $\psi(\lambda_o) = \Gamma(\frac{1}{2}(\frac{1}{2}k+s_1+s_2-1-\lambda_o))/$ $\Gamma(\frac{1}{2}(\lambda_o+s_1-s_2+1+\frac{1}{2}(a-b)))$ has all its poles at $\lambda_o = \frac{1}{2}k+s_1+s_2-1+2\sigma'$ (σ' integer ≥ 0) and all its zeroes at $\lambda_o = s_2-s_1-1-\frac{1}{2}(a-b)-2\sigma''$ (σ'' integer ≥ 0). Thus $c_+(\lambda)$ is analytic in λ_o. A similar argument works for $c_-(\lambda)$.

Q. E. D.

Theorem 6.16. *Let* $\lambda = (\lambda_o,\varepsilon)$. *Then*

(i) *the map* $T^*_{\lambda,\pm 1}: C^\infty_\lambda(O(Q)) \to C^\infty_c(\Gamma_{\pm 1})'$ *is injective at all* λ *so that* $\lambda_o \neq \frac{1}{2}k+j$ $(j \in \mathbb{Z})$.

(ii) *If* $\lambda_o = \frac{1}{2}k+j$ *with* $j \in \mathbb{Z}$ *and* $j \equiv \tilde{\varepsilon} \bmod 2$, *then* $T^*_{\lambda,+1}$ ($T^*_{\lambda,-1}$ resp.) *is injective if* $a \equiv 0 \bmod 2$ ($b \equiv 0 \bmod 2$ resp.). *If* $a \equiv 1 \bmod 2$ ($b \equiv 1 \bmod 2$ resp.), *then Kernel* $(T^*_{\lambda,1})$ (Kernel $(T^*_{\lambda,-1})$ resp.) *is an* $O(Q)$ *submodule of* $C^\infty_\lambda(O(Q))$ *which, when restricted to* K, *is the direct sum of* $[x]_a \otimes [y]_b$ ($[x']_a \otimes [y']_b$ resp.), *where* $y-x = (a+j)+(2\sigma+1)$ *as* σ *ranges over all integers* ≥ 0. ($x'-y' = b+j+(2\sigma'+1)$ *as* σ' *ranges over all integers* ≥ 0).

(iii) *If* $\lambda_o = \frac{1}{2}k+j$ *and* $j \not\equiv \tilde{\varepsilon} \bmod 2$, *then Kernel* $(T^*_{\lambda,\pm 1})$ *is an* $O(Q)$ *submodule of* $C^\infty_\lambda(O(Q))$ *which, when restricted to* K, *is the direct sum of* $[x]_a \otimes [y]_b$ *where* $x+y = j+1+2\sigma''$, *with* σ'' *ranging over all integers* > 0. *Moreover the space* $T^*_{\lambda,\pm 1}(C^\infty_\lambda(O(Q)))$ *is a finite dimensional subspace of* $C^\infty(\Gamma_{\pm 1})$.

Proof. From Proposition 6.15 we deduce that if $\tilde{f} \in \text{Kernel}\,(T^*_{\lambda,\pm 1})$, then $c_{\pm}(\lambda) \equiv 0$. Thus to determine the kernel of the map $T^*_{\lambda,\pm 1}$, it suffices to know the zeroes of the $c_{\pm}(\lambda)$ explicitly. Thus from the proof of Proposition 6.15, $c_{\pm}(\lambda) \neq 0$ if $\lambda \neq \frac{1}{2}k+j$ with $j \in \mathbb{Z}$. Thus (i) follows.

Now let $\lambda_o = \frac{1}{2}k+j$ with $j \equiv \tilde{\varepsilon} \bmod 2$. Thus $c_+(\lambda) = 0$ ($c_-(\lambda) = 0$ resp.) for $\lambda_o = s_2-s_1-1-\frac{1}{2}(a-b)-2\sigma_1$ (σ_1 integer ≥ 0) ($\lambda_o = s_1-s_2-1-\frac{1}{2}(b-a)-2\sigma_2$ (σ_2 integer ≥ 0), resp.). But this means that $a+j = s_2-s_1-1-2\sigma_1$ ($b+j = s_1-s_2-1-2\sigma_2$, resp.). Hence $j \equiv \tilde{\varepsilon} \bmod 2$ is equivalent to $a \equiv 1 \bmod 2$ ($b \equiv 1 \bmod 2$ resp.). Thus (ii) follows.

Finally if $\lambda_o = \frac{1}{2}k+j$ with $j \not\equiv \tilde{\varepsilon} \bmod 2$, then $c_+(\lambda) = 0$ for all s_1, s_2 so that $\lambda_o = \frac{1}{2}k+s_1+s_2-1+2\sigma'$ with σ' an integer < 0. Indeed each pair s_1 and s_2 has the property that the form $\lambda_o = \frac{1}{2}k+j=\frac{1}{2}k+s_1+s_2-1+2\tilde{\sigma}$ for some integer $\tilde{\sigma}$ (i.e. $j \not\equiv s_1+s_2 \bmod 2$). If $\tilde{\sigma} \geq 0$ then, from the proof of Proposition 6.15, the pole which λ_o determines in $\psi(\lambda_o)$ is cancelled off by the zero of λ_o in $\dfrac{\Upsilon(\lambda+(\frac{1}{2}(4-k),1))}{\Gamma(\lambda_o+\frac{1}{2}(4-k))}$. Moreover we note that if $\lambda_o = \frac{1}{2}k+s_1+s_2-1+2\tilde{\sigma}$ for $\tilde{\sigma} \geq 0$, then $\lambda_o+s_1-s_2+1+\frac{1}{2}(a-b) =$ $\frac{1}{2}k+s_1+s_2-1+2\tilde{\sigma}+s_1-s_2+1+\frac{1}{2}(a-b) = a+2s_1+2\tilde{\sigma} > 0$. Hence the denominator in $\psi(\lambda_o)$ does not have a zero to cancel off poles in the numerator of $\psi(\lambda_o)$ when λ_o is as above. Thus (iii) follows (a similar argument works for $c_-(\lambda)$ also.).

Q. E. D.

We recall here the notation used in Corollary 1 to Proposition 5.11.

Corollary 1 to Theorem 6.16. *In (ii) and (iii) of Theorem 6.16 we can make the following more detailed statements:*

(ii)' *Let* $\lambda_o = \frac{1}{2}k+j$ *with* $j \equiv \tilde{\varepsilon}(\lambda) \bmod 2$. *If* $a \equiv 1 \bmod 2$ ($b \equiv 1 \bmod 2$ resp.), *then the map* $(T^*_{\lambda,1})^1$ ($(T^*_{\lambda,-1})^1$ resp.) *of Corollary* 1 *to Proposition* 5.11 *is an injective* $O(Q)$ *intertwining map of* $S^o_{\lambda,1}$ to $C^\infty_c(\Gamma_{+1})'$ ($S^o_{\lambda,-1}$ to $C^\infty_c(\Gamma_{-1})'$ resp.)

(iii)' Let $\lambda_o = \frac{1}{2}k+j$ with $j \not\equiv \tilde{\varepsilon}(\lambda) \bmod 2$. Then for $a \equiv 1 \bmod 2$ ($b \equiv 1 \bmod 2$ resp.), the map $(T^*_{\lambda,1})^1$ ($(T^*_{\lambda,-1})^1$ resp.) is injective on $S^o_{\lambda,1}$ ($S^o_{\lambda,-1}$ resp.). If $a \equiv 0 \bmod 2$ ($b \equiv 0 \bmod 2$ resp.), then $S^1_{\lambda,1}$ ($S^1_{\lambda,-1}$ resp.) is an $O(Q)$ submodule of $S^o_{\lambda,1}$ ($S^o_{\lambda,-1}$ resp.) which, when restricted to K, is the direct sum of $[x]_a \otimes [y]_b$ ($[x']_a \otimes [y']_b$ resp.), where $y-x=a+j+2\tilde{\sigma}+1$ for all $\tilde{\sigma}$ integers ≥ 0 ($x'-y' = b+j+2\sigma'+1$ for all σ' integers ≥ 0). Moreover the map $(T^*_{\lambda,1})^2$ ($(T^*_{\lambda,-1})^2$ resp.) is injective on $S^1_{\lambda,1}$ ($S^1_{\lambda,-1}$ resp.).

Proof. In case (ii) of Theorem 6.16 we observe that $c_+(\lambda)$ has only simple zeroes since $\frac{\gamma(\lambda+(\frac{1}{2}(4-k),1))}{\Gamma(\lambda_o+\frac{1}{2}(4-k))}$ is nonzero when $\lambda_o = \frac{1}{2}k+j$ and $j \equiv \tilde{\varepsilon}(\lambda) \bmod 2$. Then we apply Corollary 1 to Proposition 5.11.

In case (iii) we observe that $c_+(\lambda)$ has a double zero where s_1 and s_2 satisfy the two conditions: (α) $s_1+s_2 = j+1+2\sigma$ (σ integer > 0) and (β) $s_2-s_1 = a+j+2\tilde{\sigma}+1$ ($\tilde{\sigma}$ integer ≥ 0). We then deduce that (β) is a compatible condition if and only if $a \equiv 0 \bmod 2$. We then note $c_+(\lambda)$ has in any case at most a double zero; thus we deduce our result. The case -1 is argued in a similar way.

Q. E. D.

Remark 6.7. The problem is to determine a possible intertwining map between an $O(Q)$ subrepresentation of $C^\infty_\lambda(O(Q))$ (with $\lambda = (\lambda_o,\varepsilon)$) and $U_{\pm 1}(s^2_{\mp}2s)$ (where s is as given in Theorem 6.4). First we recall that if $[s_1]_a \otimes [s_2]_b$ occurs in $U_{+1}(s^2-2s)$ ($U_{-1}(s^2+2s)$ resp.), then $s_1-s_2 = s-\frac{1}{2}(a-b)+2\mu_+$ ($s_2-s_1 = -(s-\frac{1}{2}(a-b))+2\mu_-$ resp.). Thus the first compatibility condition needed for the existence of such an intertwining map is that $s-\frac{1}{2}(a-b) \equiv \tilde{\varepsilon}(\lambda) \bmod 2$. Moreover we have that $\pi_\lambda(\omega_{O(Q)}) = \{-\lambda+\frac{1}{2}(2-k)^2\}\, I$ on $C^\infty_\lambda(O(Q))$ and that $W^{\pm}_{\xi} = -\pi_{\mathfrak{m}}(\omega_{O(Q)})$ on $\Omega_{\pm}$ resp. (this latter fact can be deduced from (6-4) and Remark 5.2). Thus the second compatibility condition needed for the possible construction of an intertwining operator is that $\lambda = \pm(s-1)$ ($\lambda = \pm(s+1)$ resp.).

<u>Theorem 6.17</u>. (A) <u>We assume</u> $s = \frac{1}{2}k+\ell$ <u>with</u> ℓ <u>an integer so that</u> $s > 1$ <u>and</u> ε_ℓ <u>belongs to</u> $\{-1,1\}$ <u>so that</u> $b+\ell \equiv \tilde{\varepsilon}_\ell$ (<u>where</u>

$$\tilde{\varepsilon}_\ell = \begin{cases} 0 & \text{if } \varepsilon_\ell = 1 \\ 1 & \text{if } \varepsilon_\ell = -1 \end{cases}) .$$

(i) <u>If</u> $a \equiv 1 \bmod 2$ <u>then for</u> $\lambda = (1-s,\varepsilon_\ell)$, $T^*_{\lambda,1}$ <u>is an</u> $O(Q)$ <u>intertwining map of</u> $C^\infty_\lambda(O(Q))$ <u>onto a dense subspace of</u> $U_{+1}(s^2-2s)$.

(ii) <u>If</u> $a \equiv 0 \bmod 2$ <u>then for</u> $\lambda = (1-s,\varepsilon_\ell)$, $(T^*_{\lambda,1})^1$ <u>is an</u> $O(Q)$ <u>intertwining map of</u> $S^o_{\lambda,1}$ <u>onto a dense subspace of</u> $U_{+1}(s^2-2s)$.

(B) <u>We assume</u> $s = \frac{1}{2}k+\ell$ <u>with</u> ℓ <u>an integer so that</u> $s < -1$ <u>and</u> ε_ℓ <u>belongs to</u> $\{-1,1\}$ <u>so that</u> $b+\ell \equiv \tilde{\varepsilon}_\ell$ $(\tilde{\varepsilon}_\ell = \begin{cases} 0 & \text{if } \varepsilon_\ell = 1 \\ 1 & \text{if } \varepsilon_\ell = -1 \end{cases})$.

(i) <u>If</u> $b \equiv 1 \bmod 2$ <u>then for</u> $\lambda = (s+1,\varepsilon_\ell)$, $T^*_{\lambda,-1}$ <u>is an</u> $O(Q)$ <u>intertwining map of</u> $C^\infty_\lambda(O(Q))$ <u>onto a dense subspace of</u> $U_{-1}(s^2+2s)$.

(ii) <u>If</u> $b \equiv 0 \bmod 2$ <u>then for</u> $\lambda = (s+1,\varepsilon_\ell)$, $(T^*_{\lambda,-1})^1$ <u>is an</u> $O(Q)$ <u>intertwining map of</u> $S^o_{\lambda,-1}$ <u>onto a dense subspace of</u> $U_{-1}(s^2+2s)$.

<u>Proof</u>. We choose r_1 and r_2 so that $r_1-r_2 = s-\frac{1}{2}(a-b)+2\mu_+$; we also choose ε_ℓ so that $b+\ell \equiv \tilde{\varepsilon}_\ell \bmod 2$. Then it follows that $r_1-r_2 \equiv \tilde{\varepsilon}_\ell \bmod 2$. Moreover if $\tilde{f}(t(\omega_1+\omega_2)) = |t|^\ell H^{r_1}(\omega_1)H^{r_2}(\omega_2) \in C^\infty_\lambda(O(Q))$ with $\lambda = (1-s,\varepsilon_\ell)$, then from Proposition 6.15 $T^*_{(-s+1,\varepsilon_\ell),1}(\tilde{f})(A\tilde{\omega}_1+B\tilde{\omega}_2)$ is $c_+((-s+1,\varepsilon_\ell))$ $A^{2-r_2-k-\ell}B^{r_2}\,{}_2F_1(\alpha_o,\beta_o,\frac{1}{2}b+r_2,B^2/A^2)H^{r_1}(\tilde{\omega}_1)H^{r_2}(\tilde{\omega}_2) = c_+((-s+1,\varepsilon_\ell))$ $(1-u)^{\frac{1}{2}(k+\ell)-1}u^{\frac{1}{2}r_2}\,{}_2F_1(\alpha_o,\beta_o,\frac{1}{2}b+r_2,u)\ H^{r_1}(\tilde{\omega}_1)\ H^{r_2}(\tilde{\omega}_2)$. Moreover

$\beta_o = \frac{1}{2}\{\frac{1}{2}(b-a)+r_2-r_1+1+s-1\} = -\mu_+$ and $\alpha_o = \frac{1}{2}(\frac{1}{2}k+r_1+r_2+s)$. But we recall from (6.13) and Theorem 6.4 that $\sqrt{1+\lambda} = s-1$, and hence for $s = \frac{1}{2}k+\ell$ (with $s > 1$), $(1-u)^{\frac{1}{2}(k+\ell)-1}u^{\frac{1}{2}r_2}\,{}_2F_1(\alpha_o,\beta_o,\frac{1}{2}b+r_2,u)\ H^{r_1}(\tilde{\omega}_1)H^{r_2}(\tilde{\omega}_2)$ defines a nonzero function belonging to $U_{+1}(s^2-2s)$ (i.e. compare to argument before (6-35)). Then we use (ii) and (iii) of Theorem 6.16. But we observe that

$\begin{Bmatrix} \text{(ii)} \\ \text{(iii)} \end{Bmatrix}$ applies if and only if $\begin{Bmatrix} a \equiv 1 \\ a \equiv 0 \end{Bmatrix}$ mod 2 . Then in case (ii) we see if x and y satisfy $x-y = s-\frac{1}{2}(a-b)+2\mu_+$, then x and y do not satisfy the condition to belong to Kernel $(T^*_{\lambda,1})$. Moreover given any isotypic component $[x]_a \otimes [y]_b$ in $C^\infty_\lambda(O(Q))$, either the pair x and y satisfy the condition for belonging to Kernel $(T^*_{\lambda,1})$ or $x-y = s-\frac{1}{2}(a-b)+2\mu_+$ for a non-negative integer μ_+ (but not both these conditions simultaneously).

In case (iii) we note if $x-y = s-\frac{1}{2}(a-b)+2\mu_+$ (μ_+ integer ≥ 0), then $[x]_a \otimes [y]_b$ belongs to Kernel $(T^*_{\lambda,1})$. Then we use (iii)' of Corollary 1 to Theorem 6.16. We observe again that any given isotypic component $[x]_a \otimes [y]_b$ in $S^o_{\lambda,1}$ must belong to $S_{\lambda,1}$ or satisfy $x-y = s-\frac{1}{2}(a-b)+2\mu_+$ (but not both conditions simultaneously). Thus (ii) of (A) follows.

Case (B) of Theorem 6.17 is argued in a similar way.

Q. E. D.

BIBLIOGRAPHY

1. BOREL, A., "Introduction aux groups arithmetiques," Hermann, Paris, 1969.

2. BOURBAKI, N., "Integration," Vol. VI, Hermann, Paris, 1963.

3. BRUHAT, F., "Sur les représentations induites des groups de Lie," Bull. Soc. Math. France, Vol. 84, 1956, pp. 97-205.

4. CHEVALLEY, C., "Theory of Lie Groups," Vol. 1, Princeton Univ. Press, Princeton, 1946.

5. DIEUDONNÉ, J., "Infinitesimal Calculus," Hermann, Paris, 1968.

6. DONAGHUE, W., "Distributions and Fourier Transforms," Academic Press, New York, 1969.

7. DUNFORD, N. and SCHWARTZ, J., "Linear Operators," Vol. 1, Interscience, New York, 1963.

8. ERDELYI, A., "Asymptotic Expansions," Dover, New York, 1956.

9. ERDELYI, A. et al., "Higher Transcendental Functions," Bateman Manuscript, Vol. I, II, McGraw Hill, New York, 1953.

10. GELFAND, I., GRAEV, M., and VILENKIN, N., "Generalized Functions," Vol. 5, Academic Press, New York, 1966.

11. GODEMENT, R., Lectures at Paris VII on Group Representations (unpublished).

12. GODEMENT, R., "Decomposition of $L^2(G/\Gamma)$ for $\Gamma = S\ell_2(Z)$," Proceedings of Symposium in Pure Mathematics, "Algebraic Groups and Discontinuous Groups," AMS, 1966, pp. 211-225.

13. HELGASON, S., "Differential Geometry and Symmetric Spaces," Academic Press, New York, 1966.

14. IGUSA, J., "Harmonic Analysis and Theta Functions," Acta Mathematica, <u>120</u>, 1968, pp. 187-222.

15. IGUSA, J., "Theta Functions," Springer Verlag, Berlin, Band 194, 1972.

16. KLEPPNER, A. and LIPSMAN, R., "Plancherel Formula for Group Extensions," Ann. Scient. Ec. Norm. Sup., Tome 5, 1972, pp. 459-516.

17. KUBOTA, T., "Topological Covering of $S\ell(2)$ over a Local Field," Jour. of Math. Soc. of Japon, 19, 1967, pp. 114-121.

18. MAURIN, K., "Allgemeine Eigenfunktionsentwicklungen, unitare Darstellungen lokalkompackter Gruppen und automorphe Funktionen," Math. Ann., 165, 1966, pp. 204-222.

19. MILLER, W., "Lie Theory and Special Functions," Academic Press, New York, 1968.

20. POULSEN, N., "On C^∞ vectors and intertwining bilinear forms," Jour. of Functional Analysis, 9, 1972, pp. 87-120.

21. SAITO, M., "Representations unitaires des groupes symplectiques," Jour. Math. Soc. of Japon, 24, 1972, pp. 232-251.

22. SCHIFFMANN, G., "Intégrales d'Entrelacement et Functions Whittaker," Bull. Soc. Math. France, 99, 1971, pp. 3-72.

23. TREVES, F., "Topological Vector Spaces," Academic Press, New York, 1967.

24. VARADARAJAN, V. S., "Geometry of Quantum Theory," Vol. 2, Van Nostrand, New York, 1970.

25. WARNER, G., "Harmonic Analysis on Semisimple Groups, I," Springer Verlag, Band 188, 1972.

26. WEIL, A., "Sur certains groupes d'operateurs unitaires," Acta Mathematica, 111, 1964, pp. 143-211.

27. GELBART, S., "Weil's Representation and the Spectrum of the Metaplectic Group," Lecture Notes in Math., Vol. 530, Springer Verlag, 1976.

28. KAZHDAN, D., "Some Applications of the Weil Representation," preprint.

29. PUKÁNSKY, L., "The Plancherel Formula for the Universal Covering Group of $S\ell_2(\mathbb{R})$," Math. Ann., 156, 1964, pp. 96-143.

30. RALLIS, S., "On a Relation Between $\widetilde{S\ell}_2$ Cusp Forms and Automorphic Forms on Orthogonal Groups," Proceedings of Symposia in Pure

Mathematics, Vol. 33, part 1, 1979, pp. 297-314.

31. RALLIS, S. and SCHIFFMANN, G., "Automorphic Forms Constructed from the Weil Representation," Amer. Jour. of Math., Vol. 100, No. 5, 1978, pp. 1049-1122.

32. RALLIS, S., "The Ideal of Relations of the Weil Representation," preprint.

33. RALLIS, S., "Eichler Commutation Relation and the Continuous Spectrum of the Weil Representation," in Non-Commutative Harmonic Analysis, Lecture Notes in Math., Vol. 728, Springer Verlag, 1979, pp. 216-243.

34. RALLIS, S., "Langlands Functoriality and the Weil Representation," preprint.

35. SALLY, P., "Analytic Continuation of the Irreducible Unitary Representations of the Universal Covering Group of $S\ell_2(\mathbb{R})$," Memoirs of AMS, 69, 1967.

36. STRICHHARTZ, R., "Harmonic Analysis on Hyperboloids," Jour. of Functional Analysis, 12, 1973, pp. 341-383.

37. HOWE, R., "θ- Series and Invariant Theory," Proceedings of Symposia in Pure Mathematics, Vol. 33, Part 1, 1979, pp. 275-286.

Stephen Rallis, Department of Mathematics, The Ohio State University, Columbus, Ohio.

Gerard Schiffmann, Institut de Recherche Mathématique Avancée, Université Louis Pasteur, Strasbourg, France.

ABCDEFGHIJ–CM–89876543210